NUMERICAL DIFFERENTIAL EQUATIONS
EQUATIONS
Theory and Technique, ODE Methods, Finite
Differences, Finite Elements and Collocation

NUMERICAL DIFFERENTIAL EQUATIONS

EQUATIONS

Theory and Technique, ODE Methods, Finite Differences, Finite Elements and Collocation

John Loustau

Hunter Collage of the City University of New York, USA

World Scientific

NEW JERSEY · LONDON · SINGAPORE · BEIJING · SHANGHAI · HONG KONG · TAIPEI · CHENNAI · TOKYO

Published by

World Scientific Publishing Co. Pte. Ltd.

5 Toh Tuck Link, Singapore 596224

USA office: 27 Warren Street, Suite 401-402, Hackensack, NJ 07601

UK office: 57 Shelton Street, Covent Garden, London WC2H 9HE

Library of Congress Cataloging-in-Publication Data
Names: Loustau, John, 1943–
Title: Numerical differential equations : theory and technique, ode methods,
    finite differences, finite elements and collocation / John Loustau.
Description: New Jersey : World Scientific, 2016.
Identifiers: LCCN 2015031782 | ISBN 9789814719490 (alk. paper)
Subjects: LCSH: Differential equations--Textbooks. | Differential equations--
    Study and teaching (Graduate)
Classification: LCC QA371 .L68 2016 | DDC 515/.35--dc23
LC record available at http://lccn.loc.gov/2015031782

British Library Cataloguing-in-Publication Data
A catalogue record for this book is available from the British Library.

We would like to thank Marina v.N. Whitman for granting us permission to use her father, John v Neumann's picture on the front cover.

In-house Editors: Tan Rok Ting/Dipasri Sardar

Typeset by Stallion Press
Email: enquiries@stallionpress.com

Printed in Singapore

To all my students

Preface

This is a book in the mathematics of numerical differential equations. I present here the basic theoretical development supporting three common methods: finite difference method (FDM), finite element method (FEM), and collocation method (CM). I include description of the basic techniques and specific examples. The examples drive the interest in the topic while the technique supports the theoretical development. Our point of view on this topic is presented in more detail in the *Foreword*.

The intended purpose of the book is to help make the current literature accessible. This is not to say that the reader can jump into any current paper. Rather, the reader can take most papers, trace back into the supporting references and find an accessible starting point that will lead forward. By including both technique and theory, then the reader will find both the mathematical and the engineering literature accessible.

Our primary concern is with simulation. Mathematical modeling is a critical first step in the process. There are many current references that teach mathematical modeling of scientific observation. Two that are associated to biology are [Wodarz and Komarova (2015)] and [van der Berg (2011)]. Other references are cited throughout Part 1. In general, it is our point of view that developing a differential equation to represent observed data is only part of the story. The full treatment must also include numerical simulation. We present this point of view more completely in the Foreword.

We have chosen not to include discontinuous Galerkin FEM (GFEM). It is well known that discontinuous GFEM better conserves basic quantities associated to a particular setting, for instance, fluid incompressibility. Nevertheless, we contend that knowledge of FEM should begin with the continuous case. With a strong foundation in the continuous case, the discontinuous can be easily understood as a variant.

I imagine that the reader is a graduate student with background in basic real and complex analysis, functional analysis, and linear algebra. The level of the background is well within standard graduate level courses in these topics. Of course background in differential equations is important.

I use *Mathematica*. At this time they are at version 10. *Mathematica* related comments may be false as later versions improve the functionality of the programming product. The particular programming platform should not be a serious issue. Students at this level know how to program and have experience writing code for numerical applications. However, *Mathematica* does provide symbolic computation capability. There are times when I do suggest that a particular algebraic computation be carried out in *Mathematica*.

There are several pathways through this material. There is, of course, the intended or sequential path. I anticipate that any one presenting this material will pick and choose the examples that he finds most interesting. Alternatively, Chapters 1, 2, and 7 yield a basic FDM course. FEM could be presented using Chapters 1, 3, 8, and 9 along with parts of 6. CM is presented in Chapters 1, 5, parts of 6, and Chapter 10.

There is material on the important area of numerical solutions to PDE with stochastic coefficients. However, I only have space to introduce some of these ideas through a single example.

None of this work would have been possible without the input of many students over a period of years. In particular, I note that each of the following students contributed directly to the examples present here. As a matter of personal pride, I remark that most have completed their education and are currently contributing to the development of technology, many as professorial faculty. In no particular order, I acknowledge. Mimi Tsuruga, Hans Gilde, Joel Dodge, Alejandro Falchettore, Saumil Patel, Samuil Jubaed, Tony Markovina, Bolanle Bob-Egbe, Yevgeniy Milman, Henry Chong, Scott Irwin, Dymtro Kedyk, John Svadlenka, Lisa Ueda, Ariel Lindorf, Maurice Lepouttre, Daniel Keegan, Amy Wang, Evan Curcio, Jason Groob, Areum Cho, Nick Crispi, Patrick Brazil, Larry Fenn, Li Qian, Gregory Jarvens, Joseph Kaneko, and Pat Fay.

The cover honors John von Neumann. He understood the effect of computers on the direction of applied mathematics. It was this vision and his research in the years following WWII that lead to the mathematics included in this book.

John Loustau
Hunter College CUNY
New York, NY
2015

Foreword

We begin with the role of numerical differential equations in science and the position the topic holds in Mathematics. First, we need to consider science and continuous mathematical models. Later, we see the role that numerical analysis plays.

Our knowledge of our environment is largely restricted to measuring change. We observe or measure changes and thereby infer knowledge to the entity that undergoes change. Mathematics first enters by providing quantifiable statements for scientific entities and their properties. The entity becomes a function, itself an idealized process, and properties such as mass, area or energy are integrals. It is by means of these mathematical constructs that we are able to quantize our abstractions. Next through observation, we identify conservation laws. Classically, these include conservation of mass or momentum. The 'no arbitrage' assumption in economics or finance is also a conservation law. No matter the source, when the conservation law is stated mathematically and then differentiated the result is zero. In other terms, conservation laws when given mathematical content give us differential equations, the equations of change. The object is a function, the change is represented as a derivative and the differential equation encapsulates our knowledge of the change. However, differential equations are largely intractable. In most cases, our knowledge of the solution is restricted to the data provided by numerical mathematics.

Writing in 1953, Braithwaite (1953) discussed this process in his book, Scientific Explanation. He used a paradigm to differentiate mathematics from science and to explain the role of mathematics in scientific development. The idea is to present science and math via a division of labor. The process is to first abstract observables as mathematical constructs and then to use conservation laws to provide the equations or relationships between the abstracted observables. In other terms, conservation laws are the means that yield mathematical models for observed reality. After the model is stated, the results of mathematical analysis are brought to bear on the problem. The expected outcome is that the differential equation is solved. With the solution at hand, the model generates data that can be used to predict observation. In the final step, if subsequent observation corresponds to prediction then the model is validated.

In 1947, John von Neumann (Grcar, 2011) knew that mathematical analysis alone could not produce the results necessary to support scientific and technological progress. The resolution of differential equations would also use numerical analysis. When the actual solution was not available then necessary information would be produced by discrete simulation. This observation must have been accepted fact for some time prior to 1947. However, the difference in the post WWII era was that computers provided the means for effective simulation.

Already at this time many of the fundamental techniques for approximating the solutions to differential equations were in place. The Runge–Kutta method for ordinary differential equations dates back to the 19$^\text{th}$ century. The mathematical foundation for this technique rested firmly on the Taylor series. For partial differential equations, finite difference method or FDM, Galerkin finite element method or GFEM, and spectral collocation were commonly applied by engineers. Indeed, this was the case as far back as the 1930s. However, little or nothing was known about the convergence and stability of these procedures. Hence, there was no means to validate much of the data that was being generated. Indeed, von Neumann recognized that convergence and stability questions were critical to the useful application of numerical analysis. In his 1947 paper with Herman

Goldstine, von Neumann set out on the path of reshaping numerical analysis as the mathematics necessary to support science and technology in the forthcoming computer age.

In terms of R. B. Braithwaite, in order to generate the data necessary to apply a mathematical model, there would need to be a second layer of theory, the stability and convergence theory. In the original setting, the conservation laws were the enabling results that interfaced between science and the continuous mathematical model. Now, there would need to be a second level of theory, the convergence and stability results, and these would be purely mathematical in nature. These results would bridge from the continuous model to the discrete model. They would validate the numerically generated data as trusted representations of the continuous mathematical model.

Early in this book, we will see an example of exactly the process just outlined. In this case, we begin with a conservation law related to conservation of energy. When this statement is rendered mathematically and differentiated, the result is the heat equation. We next apply Fourier transforms to the 1D case to arrive at a closed form solution for a specific set of boundary conditions. But beyond those specific cases, we are left with no alternative to numerical processes. In this case, the numerical technique that we choose is finite differences. As the example unfolds, the reader sees the entire process. Later in the book, we prove the Lax equivalence theorem. This result resolves the stability and convergence issue.

The purpose of this book is to present the reader with the basic techniques necessary to numerically resolve ODE and PDE and to present the underlying mathematics that supports these methods.

Historically, the stability and convergence theory has always lagged behind the technique. Even today research papers that propose to fill gaps in the convergence theory appear regularly. These gaps arise as engineers need to simulate specific cases that are beyond the current state of the theory. Often engineers introduce methods that are subsequently abstracted by mathematicians or engineers so that they may derive the convergence theorems. The theory is the off-spring of the technique. As such, the theory is only fully understood when one also understands the corresponding application.

This last statement is nicely illustrated by an example. The FEM theory is written for Sobolev spaces. The idea is that you have a PDE defined for a function on a domain. Mathematical analysis has proved the existence of a solution u for certain cases but without providing a specific representation. FEM provides a family of approximate solutions u_h for partitions of the domain. In this case, h is a parameter associated to the partition size. The convergence theorem should state that $u_h \to u$ as $h \to 0$. But then there must be a metric space that contains both u and the family u_h. Since u is the solution to a PDE, then it must be differentiable. However, the u_h are usually piecewise polynomial functions and often not differentiable. They are however weakly differentiable and Sobolev spaces are metric spaces of weakly differentiable functions. Hence, this structure is the natural space to support convergence. However, this is only natural for people who have experienced doing FEM.

But we do not begin with the numerical technique. We begin with the application. The applications are the source. They give rise to the continuous mathematical model. The numerical technique is the means to resolving the mathematical model while the theory validates the connection between method and model. Therefore, we begin with applications. Some of these applications are historical while others are current. We use them as a vehicle to methods. Later, we turn our attention to the validating theory.

It is not our intention to be encyclopedic. This is a basic introduction to the topic for a graduate student with mathematical background in linear algebra and mathematical analysis including the elements of measure theory, functional analysis and Fourier transforms. In an introduction, the writer can be narrow in scope and complete in treatment or broad in scope but somewhat superficial in treatment. We are broad in scope in that we look at several techniques. However, we are not superficial in what we do and the questions we pose. Rather, we have restricted our treatment to the standard cases. It is our contention that this introduction will provide the reader or student with sufficient background to be able to access the numerical analysis literature and thereby develop the deep understanding of those settings of particular interest. We present here the background necessary to do advanced work.

We have tried to include most of what is necessary based on a standard background in measure theory and some basic results of functional analysis. Beyond this, we have included most proofs. However, theory based on special constructs is omitted. In these cases, we point the reader to a currently available reference. We do not want the reader to get bogged down in details whose inclusion has more to do with completeness than with understanding. Our singular purpose is to introduce numerical analysis both as applied technique and as theoretical mathematics. We remain focused on this goal.

This book proceeds as follows. In Part 1, we focus on applications and techniques or methods. In Part 2, we present the theory. We begin Part 1 with techniques as seen from a list of interesting and important application areas. These include finance, mechanical engineering, civil engineering and biology. More specifically, we include fluids, traffic, environmental protection, population studies, chemotaxis and options pricing. We have chosen the list to illustrate the breadth of the topic.

We have matched techniques to applications in a manner that covers a broad list of methods. We have included methods such as Runge–Kutta and midpoint ODE methods, explicit, implicit and Crank–Nicholson FDM, trapezoid, and Adams–Bashford time stepping on top of a spatial FEM realization. We also see FEM both on rectangular and triangular partitions and three variations of collocation method, spectral collocation in one spatial dimension, Gaussian collocation or OSC in one and two dimensions and discontinuous collocation in 2D using a triangular partition. Of course each application can be approached from more than one point of view.

We have mostly stayed away from 3D. Most interesting 3D applications require more computing power than a single processor computer can provide. Currently, nearly all academic institutions have access to high performance computing. Soon we may see a high performance computer on every desktop. However, there is little conceptual difference between two or higher dimensions. Hence, we make only occasional reference to higher dimensional techniques.

In Part 1, we make special effort to lead the reader through the multi-step FEM process. It is somewhat controversial whether a student can be expected to carry out an FEM on the first go. We feel this is achievable provided the student is handed data files for the elements and nodes and provided the context is sufficiently simple. In addition, we contend that reaching this milestone is an essential step toward grasping the topic. Even if the student sees only the simplest case, everything afterward can be understood as a modification or extension.

Also in Part 1, we pay special attention to visualizing the computed results. We are convinced that this is the most effective way to communicate the results of a study. In this regard, it is critical that the researcher maintains clear focus on the purpose of the project when preparing the output.

In Part 2, we first look at FDM. We begin with the Lax Equivalence Theorem. Subsequently, we look at the special issues associated to implementations of FDM for elliptical, parabolic and hyperbolic PDE. As all transient processes are resolved via time stepping and all time stepping is resolved using finite differences, it is essential to begin the theory here.

The most complex procedure is FEM. It is the gold standard of PDE simulation techniques. We begin with a chapter that gives a detailed description of the technique. Here, we discuss the details of the geometric model, the assembly process and the application of boundary values. The convergence theorem for a class of elliptical PDE is the most mathematically demanding part of the book. This requires that we detour into Sobolev spaces. In this regard, we have included some of the basic results while others are omitted as they would lead us too far afield. We include the Lax–Milgram theorem and the Bramble–Hilbert lemma as two milestones along this path. However, we omit much of the Sobolev embedding theory.

Our last major topic is the theory of collocation method. Here, we look at three varieties, the Gaussian collocation or OSC, the spectral version and finally an FEM type that supports triangular domain partitions. The Gaussian type has excellent convergence properties but is applicable to domains that support rectangular partitions. All of the collocation methods have the advantage of not requiring numerical integration. For this reason, all three types are especially useful for PDE with random coefficients. In this regard, we expect collocation to be ever more important. In general, collocation method has the flexibility not present for FDM. On the other hand, without any need for calculating integrals, this technique is faster than FEM.

The techniques developed here may be applied to any differential equation. Nevertheless, the reader should not expect some over-reaching theory that clarifies in one blow a vast number of specific cases. Differential equations and the numerical processes associated to them are a huge topic. The supporting mathematics restricts when we may apply one method or another. Additionally, there is a basic classification system which divides the topic into subtopics. There are similarities in the development within the subtopics and there are useful things that can be said across the subtopics. Further, the goals of the theories are similar. But when you want to say something serious about any one of these cases, then the statements and arguments are specific. There are particular PDE so important that they have become a topic. For instance, there are volumes written about the Navier–Stokes equation.

Contents

Preface   vii

Foreword   xi

Part 1. Modeling and Visualization   1

1. Some Preliminaries   3

    1.1. Normed Linear Spaces   4
    1.2. A Classification of Order 2 PDEs   9
    1.3. Multivariate Polynomial Interpolation   17
    1.4. Numerical Integration for a Rectangle
        and a Triangle   29

2. Problems with Closed Form Solution   35

    2.1. Flow about an Airfoil: Harmonic Conjugates
        and Analytic Functions   37
    2.2. Temperature Distribution along a Rod, the
        Derivation and Solution of the 1D Heat Equation   47
    2.3. The Heat Equation and FDM Stability   56

3. Numerical Solutions to Steady-State Problems   71

    3.1. Ideal Fluid Flow, the Laplace Equation via FEM
        and a Rectangular Partition   73

3.2.   Waves in an Enclosed Pool, Helmholtz Equation
       via FEM and a Triangular Partition   84

4.   Population Models   93

4.1.   Predator/Prey Models, Stability Analysis   94
4.2.   Visualizing Predator/Prey Population Models
       with the Runge–Kutta Method   101
4.3.   Population Model with Herding Instinct Using
       Stochastic Collocation Method   109

5.   Transient Problems in One Spatial Dimension   115

5.1.   Options Pricing, the Black–Scholes Equation,
       and Collocation   117
5.2.   Options Pricing, the Black–Scholes Equation,
       and the 1D Heat Equation   129
5.3.   River Pollution, the Transport–diffusion Equation
       via Collocation   135
5.4.   Traffic Congestion, a Different Flow Problem   147

6.   Transient Problems in Two Spatial Dimensions   157

6.1.   Cell Chemotaxis   159
6.2.   The NSE, the Stokes Flow, and the Bernoulli
       Equation   164
6.3.   Applying FEM to the Stokes Equation,
       TR Time Stepping   169
6.4.   Herd Formation Model, FEM   177
6.5.   A Remark on Numerical Processes, Convergence,
       and Existence Theory   183

Part 2. Methods and Theory   187

7.   Finite Difference Method   189

7.1.   Convergence, Consistency, and Stability   191
7.2.   Extended Differences   200
7.3.   Difference Schemes for Hyperbolic PDE   210

7.4.	Remarks on Stability and Convergence	218
7.5.	Difference Schemes for Elliptical PDE	223
8.	**Finite Element Method, the Techniques**	**227**
8.1.	The Model 1: Elements and Nodes	230
8.2.	The Model 2: Polynomial Basis Functions	240
8.3.	Serendipity	254
8.4.	The Linear System	264
8.5.	Boundary Values, Neumann, Dirichlet, and Robin	272
9.	**Finite Element Method, the Theory**	**283**
9.1.	The Weak Derivative	285
9.2.	Sobolev Spaces	291
9.3.	Boundary Values	301
9.4.	Weak Solutions of Elliptical Equations	309
9.5.	Sobolev Embedding	317
9.6.	Polynomial Interpolation on a Sobolev Space	325
9.7.	Convergence for Finite Element Method	334
10.	**Collocation Method**	**337**
10.1.	Interpolation with Legendre Polynomials	338
10.2.	The Collocation Inner Product	343
10.3.	Convergence for Gaussian Collocation	347
10.4.	Remarks on Collocation and Superconvergence	350
Bibliography		353
Index		357

PART 1

Modeling and Visualization

Chapter 1

Some Preliminaries

Introduction

We begin with some topics that should be included but do not fit within any of the following chapters. A given reader may be familiar with parts or all of these topics. In this case, much of this material may be safely omitted. Alternatively, they may be held until needed.

The material of Section 1.1 is certainly well within the material listed in the introduction as a prerequisite. However, it is our experience that students often overlap classes with their prerequisites. Hence, it is useful to have this material here. There are two separate vector spaces that arise when doing numerical solutions to differential equations. On the one hand, the PDE is an operator defined on functions. In turn, the functions are defined on a particular domain D, where D is a subset of some real or complex n-dimensional space. The second is the space of functions themselves. For instance, this may be the space of twice differentiable functions defined on D.

In Section 2, we develop the standard classification system for second-order PDE. It is an interesting side issue that the governing equations for fluid flows change from elliptical to parabolic and then to hyperbolic depending on the Reynolds number. We make use of this in our discussion. In addition, we also introduce the concept of a well-posed problem. This idea reappears later in this text.

It is arguably true that any numerical technique for estimating the solution of a differential equation arises from some form of polynomial interpolation. However, authors often ignore multivariate polynomial interpolation, or they may mention that it is a routine extension of the single variable case. Indeed, it can be difficult to find a reference for these results. From our point of view, this theory is not so obvious that it should not be included. Our source is (Phillips, 2003), a monograph from 2003. In Section 3, we present the material for two variables. Once the reader has seen the two variable case, the way to the three or four variable extensions is apparent.

Finally, we include some material on numerical integration over rectangular or triangular regions. In this case, we are not concerned with general issues of numerical integration, Rather, we are only concerned with the cases that arise in FEM. In particular, we will only need to consider polynomial functions and then only the case where the polynomials are derived from a reference element via an affine mapping.

1.1. Normed Linear Spaces

The words vector space and linear space are synonyms. Generally, when the elements of the space (vectors) are functions, then it is common to use the term linear space or function space. All linear spaces that we consider will have real or complex scalar field. Hilbert spaces are of particular importance in the mathematics developed for FEM (Atkinson and Han, 2009; Brenner and Scott, 2008). The basic structure for FDM and collocation (Atkinson and Han, 2009; Loustau *et al.*, 2013) is the Banach space. We introduce both in this section, as well as the standard terminology associated to linear spaces.

Definition 1.1.1. Let V be a real or complex vector space. Then a *norm* is a real valued function defined on V denoted $\|v\|$ with the following properties. If v and w are in V, then

(i) $\|v\| > 0$, if $v \neq 0$,
(ii) $\|\alpha v\| = |\alpha|\|v\|$ for any scalar α,
(iii) $\|v + w\| \leq \|v\| + \|w\|$.

A vector space endowed with a norm is referred to as a *normed linear space* or *normed vector space*. If item (i) is replaced by the weaker condition, $\|v\| \geq 0$, then we call the resulting structure a *semi-norm*. Semi-norms arise naturally in numerical analysis. Note also that the ordinary or Euclidean vector length in $\mathbb{R}^n$ defines a norm on that space. Item (iii) in Definitions 1.1.1 and 1.1.2 are both referred to as the triangle inequality.

Definition 1.1.2. Let V be a linear space. A function $d : V \times V \to \mathbb{R}$ is called a *metric* on V provided for every u, v, and w in V.

(i) $d(u, v) \geq 0, d(u, u) = 0$,
(ii) $d(u, v) = d(v, u)$,
(iii) $d(u, v) + d(v, w) \geq d(u, w)$.

In turn, V is called a *metric space*.

The following result is an immediate consequence of the definitions.

Lemma 1.1.1. *Any normed linear space is a metric space.*

Proof. If V is a normed linear space, then the function $d(v, w) = \|v - w\|$ determines a metric on V. $\qquad\square$

Since convergence and continuity are metric space ideas, it makes sense to talk about convergence for sequences in normed linear spaces and continuity for functions of these spaces. Recall that a sequence u_n is called Cauchy if for every $\epsilon > 0$, there is an integer N with $d(u_m, u_n) < \epsilon$ provided $m, n \geq N$. If every Cauchy sequence in a metric space is convergent, then the space is called *complete*. Complete normed linear spaces are called *Banach spaces*.

Definition 1.1.3. Let V be a real vector space. Then a function $\sigma : V \times V \to \mathbb{R}$ is a *positive definite inner product* on V provided the following hold. For every u, v, and w in V,

(i) $\sigma(u, v) = \sigma(v, u)$,
(ii) $\sigma(\alpha u + \beta v, w) = \alpha\sigma(u, w) + \beta\sigma(v, w)$,
(iii) $\sigma(u, u) > 0$ provided $u \neq 0$.

If condition (iii) is weakened to $\sigma(u, u) \geq 0$, then we say that the form is *positive semi-definite*. In addition, we call V an *inner product space*.

Lemma 1.1.2. *Any real vector space with a positive definite inner product is a normed linear space.*

Proof. We define $\|u\| = \sigma(u, u)^{1/2}$. It is now immediate that $\|u\|$ satisfies the first two properties of a norm. To prove the third, note that for any u and v,

$$\sigma(u, v) \leq \|u\|\|v\| \tag{1.1.1}$$

(see Exercise 1). Now,

$$(\|u\| + \|v\|)^2 = \|u(\|^2 + 2\|u\|\|v\| + \|v\|^2)$$
$$\geq \|u\|^2 + 2\sigma(u, v) + \|v\|^2 = \sigma(u + v, u + v) = \|u + v\|^2,$$

and (iii) of Definition 1.1 follows.   □

Equation (1.1.1) holds in any vector space with a positive definite inner product. It is the familiar *Cauchy–Schwartz inequality*.

A *Hilbert space* is a complete normed linear space where the norm arises from an inner product in the manner just described. Note that a positive semi-definite inner product defines a semi-norm.

There are two basic examples, which occur frequently in the following chapters. In $\mathbb{R}^n$, the vector space of real n-tuples, $\sigma(u, v) = \sum_{i=1}^{n} u_i v_i$ is a positive definite inner product. In this case, the norm is given by $\|u\| = (\sum_i u_i^2)^{1/2}$. It is often referred to as the *Euclidean norm*. For the case of the space of all square (Lebesgue) integrable functions defined on an open set U, we use $\sigma(f, g) = \int_U fg$. In this case, the norm, $\|f\|_2 = (\int_U f^2)^{1/2}$ determines the well-known normed linear space $L^2(U)$ or L^2.

Additionally, there are two normed linear spaces related to L^2. $L^1(U)$ denotes the space of absolutely integrable functions on U. In this case, $\|f\|_1 = \int_U |f|$. And $L^\infty(U)$ denotes the space of all bounded functions on U. The norm $\|f\|_\infty$ is the sup of $|f|$. Note that convergence in $L^\infty(U)$ defined on the space of continuous functions is uniform convergence. From this remark, it is immediately clear that the set of continuous functions on a compact set with sup norm is a Banach space.

The L^p spaces, the pth power integrable functions with norm $\|f\|_p = (\int_U |f|^p)^{1/p}$ are also Banach spaces. These spaces will come up only briefly in the subsequent material.

For complex vector spaces, we generalize the idea of an inner product. The resulting construct is called a Hermitian form.

Definition 1.1.4. Let V be a complex vector space. Then, a function $\sigma : V \times V \to \mathbb{C}$ is a *positive definite Hermitian form* on V provided the following hold. For every $u, v,$ and w in V,

(i) $\sigma(u, v) = \overline{\sigma(v, u)}$,

(ii) $\sigma(\alpha u + \beta v, w) = \alpha \sigma(u, w) + \beta \sigma(v, w)$,

(iii) $\sigma(u, u) > 0$ provided $u \neq 0$.

Note that by (i), $\sigma(u, u)$ is always real. Also by (ii), $\sigma(u, \alpha v) = \overline{\sigma(\alpha v, u)} = \overline{\alpha}\ \overline{\sigma(v, u)} = \overline{\alpha}\ \sigma(u, v)$. As with positive definite inner products, positive definite Hermitian forms induce a norm and a metric on the space V. As in the real case, the norm is defined as $\sigma(u, u) = \|u\|^2$.

We next turn to the linear operators of the normed linear space.

Definition 1.1.5. Let U and V be normed linear spaces, then a linear transformation L is bounded provided there is a constant M such that for any u in U, $\|L(u)\| \leq M\|u\|$. The least M is written $\|L\|$ and called the *linear transformation norm* or *operator norm* of L.

It is easy to prove that a bounded linear transformation of a normed linear space is uniformly continuous (see Exercise 10). The following results are standard.

Theorem 1.1.1. *The set of bounded linear transformations of a normed linear spaces is a normed linear space with respect to the linear transformation norm. Moreover, if $U = V$, then for any bounded linear transformations L and N, $\|LN\| \leq \|L\|\|N\|$.*

Proof. The proof is left as an exercise. $\qquad\square$

Theorem 1.1.2. *If U and V are normed linear spaces and U is finite dimensional, then every linear transform $L: U \to V$ is bounded.*

Proof. Take u a unit vector in U and a basis $u_1, \ldots, u_n$ of unit vectors in U. We write $u = \sum_i \xi_i u_i$ with each $|\xi_i| \leq 1$ and calculate

$$\|L(u)\| = \left\| \sum_{i=1}^{n} \xi_i L(u_i) \right\| \leq nC,$$

where $C = \max_i \|L(u_i)\|$. It now follows for any u in U and $v = u/\|u\|$, $\|L(u)\| = \|u\| \|L(v)\| \leq nC\|u\|$. $\qquad\qquad\square$

We mention two results that will come up later. The Riesz representation theorem holds in Hilbert space. In particular, if f is a bounded linear functional defined on a Hilbert Space H, then there is an element u in H with $f(v) = \sigma(u, v)$ for every v in H. Secondly, any normed linear space V is a dense subspace of a Banach space $\widehat{V}$. Furthermore, if L is a bounded linear transformation of V, then L extends to a bounded linear transformation of $\widehat{V}$. See Rudin (1986) for the above theorems.

Exercises

1. Suppose that V is a linear space with a positive definite inner product space, that σ is the inner product and that $\|u\|$ is given by $\sigma(u, v)^{1/2}$. Prove the Cauchy–Schwartz inequality,

$$|\sigma(u, v)| \leq \|u\| \|v\|$$

   for any u and v in V. (*Hint*: Expand $0 \leq \sigma(\alpha u - \beta v, \alpha u - \beta v)$, and set $\alpha = \|v\|$ and $\beta = \|u\|$.)
2. Suppose that V is a linear space of square integrable real valued functions on U contained in $\mathbb{R}^n$. Prove $\sigma(f, g) = \int_U fg$ defines a positive definite inner product on V.
3. Prove that a Hermitian form on complex vector spaces give rise to a normed linear spaces.
4. Prove that $L^1(U)$ is a normed linear space.
5. Consider the continuous functions $C^0[U] = V$. Prove that $L^\infty(U)$ induces a norm on V.

6. With the notation of Problem 5, prove that a sequence in V with $L^\infty(U)$ converges if and only if the sequence converges uniformly.

7. Prove that for compact $U, C^0(U)$ with sup norm is a Banach space.

8. Let V be a linear space of functions defined on D contained in $\mathbb{R}^n$. For any x in $\mathbb{R}^n$, define $F_x(f) = f(x)$. Prove that F_x is a linear transformation of V.

9. Continuing with the notation of 8, suppose that there are points $x_1, \ldots, x_n$ of D and functions f_j, $j = 1, \ldots, n$, in V with $F_{x_i}(f_j) = \delta_{i,j}$, (*the Kronecker delta*) for every i and j. Prove that the $f_j, j = 1, \ldots, n$ are linearly independent. In this case, the x_i are called dual to the f_j.

10. Prove that a bounded linear transformation of normed linear space is uniformly continuous.

11. Prove Theorem 1.1.1.

1.2. A Classification of Order 2 PDEs

In this section, we present the classification system for second-order PDE. The theoretical mathematics associated to the PDE differs greatly from one class to the next. Hence, each PDE class is in a sense its own topic. It is not surprising that the numerical simulation techniques also vary by PDE class. As motivation for the three classes, we use an example from computational fluid dynamics (CFD).

We begin the section with a brief discussion of the concept, well-posed. A differential equation represents a mathematical rendering of an observed event. In general terms, the mathematical rendering is well posed provided it faithfully represents the reality it is intended to model. To be useful, we need a concept that can be verified. For this purpose, we present the characterization due to Jacques Hadamard (1865–1963). A differential equation (with boundary values and initial conditions) is well posed provided it has a unique solution that varies continuously with problem data. Later, this concept will reappear for each of the classes of numerical process that we develop. Each time we recast the concept for the specific setting.

In a very general context, a differential equation may be considered as arising from a transformation, T defined on a function space, V and taking values in another space, W. An element u in V is a solution to T, provided $T(u) = 0$. For instance, in the case of the Laplace equation $\nabla^2 u = 0$, T is the Laplace operator ∇^2. For the Helmholtz equation, $\nabla^2 u + \lambda u = 0$, T is $\nabla^2 + \lambda I$, where I denotes the identity transformation. In both cases, T involves differentiation. For the time being, we suppose that elements of V are sufficiently differentiable so that T is defined on all of V. Later, we will allow for weak differentiability.

At this level of abstraction, the only difference between a PDE and an ODE is whether the elements of V are functions of one or more independent variables. For the most part, we will use PDE as the generic term and use ODE when we want to develop specific results for functions of a single variable.

A differential equation is called *linear* if T is a geometric transformation of the underlying function space V. In particular, T is geometric, if there exists an element f in W with $T(u) = L(u) - f$, where L belongs to $Hom[V, W]$, the space of linear transformations from V to W. If the transformation L is non-singular, then the geometric transformation is often referred to as an affine mapping. The differential equation T is called homogeneous provided f is zero. Since $T = L$ when T is homogeneous, then we may call L the homogeneous part of T. We will see that f is related to the boundary values. Furthermore, it is immediate that when L is linear, then the set of u in V for which $L(u) = 0$ is a subspace of V, the kernel of L. Hence, when f belongs to V, then the solution space of T is a coset of the kernel of L. If we denote the translation by f via t_f, then the PDE may be written in compact form as $t_f \circ L$.

The order of a differential equation is the highest order derivative that occurs in the equation. For the most part, we are concerned with *second-order* equations. However, equations of other orders do arise. The classification of second-order PDE into *elliptical, parabolic,* and *hyperbolic* equations is useful both from the theoretical and practical standpoints. The practitioner associates heuristic qualities to the solution depending on the type of equation. The mathematical theory associated to each type is distinct and separate.

We begin by considering a simple second-order PDE with two independent variables, x and y, and a scalar valued dependent variable u.

$$A\frac{\partial^2 u}{\partial x^2} + 2B\frac{\partial^2 u}{\partial x \partial y} + C\frac{\partial^2 u}{\partial y^2} + D\frac{\partial u}{\partial x} + E\frac{\partial u}{\partial y} + Fu + G = 0. \quad (1.2.1)$$

For the moment, we assume that the coefficients are constant. We think of u as a surface over a domain D in the xy-plane. We want to consider the impact on (1.2.1) should the second partials of u be discontinuous.

Let $\gamma(s) = (x(s), y(s))$ be a rectifiable curve in D, then $u \circ \gamma$ is a function from $\mathbb{R}$ to $\mathbb{R}$. From the chain rule, the first coordinate of the tangent vector to the graph is

$$(u \circ \gamma)_x = \frac{d}{ds}(u \circ \gamma) = \frac{\partial u}{\partial x}\frac{dx}{ds} = \frac{\partial u}{\partial x}x'.$$

Hence, the tangent vector is given by

$$\begin{pmatrix} (u \circ \gamma)_x \\ (u \circ \gamma)_y \end{pmatrix} = \begin{pmatrix} \dfrac{du}{dx}x' \\ \dfrac{du}{dy}y' \end{pmatrix}. \quad (1.2.2)$$

This expression describes a curve in $\mathbb{R}^2$ parameterized by s. Hence, the second derivative or acceleration vector (provided s is time) for $u \circ \gamma$ is the tangent vector to this arc. Therefore, again using the chain rule, the acceleration vector is

$$\begin{pmatrix} (u \circ \gamma)_{xx} \\ (u \circ \gamma)_{yy} \end{pmatrix} = \begin{pmatrix} \dfrac{d}{dx}(u \circ \gamma)_x \\ \dfrac{d}{dy}(u \circ \gamma)_y \end{pmatrix}$$

$$= \begin{pmatrix} (\nabla u_x) \cdot \gamma' \\ (\nabla u_y) \cdot \gamma' \end{pmatrix} = \begin{pmatrix} \dfrac{\partial^2 u}{\partial x^2} & \dfrac{\partial^2 u}{\partial x \partial y} \\ \dfrac{\partial^2 u}{\partial y \partial x} & \dfrac{\partial^2 u}{\partial y^2} \end{pmatrix}\begin{pmatrix} x' \\ y' \end{pmatrix}. \quad (1.2.3)$$

Recall that if $u \circ \gamma$ is C^2, twice continuously differentiable, then $\partial^2 u / \partial x \partial y = \partial^2 u / \partial y \partial x$. Hence, for any value of s for which $u(\gamma)$ is C^2, then (1.2.3) yields three linear relations in u_{xx}, u_{xy} and u_{yy}. Setting

$$H = - \left(D \frac{\partial u}{\partial x} + E \frac{\partial u}{\partial y} + Fu + G \right),$$

we state the relation as

$$\begin{pmatrix} A & 2B & C \\ x' & y' & 0 \\ 0 & x' & y' \end{pmatrix} \begin{pmatrix} \dfrac{\partial^2 u}{\partial x^2} \\ \dfrac{\partial^2 u}{\partial x \partial y} \\ \dfrac{\partial^2 u}{\partial y^2} \end{pmatrix} = \begin{pmatrix} H \\ (u(\gamma))_{xx} \\ (u(\gamma))_{yy} \end{pmatrix}. \qquad (1.2.4)$$

Suppose that there is a discontinuity of the second partials at s_0 and that the discontinuity is isolated in the sense that the second partials are continuous in any punctured neighborhood of s_0. A discontinuity in the second partial of u at a location (x, y) infers a discontinuity in the acceleration vector for a curve $u(\gamma)$ passing through $(x, y) = (x(s_0), y(s_0))$. In particular, the acceleration vector computed as $s \to s_0$ from below must be distinct from the vector computed as s approaches from above. Since the matrix on the left-hand side of (1.2.4) is a function of the parameter s, it follows that the matrix (1.2.4) at s_0 must be singular. Computing the determinant and setting it to zero yields the following quadratic relation in the slope of γ:

$$A \left(\frac{y'}{x'} \right)^2 - 2B \left(\frac{y'}{x'} \right) + C = 0. \qquad (1.2.5)$$

The solution to this equation depends upon the discriminant $4(B^2 - AC)$. We now borrow the terminology from analytic geometry and call (1.2.1) *elliptical* if the discriminant is negative, *parabolic* if it is zero and *hyperbolic* if it is positive. Restated in terms of (1.2.4), Eq. (1.2.5) has no real roots in the elliptical case. Hence, the second partials are continuous. In the parabolic case, there is a single direction along which the second partials are discontinuous and in the hyperbolic case, there are two independent directions.

The formal definitions of elliptical, parabolic and hyperbolic follow. These terms will be defined in considerably greater generality than above. But first we want to see how this works out in a specific case from fluid dynamics.

Recall that the speed of sound in air is about 1700 ft/sec. This number is denoted $M1$ or $Mach - 1$ or simply M. Consider an object traveling along a linear path at a rate of 1000 ft/sec. Since the rate is less than $M1$, the sound of the object proceeds it (1000 ft/sec is about $0.588M$). The sound emitted at any point in time causes a disturbance in the air along a circle that is centered at the object location. For a fixed time, the circle is called the event horizon. In Figure 1.2.1, the object is at point A at time $t = 0$ and at D when $t = 1$. The circles centered at B and C show the event horizon for

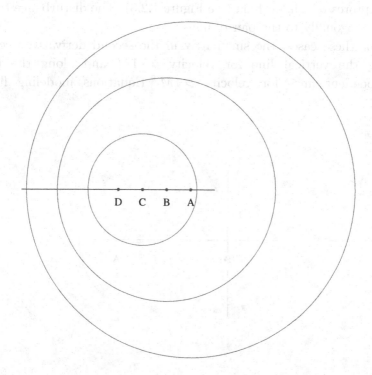

Figure 1.2.1.   Sound horizon at $t = 1$ for an object traveling at $0.588M$.

the sound emitted at $t = 1/3$ and $t = 2/3$, respectively. Note that the several event horizons remain inside the initial event horizon. An observer located at D will experience steady sound from the moving object. There is a one-to-one correspondence between the sound heard at any time and the sound emitted. Of course, the pitch will rise as the object approaches due to the Doppler effect. (Note the compression ahead of the moving object in Figure 1.2.1.)

Next, we suppose that the object travels at $1M$ then the sound horizons are tangential at D (see Figure 1.2.2). The coincidence of the several sound horizons causes the boom which propagates outward along the tangent (note the vertical line in the figure). In this case, there is no longer a one-to-one correspondence between sound emitted and sound heard.

Finally, we look at the case with velocity $= 2M$ after 0.25 s. Now the sound horizon occurs along a pair of lines which are the asymptotes to a hyperbola (see Figure 1.2.3). The disturbance travels orthogonally to the pair of lines.

For these cases, the singularity in the second derivative occurs along the vertical line for velocity $= 1M$ and along the two independent lines for velocity $> 1M$. Equations modeling fluid

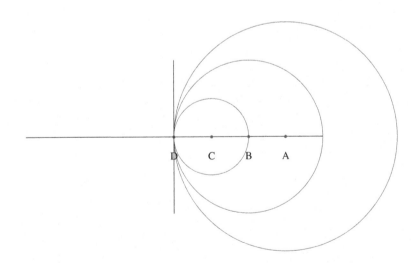

Figure 1.2.2.   Sound horizon at $t = 0.588$ for an object traveling at $1M$.

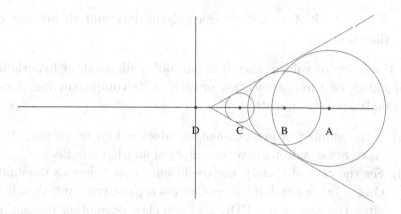

Figure 1.2.3. Sound horizon at $t = 0.25$ for an object traveling at $2M$.

dynamics depend on the Reynolds number. But if the scale, viscosity, and density are fixed, then the Reynolds number depends on velocity. Subsonic flow is elliptical and supersonic flow is hyperbolic. Parabolic equations are used for transitional flow.

We now turn to the formal definitions. For this purpose, we need to restate the concept in terms which can be extended to a real vector space of dimension $n < \infty$. In the case of Eq. (1.2.1), we called the PDE elliptical, parabolic, or hyperbolic depending on whether $B^2 - AC$ is respectively negative, zero or positive. In turn, the eigenvalues of the matrix of coefficients for the second-order terms have the same sign, at least one is zero, or have different signs. The advantage of this last assertion is that it is easily extended to the case of a PDE in n independent variables and with variable coefficients.

Definition 1.2.1. Consider the following second-order PDE for a scalar valued function u in n independent variables, $x = (x_1, \ldots, x_n)$.

$$\sum_{i,j=1}^{n} A_{ij} \frac{\partial^2 u}{\partial x_i \partial x_j} - F\left(x; u, \frac{\partial u}{\partial x_1}, \ldots, \frac{\partial u}{\partial x_n}\right) = 0, \qquad (1.2.6)$$

where for each x, the matrix $A(x) = A_{ij}(x)$ is symmetric. Then, the PDE is

(a) *Parabolic* if $A(x)$ has at least one zero eigenvalue,
(b) *Elliptical* if $A(x)$ has non-zero eigenvalues all of the same sign,

(c) *Hyperbolic* if $A(x)$ has non-zero eigenvalues and all but one of the same sign.

If u is vector valued, then it is parabolic, elliptical, or hyperbolic depending on the classification of each of its component functions. We make some comments.

(A) If the number of independent variables is four or greater, then there are equations between elliptical and hyperbolic.

(B) For the case of (1.2.4), the coefficients were taken as constant. On the other hand, if the coefficients are not constant, then it is often the case that a PDE is in one class or another depending on the spatial variable.

(C) The steady-state Laplace, Poisson, and Helmholtz equations are elliptical. But the transient form of each is parabolic. For instance, consider the heat equation in two spatial dimensions. In the steady-state formulation, we have $\nabla^2 u = 0$. It is certainly elliptical, whereas, the transient form, $u_t = \alpha \nabla_x^2 u$, is parabolic since now u is a function of three independent variables. Here, we appeal to the fact that any 3×3 matrix has a real eigenvalue.

   We will see later that the transient form equation is often handled numerically by separating the time variable from the spacial variables. In particular, you fix t and then build the discrete form of the resulting equation via any one of the techniques presented below. Finally, you use FDM to step through the time sequence.

   Alternatively, the time variable is separated from the spatial variables, the spatial variables are handled via FDM or FEM or some other method and the time variables use an ODE technique such as Runge–Kutta (see Chapter 4). When the spatial variables are resolved with FDM, then the technique is called the Method of Lines.

(D) The second-order wave equation, $(1/c^2)u_{tt} = \nabla_x^2 u$ is hyperbolic.

(E) There is also a classification system for first-order PDE. We omit it here as our concern lies primarily with second-order cases.

(F) A PDE with variable coefficients may be elliptical, yet the eigenvalues, as a function of x, may not be bounded away from zero.

An elliptical PDE with this case excluded is called uniformly elliptical. We see this concept in Chapter 9.

Exercises

1. Prove the assertion that the eigenvalues of

$$\begin{pmatrix} A & B \\ B & C \end{pmatrix}$$

have the same sign if and only if $B^2 - AC < 0$. Prove that the matrix has a zero eigenvalue if and only if $B^2 - AC = 0$.

2. Verify that the transient heat equations in two spatial variables are parabolic (see Section 2.2).

3. Determine the classification of the Black–Scholes equation for pricing futures (see Section 5.1).

4. Equation (1.2.1) is sometimes referred to as the diffusion transport equation. $A, B,$ and C are the diffusion coefficients, D and E are the transport coefficients, F is the decay term, and G is the source. Compare the one variable diffusion transport equation and the Black–Scholes equation (see Section 5.1). Interpret the terms of Black–Scholes in the light of the interpretation of terms of the diffusion transport equation. We will see an application of this in Section 5.3.

1.3. Multivariate Polynomial Interpolation

We have already stated that polynomial interpolation is the central theme for numerical solutions to PDE. To develop interpolation in two variables, we introduce divided differences. This technique is classical. It goes back at least as far as Newton. We begin with some general ideas that lead to results essential to manipulating divided differences. After these preliminaries, we state and prove the basic theorem for this section, the error estimate for two variable polynomial interpolation. At the end of the section, we turn to the problem of computing the interpolating polynomial. In this context, we introduce the two variable Vandermonde matrix and derive sufficient conditions for non-singularity.

Before starting, it is important to note that it only makes sense to interpolate continuous functions. This will cause problems when the underlying space is L^2.

We begin considering the $n+1$ point polynomial interpolation of a real valued function f defined on a real interval $[a, b]$. We designate the interpolation points by $x_0, \ldots, x_n$ and the interpolating polynomial as p. Note that since the polynomials $\prod_{j=0}^{k-1}(x - x_j), k \leq n$ form a basis for the space of polynomials of degree no greater than n, then p may be written in the form $\sum_{k=0}^{n} c_k \prod_{j=0}^{k-1}(x - x_j)$. This is called the *Newton form of the interpolating polynomial*.

Definition 1.3.1. Given $x_0, \ldots, x_n \in [a, b]$ and f a continuous function, we define the kth *divided difference*, $[x_0, \ldots, x_k]f = c_k$, where c_k is the kth coefficient of the Newton form of the interpolating polynomial.

Note that the divided difference, $[x_0, \ldots, x_k]$ is a linear operator on the Banach space $C^0[a, b]$. It is immediate from the definition that $[x_0]f = f(x_0)$. Also, $[x_0, x_1]f = (f(x_1) - f(x_0))/(x_1 - x_0)$. These remarks are also a consequence of the next theorem.

Theorem 1.3.1. *For each k and each f in $C^0[a, b]$,*

$$[x_0, \ldots, x_k]f = \sum_{i=0}^{k} f(x_i) \prod_{j \neq i, j=0}^{k} \frac{1}{(x_i - x_j)}. \qquad (1.3.1)$$

Proof. We write the Newton form using divided difference notation.

$$p(x) = [x_0]f + (x - x_0)[x_0, x_1]f + (x - x_0)(x - x_1)[x_0, x_1, x_2]f$$

$$+ \cdots + (x - x_0)(x - x_1), \ldots, (x - x_{n-1})[x_0, x_1, x_2, \ldots, x_n]f. \qquad (1.3.2)$$

Now the coefficient of x^n in (1.3.2) is $[x_0, x_1, x_2, \ldots, x_n]f$. On the other hand, the ith Lagrange polynomial is $\prod_{j \neq i, j=0}^{n}(x - x_j)/(x_i - x_j)$. Hence, the coefficient of x^n in the ith Lagrange polynomial is $\prod_{j \neq i, j=0}^{n} 1/(x_i - x_j)$ and the coefficient of x^n in the Lagrange form of p is $\sum_{i=0}^{n} f(x_i) \prod_{j \neq i, j=0}^{n} 1/(x_i - x_j)$. Equating the two forms of this coefficient yields (1.3.1). $\qquad \square$

Corollary 1.3.1. *For any permutation $\sigma \in S_{n+1}$, $[x_{\sigma(0)}, x_{\sigma(1)}, \ldots,$ $x_{\sigma(n)}]f = [x_0, x_1, \ldots, x_n]f$.*

Proof. This result is an immediate consequence of (1.3.1). □

The next theorem is often referred to as the recursion property for divided differences. With this result, high order divided differences may be calculated from the lower order terms. This result is useful in practice and in induction arguments.

Theorem 1.3.2. *Given $x_0, x_1, \ldots, x_{n-1}, x_n$ in $[a, b]$,*

$$[x_0, \ldots, x_n]f = \frac{1}{x_n - x_0}([x_1, \ldots, x_n]f - [x_0, \ldots, x_{n-1}]f). \quad (1.3.3)$$

Proof. We start with the right-hand side of (1.3.3) and expand using (1.3.1)

$$\frac{1}{x_n - x_0}\big[[x_1, \ldots, x_n]f - [x_0, \ldots, x_{n-1}]f\big]$$

$$= \frac{1}{x_n - x_0}\left[-\sum_{i=0}^{n-1} f(x_i) \prod_{j \neq i} \frac{1}{x_i - x_j} + \sum_{i=1}^{n} f(x_i) \prod_{j \neq i} \frac{1}{x_i - x_j}\right]$$

$$= \frac{1}{x_n - x_0}\left[-f(x_0) \prod_{j=1}^{n-1} \frac{1}{x_0 - x_j} + f(x_1)\left[\prod_{j \neq 1, j=2}^{n} \frac{1}{x_1 - x_j}\right.\right.$$

$$\left.\left.- \prod_{j \neq 1, j=0}^{n-1} \frac{1}{x_1 - x_j}\right] + \cdots + f(x_n) \prod_{j=1}^{n-1} \frac{1}{x_n - x_j}\right]$$

$$= f(x_0) \prod_{j=1}^{n} \frac{1}{x_0 - x_j}$$

$$+ f(x_1) \frac{1}{x_n - x_0}\left[\prod_{j \neq 1, j=0}^{n} \frac{x_1 - x_0}{x_1 - x_j} - \prod_{j \neq 1, j=0}^{n} \frac{x_1 - x_n}{x_1 - x_j}\right]$$

$$+ \cdots + f(x_n) \prod_{j=0}^{n-1} \frac{1}{x_n - x_j}$$

$$= f(x_0) \prod_{j=1}^{n} \frac{1}{x_0 - x_j} + f(x_1) \frac{1}{x_n - x_0} \prod_{j \neq 1, j=0}^{n} \frac{x_n - x_0}{x_1 - x_j}$$

$$+ \cdots + f(x_n) \prod_{j=0}^{n-1} \frac{1}{x_n - x_j}$$

$$= f(x_0) \prod_{j=1}^{n} \frac{1}{x_0 - x_j} + f(x_1) \prod_{j \neq 1, j=0}^{n} \frac{1}{x_1 - x_j}$$

$$+ \cdots + f(x_n) \prod_{j=0}^{n-1} \frac{1}{x_n - x_j}$$

$$= [x_0, \ldots, x_n] f,$$

by (1.3.1).   □

Next, we express f via its Newton form polynomial interpolation with error term. By equating the error term to the standard one variable polynomial interpolation error expression, we arrive at an expression for the nth divided difference in terms of the $(n+1)$th derivative of f.

Theorem 1.3.3. *Let f be a continuous function on $[a, b]$, then given $x_0, \ldots, x_n$ and x in the interval,*

$$f(x) = [x_0]f + (x - x_0)[x_0, x_1]f + (x - x_0)(x - x_1)[x_0, x_1, x_2]f$$

$$+ \cdots + (x - x_0)(x - x_1) \ldots (x - x_{n-1})[x_0, x_1, x_2, \ldots, x_n]f$$

$$+ E_n, \tag{1.3.4}$$

where

$$E_n = \prod_{i=0}^{n} (x - x_i)[x_0, x_1, x_2, \ldots, x_n, x]f. \tag{1.3.5}$$

Proof. We use (1.3.3) to derive

$$f(x) = [x]f = [x_0]f + (x - x_0)[x_0, x]f.$$

Repeating the same process with the last term on the right-hand side,

$$f(x) = [x]f = [x_0]f + (x - x_0)[x_0, x_1]f$$
$$+ (x - x_0)(x - x_1)[x_0, x_1, x]f.$$

Continuing, we arrive at (1.3.4) and (1.3.5). □

Corollary 1.3.2. *If f is $n+1$ times differentiable, then there exists ξ in $[a, b]$ with*

$$E_n = \frac{f^{(n+1)}(\xi)}{(n+1)!} \prod_{i=0}^{n} (x - x_i).$$

In particular,

$$[x_0, \ldots, x_n, x]f = \frac{f^{(n+1)}(\xi)}{(n+1)!}.$$

Proof. The result is an immediate consequence of Theorem 1.3.3 and the standard development of one variable polynomial interpolation (see for instance, Atkinson (1989). □

Our goal is to derive the two variable polynomial interpolation with error term. In the one variable case, the interpolation points needed to be only distinct. For a degree n interpolating polynomial, we needed $n+1$ interpolating points. For a polynomial that is degree n in two variables, there are $(n+1)(n+2)/2$ independent coefficients, which is also the number of points with integer entries enclosed in the right triangle with vertices $(0,0), (n,0)$, and $(0,n)$. This is the reasoning behind the terminology we introduce in the following definition.

Definition 1.3.2. Given n positive, the set $S_n^{\triangle} = \{(x_i, y_i) : i, j \geq 0, i + j \leq n\}$ is called a *triangular set of points*.

A triangular set always has $(n+1)(n+2)/2$ points. There is no reason that the points lie in any particular triangle. The reference is to the standard means of writing monomials $x^i y^j$ in a triangular array. In particular, we write $x^i y^j, i + j = n$, for $1, \ldots, n$ as

$$1$$
$$x^1, y^1$$
$$x^2, x^1 y^1, y^2$$
$$x^3, x^2 y^1, x^1 y^2, y^3$$

and so forth.

We now prove our main result for divided differences. A proof of the theorem along with related material may be found in Phillips (2003). In the statement of the theorem, we use the following notation, $\pi_0(x) = 1$, $\pi_m(x) = \prod_{k=0}^{m-1} x - x_k, m > 0$.

Theorem 1.3.4. *Let f be a real valued function defined on a subset of $\mathbb{R}^2$ including a triangular set S_n^Δ. For the points in S_n^Δ, we define*

(i) $p_0(x, y) = [x_0][y_0]f = f(x_0, y_0)$.

(ii) $p_m(x, y) = p_{m-1}(x, y)$
$\quad + \sum_{k=0}^{m} \pi_k(x) \pi_{m-k}(y)[x_0, \ldots, x_k][y_0, \ldots, y_{m-k}]f$.

(iii) $r_m(x, y) = \sum_{k=0}^{m} \pi_{k+1}(x) \pi_{m-k}(y)[x, x_0, \ldots, x_k][y_0, \ldots, y_{m-k}]f$
$\quad + \pi_{m+1}(y)[x][y_0, \ldots, y_m]f$.

Then for any $0 \le m \le n$,

(a) $f(x, y) = p_m(x, y) + r_m(x, y)$,

(b) p_m *interpolates f on S_n^Δ.*

Proof. We begin by verifying property a for each m. We proceed by induction. For $m = 0$,

$$p_0(x, y) + r_0(x, y)$$
$$= f(x_0, y_0) + r_0(x, y)$$
$$= f(x_0, y_0) + (x - x_0)[x_0][y_0]f + (y - y_0)[x][y, y_0]f.$$

By (1.3.3),

$$(x - x_0)[x, x_0][y_0]f = -[x_0][y_0]f + [x][y_0]f,$$
$$(y - y_0)[x][y, y_0]f = -[x][y_0]f + [x][y]f.$$

Therefore,

$$p_0(x,y) + r_0(x,y)$$

$$= f(x_0, y_0) - [x_0][y_0]f + [x][y_0]f - [x][y_0]f + [x][y]f$$

$$= f(x,y).$$

We have verified the identity for this case.

Suppose now that a holds for m. We set

$$S_1 = \sum_{k=0}^{m} \pi_{k+1}(x)\pi_{m-k}(y)[x_0, \ldots, x_{k+1}][y_0, \ldots, y_{m-k}]f,$$

$$S_2 = \sum_{k=0}^{m} \pi_{k+2}(x)\pi_{m-k}(y)[x, x_0, \ldots, x_{k+1}][y_0, \ldots, y_{m-k}]f$$

and compute using (1.3.3)

$$S_1 + S_2 = \sum_{k=0}^{m} \pi_{k+1}(x)\pi_{m-k}(y)[x_0, \ldots, x_{k+1}][y_0, \ldots, y_{m-k}]f$$

$$+ \sum_{k=0}^{m} \pi_{k+2}(x)\pi_{m-k}(y)[x, x_0, \ldots, x_{k+1}][y_0, \ldots, y_{m-k}]f$$

$$= \sum_{k=0}^{m} \pi_{k+1}(x)\pi_{m-k}(y)[x_0, \ldots, x_{k+1}][y_0, \ldots, y_{m-k}]f$$

$$+ \sum_{k=0}^{m} \pi_{k+1}(x)\pi_{m-k}(y)(x - x_{k+1})[x, x_0, \ldots, x_{k+1}]$$

$$\times [y_0, \ldots, y_{m-k}]f$$

$$= \sum_{k=0}^{m} \pi_{k+1}(x)\pi_{m-k}(y)[x_0, \ldots, x_{k+1}][y_0, \ldots, y_{m-k}]f$$

$$+ \sum_{k=0}^{m} \pi_{k+1}(x)\pi_{m-k}(y)[x, x_0, \ldots, x_k][y_0, \ldots, y_{m-k}]f$$

$$- \sum_{k=0}^{m} \pi_{k+1}(x)\pi_{m-k}(y)[x_0, \ldots, x_{k+1}][y_0, \ldots, y_{m-k}]f$$

$$= \sum_{k=0}^{m} \pi_{k+1}(x)\pi_{m-k}(y)[x, x_0, \ldots, x_k][y_0, \ldots, y_{m-k}]f,$$

which is the left-hand term in the expression for r_m. Continuing, we define three terms whose sum is the right-hand term in the expression for r_m.

$$T_1 = \pi_{m+1}(y)[x_0][y - 0, \ldots, y_{m+1}]f,$$
$$T_2 = \pi_1(x)\pi_{m+1}(y)[x, x_0][y_0, \ldots, y_{m+1}]f,$$
$$T_3 = \pi_{m+2}(y)[x][y, y_0, \ldots, y_{m+1}]f.$$

Again, we use (1.3.3).

$$\begin{aligned}
T_1 + T_2 + T_3 &= \pi_{m+1}(y)[x_0][y_0, \ldots, y_{m+1}]f \\
&\quad + \pi_1(x)\pi_{m+1}(y)[x, x_0][y_0, \ldots, y_{m+1}]f \\
&\quad + \pi_{m+2}(y)[x][y, y_0, \ldots, y_{m+1}]f \\
&= \pi_{m+1}(y)[x_0][y_0, \ldots, y_{m+1}]f \\
&\quad + \pi_1(x)\pi_{m+1}(y)[x, x_0][y_0, \ldots, y_{m+1}]f \\
&\quad + \pi_{m+1}(y)(y - y_{m+1})[x][y, y_0, \ldots, y_{m+1}]f \\
&= \pi_{m+1}(y)[x_0][y_0, \ldots, y_{m+1}]f \\
&\quad + \pi_{m+1}(y)(x - x_0)[x, x_0][y_0, \ldots, y_{m+1}]f \\
&\quad + \pi_{m+1}(y)(y - y_{m+1})[x][y, y_0, \ldots, y_{m+1}]f \\
&= \pi_{m+1}(y)[x_0][y_0, \ldots, y_{m+1}]f \\
&\quad + \pi_{m+1}(y)[x][y_0, \ldots, y_{m+1}]f \\
&\quad + \pi_{m+1}(y)(y - y_{m+2})[x][y, y_0, \ldots, y_{m+1}]f \\
&= \pi_{m+1}(y)[x_0][y_0, \ldots, y_{m+1}]f \\
&\quad + \pi_{m+1}(y)[x][y_0, \ldots, y_{m+1}]f \\
&\quad - \pi_{m+1}(y)[x_0][y_0, \ldots, y_{m+1}]f \\
&\quad + \pi_{m+1}(y)(y - y_{m+1})[x][y, y_0, \ldots, y_{m+1}]f \\
&= \pi_{m+1}(y)[x][y_0, \ldots, y_{m+1}]f \\
&\quad + \pi_{m+1}(y)(y - y_{m+1})[x][y, y_0, \ldots, y_{m+1}]f
\end{aligned}$$

$$= \pi_{m+1}(y)[x][y_0, \ldots, y_{m+1}]f$$
$$+ \pi_{m+1}(y)[x][y, y_0, \ldots, y_m]f$$
$$- \pi_{m+1}(y)[x][y_0, \ldots, y_{m+1}]f$$
$$= \pi_{m+1}(y)[x][y, y_0, \ldots, y_m]f.$$

Hence, $S_1 + S_2 + T_1 + T_2 + T_3 = r_m$. Moreover, replacing k with $k-1$ in S_1, we have

$$S_1 + T_1 = \sum_{k=0}^{m} \pi_{k+1}(x)\pi_{m-k}(y)[x_0, \ldots, x_{k+1}][y_0, \ldots, y_{m-k}]f$$

$$+ \pi_{m+1}(y)[x_0][y_0, \ldots, y_{m+1}]f$$

$$= \sum_{k=1}^{m+1} \pi_k(x)\pi_{m+1-k}(y)[x_0, \ldots, x_k][y_0, \ldots, y_{m+1-k}]f$$

$$+ \pi_{m+1}(y)[x_0][y_0, \ldots, y_{m+1}]f$$

$$= \sum_{k=0}^{m+1} \pi_k(x)\pi_{m+1-k}(y)[x_0, \ldots, x_k][y_0, \ldots, y_{m+1-k}]f$$

$$= p_{m+1} - p_m.$$

In turn, setting k to $k-1$ in S_2 and adding $T_2 + T_3$ yields

$$\sum_{k=0}^{m+1} \pi_{k+1}(x)\pi_{m-k+1}(y)[x, x_0, \ldots, x_k][y_0, \ldots, y_{m+1-k}]f$$

$$+ \pi_1(x)\pi_{m+1}(y)[x, x_0][y_0, \ldots, y_{m+1}]f$$

$$+ \pi_{m+2}(y)[x][y, y_0, \ldots, y_{m+1}]f$$

$$\sum_{k=1}^{m+1} \pi_{k+1}(x)\pi_{m-k+1}(y)[x, x_0, \ldots, x_k][y_0, \ldots, y_{m+1-k}]f$$

$$+ \pi_{m+2}(y)[x][y, y_0, \ldots, y_{m+1}]f = r_{m+1}.$$

Now, by induction

$$f = p_m + r_m = p_m + S_1 + S_2 + T_1 + T_2 + T_3$$
$$= p_m + S_1 + T_1 + S_2 + T_2 + T_3 = p_{m+1} + r_{m+1}.$$

This completes the verification of a.

To show that p_m interpolates f with respect to S_n^Δ, it suffices to verify that $r_m(x_i, y_i) = 0$ for each (x_i, y_j) in S_n^Δ. Looking at the kth summand of r_m, if $i \leq k$, then $\pi_{k+1}(x_i) = 0$. If, on the other hand, $i > k$, then $i + j \leq m$, implies that $j < (i - k) + j \leq m - k$. Hence, $\pi_{m-k}(y_j) = 0$. This completes the proof of the theorem. □

Next, we recast the remainder or error term into a more familiar form provided f is sufficiently differentiable. As the notation is somewhat cumbersome, we develop it now. By Corollary 1.3.2,

$$[x, x_0, \ldots, x_k][y_0, \ldots, y_{m-k}]f = \frac{\partial^{k+1}}{\partial x^{k+1}} \frac{\partial^{m-k}}{\partial y^{m-k}} f(\xi_{k,x}, \eta_{m-k})$$

$$[x][y_0, \ldots, y_m]f = \frac{\partial^m}{\partial y^m} f(\xi_x, y_{m,y}),$$

where $(\xi_{k,x}, \eta_{m-k})$ and $(\xi_x, y_{m,y})$ lie in the interior to the convex hull spanned by the points of S_n^Δ.

Corollary 1.3.3. *If f is sufficiently differentiable, then*

$$r_m(x, y) = \sum_{k=0}^{m} \prod_{i=0}^{k} (x - x_i)(y - y_{m-(i-1)})$$

$$\times \frac{1}{(k+1)!(m-k)!} \frac{\partial^{k+1}}{\partial x^{k+1}} \frac{\partial^{m-k}}{\partial y^{m-1}} f(\xi_{k,x}, \eta_{m-k})$$

$$+ (y - y_0) \ldots (y - y_m) \frac{1}{(m+1)!} \frac{\partial}{\partial x} \frac{\partial^m}{\partial y^m} f(\xi_x, y_{m,y}).$$

$$(1.3.6)$$

It is immediate that Corollary 1.3.3 is a direct generalization of the error term for one variable interpolation. The final task is to develop a means to compute the interpolating polynomial from the triangular set of points. To do this, we need to extend the idea of a Vandermonde matrix to the two variable case. Suppose that we want the degree 2 polynomial interpolation in two variables. There are six monomials of degree less than or equal to two: $1, x^1, y^1, x^2, x^1 y^1, y^2$. Take a six element triangular set $(x_0, y_0), (x_1, y_0), (x_2, y_0), (x_0, y_1), (x_1, y_1), (x_0, y_2)$

and then evaluate the monomials at the points. The result is the 6×6 degree 2 *Vandermonde matrix.*

$$A = \begin{pmatrix} 1 & x_0^1 & y_0^1 & x_0^2 & x_0^1 y_0^1 & y_0^2 \\ 1 & x_1^1 & y_0^1 & x_1^2 & x_1^1 y_0^1 & y_0^2 \\ 1 & x_2^1 & y_0^1 & x_2^2 & x_2^1 y_0^1 & y_0^2 \\ 1 & x_0^1 & y_1^1 & x_0^2 & x_0^1 y_1^1 & y_1^2 \\ 1 & x_1^1 & y_1^1 & x_1^2 & x_1^1 y_1^1 & y_1^2 \\ 1 & x_0^1 & y_2^1 & x_0^2 & x_0^1 y_2^1 & y_2^2 \end{pmatrix}.$$

We claim that the matrix A is non-singular. We begin by considering the x_i and y_i as independent indeterminates. Hence, the determinant of the matrix becomes a polynomial in these variables. If we subtract row 2 from row 1 and expand the determinant by cofactors on row 1, we see that $(x_0 - x_1)$ divides the determinant. Repeating the process for rows 3 and 1, we have that $(x_0 - x_1)(x_0 - x_2)$ divides the determinant. Continuing the process, we eventually have that

$$(x_1 - x_2)(x_0 - x_2)(x_0 - x_1)^2(y_1 - y_2)(y_0 - y_2)(y_0 - y_1)^2$$

divides the determinant. Next, we use a degree argument to prove that the determinant is

$$\alpha(x_1 - x_2)(x_0 - x_2)(x_0 - x_1)^2(y_1 - y_2)(y_0 - y_2)(y_0 - y_1)^2 \quad (1.3.7)$$

for real α. It now follows that the determinant is not zero on any triangular set of points.

Theorem 1.3.5. *The degree 2, two variable Vandermonde matrix determined by a triangular set of points is non-singular.*

Similar results hold for degree 1, degree 3 and so forth (see Exercises 3–5).

Theorem 1.3.6. *Suppose we are interpolating using a triangular set associated to polynomials of degree no larger than n. Let A be the Vandermonde matrix for the triangular set and e_k be the kth standard* *busis vector. Then vector a satisfying the linear system $Aa = e_k$ determines a polynomial dual to the kth point.*

Proof. With the given notation, a has $(n+1)(n+2)/2$ entries with an entry associated to each of the monomials. Hence, we may identify the entries of a as $a_{i,j}$ with $i+j \leq n$. Next, we write $p_k = \sum_{i+j\leq n} a_{i,j}x^i y^j$. Let (x_s, y_t) be the kth point in the triangular set, then since the coefficient vector of p_k is the solution to $Aa = e_k$, it follows that $p_k(x_i, y_j) = \delta_{s,i}\delta_{t,j}$, where $\delta_{s,i}$ denotes the Kronecker delta. $\square$

Corollary 1.3.4. *The polynomials $p_{i,j}, i + j \leq n$ form a basis for the degree no larger than n polynomials in two variables.*

Exercises

1. Prove Corollary 1.3.4.
2. Given a triangular set of points $(0,0)$, $(1,0)$, $(2,0)$, $(0,1)$, $(1,1)$, $(0,2)$ derive the six polynomials of degree no larger than three dual to these points.
3. Prove that the degree 1 Vandermonde matrix is non-singular.
4. Derive an expression analogous to (1.3.7) for the degree 3 Vandermonde matrix. Prove that the matrix is non-singular on a triangular set of points.
5. Derive an expression analogous to (1.3.7) for the degree n Vandermonde matrix. Prove that the matrix is non-singular on a triangular set of points.
6. Extend the definition for two point Gaussian quadrature to a rectangle $[a, b] \times [c, d]$. Suppose that f is defined on the rectangle and set

$$\int_b^d \int_a^c f\,dy\,dx = \frac{1}{4}\sum_{i,j=1}^2 f(x_i, y_j),$$

where x_i, x_2 are the x-direction Gaussian points and y_1, y_2 are the corresponding points in the y-direction. Prove that the expression for the integral is exact for polynomials of degree no larger than 2 in either variable.

1.4. Numerical Integration for a Rectangle and a Triangle

In this section, we consider two cases of numerical integration that arise in FEM. There are opportunities to use these techniques in Chapters 3 and 6. FEM always requires integration and these integrals need to be carried out hundreds or thousands of times. Hence, it is imperative that we have an efficient procedure to calculate integrals of the type that occur in FEM. The problem is that a general purpose mathematics package that is available in a 4GL such a *Mathematica* is designed to handle all cases. Therefore, it may be too slow for the particular cases that interest us. In response, we develop two simple ideas that will be very efficient for our purpose. To illustrate these ideas, we restrict our attention to low degree polynomials.

Of course there are commercially available FEM packages as stand alone applications and as part of a product life cycle system. In addition, there are open source FEM systems especially for particular equations such as the Navier–Stokes equation. These programs do well for the cases envisioned by their designers. However, in any research setting, we will always push the envelope to its limits and will always need researchers capable of modifying our existing systems.

For a PDE with constant coefficients in $\mathbb{R}^2$, there are four integrals of concern.

$$\int_E \varphi_1 \varphi_2, \int_E \frac{\partial \varphi_1}{\partial x} \varphi_2, \int_E \frac{\partial \varphi_1}{\partial y} \varphi_2, \int_E \nabla \varphi_1 \cdot \nabla \varphi_2, \qquad (1.4.1)$$

where E is a rectangle or a triangle and the φ_i are polynomials in two variables. The same basic ideas apply to PDE with polynomial coefficients.

In the first case, we consider the integral over a rectangle. In this setting, we have a rectangle with lower left vertex (α, β) and upper right vertex (γ, δ). The four Lagrange polynomials that are dual to the vertices are

$$N_{\alpha,\beta}(x,y) = \frac{\gamma-x}{\gamma-\alpha}\frac{\delta-y}{\delta-\beta}, \quad N_{\gamma,\beta}(x,y) = \frac{x-\alpha}{\gamma-\alpha}\frac{\delta-y}{\delta-\beta},$$

$$N_{\gamma,\delta}(x,y) = \frac{x-\alpha}{\gamma-\alpha}\frac{y-\beta}{\delta-\beta}, \quad N_{\alpha,\delta}(x,y) = \frac{\gamma-x}{\gamma-\alpha}\frac{y-\beta}{\delta-\beta}. \qquad (1.4.2)$$

We want an extremely fast means to calculate the integrals in (1.4.1), where the φ_i are Lagrange polynomials.

In this case, we suggest Gaussian quadrature. Since the polynomials are no more than degree 2 in each variable, then two point quadrature over a rectangle will exactly compute the integral of each case listed in (1.4.1). In particular,

$$\int_{\alpha}^{\gamma}\int_{\beta}^{\delta} N_{\alpha,\beta}(x,y)N_{\gamma,\beta}(x,y)dydx = \sum_{i,j=1}^{2} N_{\alpha,\beta}(\xi_i,\eta_j)N_{\gamma,\beta}(\xi_i,\eta_j),$$

where $\xi_i = (\gamma-\alpha)(\zeta_i+1)/2+\alpha, \eta_i = (\delta-\beta)(\zeta_i+1)/2+\beta, \eta_i = \pm\sqrt{1/3}$. For the case of Hermite cubics, we would use 3 point quadrature.

In the case of triangles, there are numerical integration techniques similar to Gaussian quadrature (see Lyness and Jespersen (1975)). It is our experience that the results from these procedures are not satisfactory for FEM. Instead, consider two alternatives.

First, there is a transformation from a rectangle to a triangle which is bijective on the interior of the two domains. For instance, for the standard rectangle $[-1,1] \times [-1,1]$ and the reference triangle with vertices $(0,0),(1,0),(0,1)$, the transformations are

$$\Phi(x,y) = (\xi,\eta) = \left(\frac{(1+x)(1-y)}{4}, \frac{1+y}{2}\right)$$

and

$$\Phi^{-1}(\xi,\eta) = \left(\frac{2\xi}{1-\eta}, 2\eta-1\right).$$

In addition, given a triangle E with vertices $(\alpha,\beta),(\gamma,\delta),(\mu,\nu)$, the basic affine mapping $A(x,y) = T(x,y)+(\alpha,\beta)$, where T is the linear transformation given by the matrix,

$$\begin{pmatrix} \gamma - \alpha & \mu - \alpha \\ \delta - \beta & \nu - \beta \end{pmatrix},$$

will map the reference triangle to the triangle E. Hence, by combining A and Φ, we have a bijection from the standard rectangle to any triangle. Therefore, employing the usual varible change procedure [Marsden and Tromba (2003)], we are able rewrite

$$\int_E f(\xi, \eta) dA = \int_{-1}^{1} \int_{-1}^{1} f \circ A \circ \Phi(x, y) J(A \circ \Phi) dx dy.$$

Now, we employ the rectangle quadrature techniques to the integral on the right-hand side.

The procedure just described is so commonly implemented that it could be described as the standard. In the remainder of this section, we describe a procedure that is actually faster. It requires the development of an inhouse infrastructure that includes the common cases. However, it would not be practical for spectral methods with large degree polynomials.

We next look at an altenative. Consider again a triangle E with vertices (α, β), (γ, δ), (μ, ν). Additionally, we have the reference triangle R with vertices $(0,0), (1,0)$, and $(0,1)$ and the affine map $A(x, y) = T(x, y) + (\alpha, \beta)$. If p is a polynomial defined on R, then $\varphi = p \circ A^{-1}$ is a polynomial on the element triangle. If we start with the three degree one basis polynomials p_i dual to the vertices of R, then each $\varphi_i = p_i \circ A^{-1}$ is dual to the vertices of E. Looking back at (1.4.1), this choice results in a total of 24 distinct integrals. Note that each of the 24 can be resolved as rational functions in the six parameters $\alpha, \beta, \gamma, \delta, \mu, \nu$. For instance,

$$p_1(x, y) = 1 - x - y, \quad p_3(x, y) = y$$

are dual to the first and third vertices of R. The corresponding polynomials on E are

$$\varphi_1(x, y) = \frac{1}{det(A)}[(\delta - \gamma)x + (\mu - \gamma)y + \gamma\nu - \delta\mu],$$

$$\varphi_3(x, y) = \frac{1}{det(A)}[(\beta - \delta)x + (\gamma - \alpha)y + \alpha\delta - \beta\gamma].$$

In the first case, the change of variables yields

$$\int_E \varphi_1(x,y)\varphi_3(x,y) = \int_E p_1 A^{-1}(x,y)p_3 A^{-1}(x,y)dxdy$$

$$= |det(T)| \int_0^1 \int_0^{1-y} p_1 p_3 dxdy = \frac{|det(T)|}{24}.$$

For the second case, we start with

$$\frac{\partial \varphi_1}{\partial x} = \frac{\delta}{|det(T)|}.$$

Hence,

$$\int_E \frac{\partial \varphi_1}{\partial x}(x,y)\varphi_3(x,y)dxdy = \int_0^1 \int_0^{1-y} \frac{\partial \varphi_1}{\partial x} A(x,y)p_3(x,y)J(A)dxdy$$

$$= \int_0^1 \int_0^{y-1} y(\nu-\delta)dxdy = \frac{\nu-\delta}{6}.$$

As the third case is similar, we move onto the fourth.

$$\int_E \nabla\varphi_1(x,y)\cdot\nabla\varphi_3(x,y)dxdy = \frac{(\alpha-\gamma)(\gamma-\mu))+(\beta-\delta)(\gamma-\nu)}{2|det(T)|}.$$

$$(1.4.3)$$

Suppose we select a representative triangle with vertices (0.06, −0.3), (0.46, 0.02), (−0.02, 0.3) and compute the integral as on the left-hand side of (1.4.3), and alternatively we calculate the integral by directly evaluating the expression on the right-hand side. The resulting CPU time will vary by triangle and computer. From our test of this case, we found that the right-hand side was more than 780 times faster.

For higher degree polynomials, the differences are even greater. Consider the degree 2 model of polynomials dual to the three vertices and three edge midpoints. Again, we start with polynomials dual to the first and third vertices.

$$p_1(x,y) = 1 - 3x + 2x^2 - 3y + 6y^2, \quad p_3(x,y) = -y + 2y^2.$$

For the first integral in (1.4.1), the fast alternative is 30,000 times faster.

Certainly, we are suggesting a very tedious alternative to the usual integral. But once done and rendered as a look up table that can be copied into any program, there are only benefits. To generate such a table for degrees 1, 2, or 3 polynomials is a doable task as a semi-automated procedure. For polynomials of the size encountered with spectral methods, the process must be fully automated. Fortunately, *Mathematica* provides a symbolic programming platform that would make this task possible.

Exercises

1. For the degree 2 polynomials, p_1 and p_3 use Mathematica to compute φ_1 and φ_3 for a general triangle with vertices $(\alpha, \beta), (\gamma, \delta), (\mu, \nu)$. Next, compute $\nabla\varphi_1 \cdot \nabla\varphi_3$ and then using variable change operators, compute the integral of $\nabla\varphi_1 \cdot \nabla\varphi_3$ over a general triangle as a rational function of the triangle vertices.

2. Select the element triangle with vertices at $(1, 1), (1.1, 1.2)$, and $(1.07, 1.04)$. Compute the integral of $\nabla\varphi_1 \cdot \nabla\varphi_3$ using standard integral procedures and using the method introduced above and completed in Exercise 1. In effect, test the effectiveness of the fast procedure for this case.

Chapter 2

Problems with Closed Form Solution

Introduction

In this chapter, we present two problems which give rise to the same differential equation. One represents heat diffusion and the other fluid flow about an Joukowski airfoil. The common underlying differential equation is $\partial u / \partial t = \alpha \nabla^2 u$, where $u(t, x)$ is a scalar valued function of a spatial independent variable x and a time variable t. In each case, there is a closed form solution for certain cases.

Settings where a differential equation has a known solution are rare. These cases are important to the numerical analyst as they provide a means to measure the correctness of the numerical solution. Of course, the estimation procedures are intended for cases without known solution. Furthermore, the numerical procedures are verified by mathematical theory. Nevertheless, comparing the actual to the estimated does help to develop intuition. We do this in this chapter by implementing three primary FDM techniques, explicit, implicit and Crank–Nicolson and measuring their output against the actual data.

The first application, flow of an ideal fluid about an airfoil, provides several additional opportunities. Mathematica has exceptional symbolic computational capabilities. In that regard, it is special as a mathematical 4GL. In this chapter, we have the opportunity to exercise these capabilities. Additionally, drawing the flow streamlines

is an interesting visualization problem. However, the techniques
developed here require that the flow field is unbounded and the object
is smooth. Later, these limitations lead us to introduce finite element
method as a preferred technique for fluid dynamics. At that time, we
will find the experience with this case to be helpful.

The heat flow problem in one dimension has an important consequence. It is equivalent to the Black–Scholes equation for pricing
stock market options. Hence, the work we do now will be applicable to options pricing in Chapter 4. Further, the heat equation is an
excellent vehicle for introducing the FDM. In this regard, we see that
the method produces reasonable or unreasonable results depending
on the size of the time step. This leads us to the idea of stability
and Neumann stability analysis. And the latter leads in turn to the
discrete Fourier transform. This version of the Fourier transform is
the primary tool for dealing with stability. Nevertheless, there are
other procedures. These are usually based on the matrix norm.

Nearly all transient processes handle the time variable via FDM.
In this chapter, we also use FDM for the spatial variable. However, no
matter how the spatial variable is realized, time stepping is usually
FDM. Hence, the issues that arise here will come up repeatedly in
subsequent chapters.

The mathematical foundation for the material presented in this
chapter lies mainly in complex analysis. We do not develop this supporting theory. Rather, we do identify specific theorems necessary to
support the application and point the reader to references.

In Section 1, we present ideal fluid flow and the Laplace equation. In Section 2, we derive the heat equation from the conservation
law for heat energy and the temperature. We next apply the Fourier
transform to the equation to arrive at the Green's function. For the
one-dimensional case, a complete solution is possible in restricted
cases. In addition, we develop FDM and stability in one dimension.
In Section 3, we introduce FDM for the multidimensional heat equation and extend the one-dimensional stability theory. In this section, we begin the discussion of consistency and convergence. These
concepts reappear when we do the Lax Equivalence Theorem in
Chapter 7.

2.1. Flow about an Airfoil: Harmonic Conjugates and Analytic Functions

As our first detailed example, we consider the following problem. Describe the flow about an Joukowski airfoil. The mathematical model for this problem will take us into complex analysis. We begin by developing the terminology, stating the necessary theorems and narrowing down the question, *describe the flow*.

The first step is to draw an airfoil. We begin with the unit circle, $\delta(t) = (\cos(t), \sin(t))$ where t is the interval $[0, 2\pi]$. Next, we offset the circle so that its center is in the second quadrant, $\beta(t) = (\cos(t) + \Delta x, \sin(t) + \Delta y)$. For instance, we take $\Delta x = -0.1$ and $\Delta y = 0.05$. Additionally, we expand the translated circle so that it passes through the point $(1, 0)$ on the x-axis. The resulting curve is given by $\alpha(t) = d\delta(t) + (\Delta x, \Delta y)$ where d is the distance from $(1, 0)$ to $(\Delta x, \Delta y)$. Figure 2.1.1 shows the unit circle and the circle $\alpha(t)$.

If we identify $\mathbb{R}^2$ with the complex plane $\mathbb{C}$ via the correspondence $(x, y) = x + \mathbf{i}y$, then we can write α as a function taking complex values, $\alpha(t) = (d\cos(t) + \Delta x) + \mathbf{i}(d\sin(t) + \Delta y)$. Next, we compose α with the complex function $f(z) = z + 1/z$ to get $\gamma(t) = f(\alpha(t))$ and plot (see Figure 1.1.2). The resulting shape is called a *first-order Joukowski airfoil*. Changes to Δx and Δy will change the shape of the airfoil. Hence, these values are called shape parameters and the first-order airfoil is considered a 2-parameter shape. There is another and more complicated construction due also to Joukowski which depends on 5 shape parameters. This is called the *second-order Joukowski airfoil* (see Jones (1990)).

Now, we restate the problem. We want to describe the flow about the first-order airfoil given by $\gamma(t)$. But what do we mean by describe 'the flow'? In other words, we need first to state the actual physical problem and then carefully translate that into a mathematical problem that we can resolve. Now, we are looking at a cross-section of an aircraft wing. In other words, the setting is the flow of air about a wing. Next, at each location, the air particles should have well-defined angular velocity. In particular, the setting determines a vector field. Hence, describing the flow should require that we describe this vector

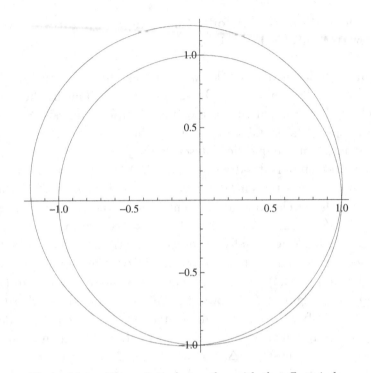

Figure 2.1.1.   The unit circle together with the offset circle.

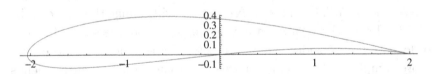

Figure 2.1.2.   The Joukowski airfoil with shape parameters −0.1 and 0.05.

field. Further, we might like to know the path of a particle moving through the system. Such a path should have tangent everywhere parallel to the underlying vector field. These paths are called streamlines. We would like to show the streamlines. Finally, we would like to be able to calculate the lift and drag. That is, determine if the shape would fly.

What assumptions would be reasonable? We will first assume that the flow is laminar.That is, that the flow occurs in parallel planes orthogonal to the axis of the wing. This assumption amounts to reducing the problem from 3D to 2D. We assume the fluid is inviscid, that particle motion is not affected by neighboring particles.

Now, we write the vector field as a function $u : D \to \mathbb{R}^2$, $u = u(x,y) = (u_1(x,y), u_2(x,y))$, where D denotes the region of the plane exterior to γ and including γ. Next, we should assume that the vector field is continuous. This amounts to the statement that there are no abrupt changes in the vector field as we move about the region, D. In addition, we assume that the flow is steady. This means that the flow far from the airfoil should approach a constant value called the ambient flow. The vectors for the ambient flow would be parallel to the x-axis and have constant length.

We need two more assumptions, which are more subtle than the laminar, continuity and steady-state assumptions. First, even though we know that air is compressible, it is incompressible under the temperatures and pressures experienced during normal flight. We can express this experimentally verified fact via a line integral. Suppose that δ is a smooth, connected curve in D whose interior also lies in D, $\delta(t) = (\delta_1(t), \delta_2(t))$, $t \in [a, b]$ with tangent vector $\delta'(t)$ and normal $n(t)$. Incompressibility may be expressed via the following identity:

$$0 = \int_a^b u \circ \delta(t) \cdot n(t) \|\delta'(t)\| dt = \int_0^1 u \circ \delta(s) \cdot n ds. \qquad (2.1.1)$$

The second, more compact notation arises by parameterizing the curve via arc length. The next assumption is called irrotational and is also defined as a line integral.

$$0 = \int_a^b u \circ \delta(t) \cdot \delta'(t) \|\delta'(t)\| dt = \int_0^1 u \circ \delta(s) \cdot \delta'(s) ds. \qquad (2.1.2)$$

Intuitively, irrotational may be described as follows. The expression $(u \circ \delta)$ denotes the vector field restricted to the curve. Now, the dot product gives the cosine of the angle between the tangent to δ and the vector field restricted to δ. The statement that the integral along δ is zero means that the angle between the vectors does not rotate through 2π radians as you move along the curve. Alternatively, Eqs. (2.1.1) and (2.1.2) may be derived by assuming that u is

divergence free and curl free. Most multivariate calculus books will
have these derivations. Before continuing, note that divergence free
and curl free are conservation laws.

We are now ready to advance the mathematics. First, select the
point on the airfoil where the curve meets the positive y-axis and
extend a half line vertically along the y-axis. Next, we fix a point
(x_0, y_0) in D and define for (x, y) in D a real valued function φ by
means of

$$\varphi(x, y) = \int_\delta u \circ \delta(s) \cdot \delta'(s) ds, \qquad (2.1.3)$$

where δ represents a smooth curve in D connecting (x_0, y_0) to (x, y)
and not crossing the half line. Note that (2.1.2) shows that the definition of φ is independent of the choice of curve δ. Indeed, given
any other curve $\bar{\delta}$ connecting the two points and not crossing the
half line, set α equal to the curve formed by a point traveling from
(x_0, y_0) to (x, y) along δ and then returning to (x_0, y_0) along $\bar{\delta}$. Since
the interior of the airfoil is not included inside α, (2.1.2) states
that $0 = \int_0^1 u \circ \alpha(t) \cdot \alpha' dt$. In particular, $\int_\delta u \circ \delta(s) \cdot \delta'(s) ds = \int_{\bar{\delta}} u \circ \bar{\delta}(s) \cdot \bar{\delta}'(s) ds$.

Now exactly as in calculus, we can prove (see the Fundamental
Theorem of Calculus in most any first semester calculus book) that
φ has partial derivatives and that

$$\frac{\partial \varphi}{\partial x} = u_1, \quad \frac{\partial \varphi}{\partial y} = u_2. \qquad (2.1.4)$$

Note that the partial derivatives of φ are the components of the
vector field. Similarly, we define another real valued function ψ,

$$\psi(x, y) = \int_0^1 u \circ \delta \cdot n ds. \qquad (2.1.5)$$

Again, (2.1.5) determines a well-defined function and now

$$\frac{\partial \psi}{\partial x} = -u_2, \quad \frac{\partial \psi}{\partial y} = u_1. \qquad (2.1.6)$$

Next, we define a function from the complex plane to the complex
plane via $F(z) = \varphi(x, y) - i\psi(x, y)$, where $z = x + iy$. In this situation,

we call φ the real part of F (denoted $\operatorname{Re}[F]$) and ψ the imaginary part of F (denoted $\operatorname{Im}[F]$). Formulas (2.1.5) and (2.1.6) show that the real and imaginary parts of F have partial derivatives which satisfy the Cauchy Riemann equations. Further, since the vector field is continuous, then φ and ψ have continuous partial derivatives. Hence, (by theorems of Complex Analysis) F is complex differentiable (analytic) on D and u has continuous partial derivatives on D. Furthermore, the gradient of φ satisfies $\nabla\varphi = u$.

But even more is true. Again from (2.1.5) and (2.1.6), $\nabla\varphi$ is orthogonal to $\nabla\psi$. We know from calculus that the level curves for ψ ($\psi(x,y) = constant$) are orthogonal to the gradient $\nabla\psi$. Hence, the level curves for ψ are parallel to $\nabla\varphi = u$. Since the streamlines for u are the level curves for ψ, then the question of determining the flow about the airfoil leads us to determining F. Indeed, the flow field is the gradient of the real part of F and the streamlines are the level curves of the imaginary part of F. The technical term for F is the *complex potential of the flow*.

Before moving on, we make the observation that with (2.1.5) and (2.1.6), we can prove that φ and ψ satisfy the *Laplace equation*, $\partial^2\varphi/\partial x^2 + \partial^2\varphi/\partial y^2 = 0$. Real valued functions which satisfy Laplace and combine to form an analytic function are called harmonic pairs. The Laplace equation will not be needed here, but will play an important role when we return to ideal fluid flow in a later chapter.

The theory of analytic functions provides the answer to describing the flow about the airfoil. First, we should expect that the surface of the airfoil be a streamline. Stated otherwise, the flow at the airfoil should be tangent to the airfoil. Hence, the imaginary part of F should be constant on γ. Recall that γ is the result of first applying the function $h(x,y) = d(x,y) + (\Delta x, \Delta y)$ to the unit circle and then applying $f(x,y) = (x + x/(x^2 + y^2), y - y/(x^2 + y^2))$ to the result. Hence, $h^{-1}f^{-1}$ maps the airfoil to the unit circle. In addition, $h^{-1}f^{-1}$ maps the exterior of the airfoil to the exterior of the unit circle. Furthermore, the imaginary part of f, $\operatorname{Im}(f) = y - y/(x^2 + y^2)$ when applied to the unit circle is zero. So, we have it. Set $F = f(h^{-1})f^{-1}$.

By theorems of Complex Analysis, there is a unique analytic function, $F(z) = \varphi(x, y) - \mathbf{i}\psi(x, y)$ on D with ψ constant on the airfoil, γ, provided we stipulate the boundary values, the constant value ψ takes on γ, the minimum value of $|\varphi|$ on D (which in fact occurs on γ) and the value of $\lim_{z \to \infty} u(z)$, the ambient flow. Therefore, the function just demonstrated is in effect the only complex potential for the flow.

At this point, we pause to consider what we have done. We began with a question which we restated mathematically. Then, the mathematical statements were developed using known mathematical theory to yield a complex potential function $F = \varphi + \mathbf{i}\psi$ with the flow field equal to $\nabla\varphi$ and streamlines given by $\psi = constant$. Hence, F is a mathematical model for the physical problem. However, the model does not exactly conform to the physical setting. This comes up later. Right now, we have the problem of writing down φ and ψ so that we can calculate the vector field and the streamlines. And this is no easy matter.

If $w = f(z) = z + 1/z$, then $wz = z^2 + 1$ or $0 = z^2 - wz + 1$, and solving for z and changing the variable name yields

$$f^{-1}(z) = \frac{z \pm \sqrt{z^2 - 4}}{2}. \tag{2.1.7}$$

It is easier to see that setting $\Delta z = (\Delta x, \Delta y)$,

$$h^{-1}(z) = \frac{z - \Delta z}{d}.$$

This gives us a formula for F (using the positive branch on the square root)

$$F(z) = \frac{z + \sqrt{z^2 - 4} - 2\Delta z}{2d} + \frac{2d}{z + \sqrt{z^2 - 4} - 2\Delta z}.$$

We must now try to find the real and imaginary parts of F. Actually, this is an important question, important enough that two mathematicians, Theodorsen and Garrick, working for NACA, the National Advisory Committee for Aerodynamics (the predecessor US government agency to NASA) worked this out in the 1930s. They did this using some very elaborate trigonometric arguments (see Theodorsen (1932), Theodorsen *et al.* (1934)). Fortunately, *Mathematica* will compute φ and ψ for us. Applying the function ComplexExpand to

F will yield the desired representation for F decomposed into a real and imaginary part. We leave this as an exercise for the student. Even though the expression is ugly, you can use *Mathematica* to compute $\text{Im}(F) = constant$. And subsequenly display the streamlines. Unfortunately, the expression will not differentiate. See Exercise 4 to compute the flow field.

However, the plot of $\text{Im}(F) = constant$ does not give the streamlines that we expect. First of all, there are two functions F, one for the positive square root in (2.1.7) and one for the negative root. Using a positive constant, one choice in (2.1.7) will plot the left-hand side of a streamline above the x-axis, while the other choice will draw the right-hand side of the same streamline. Setting the constant to a negative yields a streamline below the airfoil.

From Complex Analysis, we know that when using the square root function, you must introduce a cut in the complex plane and not allow curves to cross the cut. In this case, the cut lies on the y-axis. The streamlines on either side of the cut do not meet at a point on the y-axis, rather the two branches that we described meet at ∞. As they cross the y-axis, they actually diverge and describe something that bares no resemblance to the streamline we expect. In particular, the mathematical model diverges significantly from the physics. In Figure 2.1.3, we demonstrate this by extending a left side streamline extended past the y-axis.

This brings us to visualization. We need to show a picture of streamlines that corresponds to physics, not the mathematical model. We need to 'fix' the picture. We fix the picture by drawing the streamline for values $x < -\epsilon$ and $x > \epsilon$ for a very small $\epsilon > 0$. Then, we connect the two separate branches with a straight line segment.

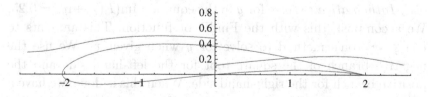

Figure 2.1.3. The left side streamline for $y = 0.5$ extended past $x = 0.2$.

We have mentioned lift and drag but have not defined them. If (α, β) is a tangent vector on the airfoil, $u(\gamma(t)) = (\alpha, \beta)$, then $(-\beta, \alpha)$ is the normal at the location and the lift is given by α, the y-component of the normal and the drag by $-\beta$, the x-coordinate at the location. The lift for the airfoil is the sum (integral) of the lift along the airfoil. The lack of symmetry ensures that the lift for the airfoil is positive. However, these sums should only be carried out for about 60% of the airfoil. After that, the flow has separated creating a vacuum between the flow and airfoil which fills as an area of turbulence. This fact increases the lift and the drag since the rear of the airfoil generates downward force (negative lift) and forward force (negative drag). Determine the 60% point on the upper and lower surface of the shape by locating a point 60% along the line joining the nose and the tail of the airfoil (diameter). Then, construct the perpendicular to the diameter at that location and identify where this line intersects the airfoil.

The flow separation is caused by inertia. The Laplace equation does not model inertia as this effect is related to viscosity, the tendency of fluid particles to affect the flow of neighboring particles. For a fluid flow model that includes the effect, see the Navier–Stokes equation in Chapter 6.

Before concluding this section, we remark that there is another way to locate points along the streamline without resolving F in terms of its real and imaginary parts. This procedure is important as the second-order Joukowski airfoil gives rise to a function with a square root within a square root. In this case, Mathematica fails to give us the decomposition.

Suppose we are looking for points along the streamline $\text{Im}[F(x + iy)] = 0.25$ with $-3 < x < 3$. First, construct a sequence $-3 = x_0 < x_1 < \cdots < x_{300} = 3$, where $x_j - x_{j-1} = 0.01$. Now, for each x_j, we can use *Mathematica* to solve for y in the equation $\text{Im}(F(x_j + iy) = 0.25$. We accomplish this with the FindRoot function. This amounts to using Newton's method to solve for y when given x_j. We use the negative branch of the square root for the left-hand side and the positive branch for the right-hand side. When this is done, we have a sequence of points along the streamline. We complete the streamline

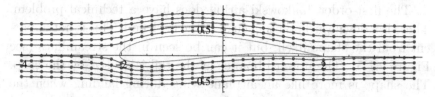

Figure 2.1.4. Airfoil with streamlines rendered with B-splines with guide points.

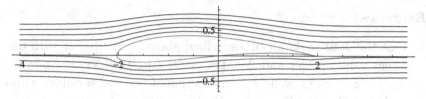

Figure 2.1.5. Airfoil with streamlines rendered with B-splines without guide points.

by using the calculated points as guide points for a B-Spline. As in the earlier discussion, we will do no calculation at $x_j = 0$. Instead, we connect the left hand B-Spline to the right hand B-Spline with a line segment. Now, we know that the B-Spline does not pass through the calculated points. However, it passes near enough that the resulting curve works very well as a visualization. Figure 2.1.4 shows the streamlines around a first-order Joukowski airfoil. It was constructed using the technique just described. You need to look very closely to see that the B-Splines do not pass through the computed streamline points.

Figure 2.1.5 is the same diagram, this time we only show the B-splines.

We turn to techniques to calculate the flow field, tangents, normals, lift and drag for the first-order Joukowski airfoil. In the case of the flow field, the expression provided by Mathematica for $\varphi = Re[F]$ will not differentiate. In Exercise 4, we show how to use finite differences to display the flow field. We may use similar techniques to get flow vectors on the surface of the airfoil. In turn, we calculate the normals, then the lift and drag vectors.

The first-order Joukowski airfoil does have a technical problem. There is a downward pointing cusp at the back end. This is not apparent from the shape itself, but it can be seen in the streamlines (see Figure 2.1.5). The upward bounce at the tail is caused by the cusp. The shape is more like an airplane wing during landing when the flaps are extended and pointed downward. This problem is resolved with the second-order airfoil.

Exercises

1. Plot the Joukowski airfoil described above. Experiment with different values of Δx and Δy. Try $\Delta x = \Delta y$. Try $|\Delta x| < \Delta y$.
2. Setting $z = x + iy$, write F as a function of two real variables x and y. Apply ComplexExpand to Re$[F]$ and Im$[F]$ to get expressions for φ and ψ. Do this for each branch of the square root.
3. Use ContourPlot to draw a streamline (level curves of ψ) above and below the airfoil. Include the airfoil in your image. Keep in mind that you will have to fix the streamlines near the y-axis.
4. Draw the vector field.

   a. Make a rectangular mesh (list) of points (x, y) lying outside the airfoil and inside the rectangle with vertices $(-3, -3)$, $(3, -3)$, $(3, 3)$, and $(-3, 3)$. So that the image is not too busy, we want the distance between horizontally and vertically adjacent points to be 0.2. Since f^{-1} maps the outside of the airfoil to the outside of the offset circle, then select the pairs (x, y) with $|f^{-1}(x + iy) - (\Delta x + i\Delta y)| \geq d$.
   b. Approximate $\nabla \varphi$ via the extended central difference finite. For each (x, y) in the list, compute

   $$\nabla \varphi(x, y) = \nabla \varphi(x, y)$$
   $$= \frac{1}{0.1}(-\varphi(x + 0.2, y) + 8\varphi(x + 0.1, y)$$
   $$- 8\varphi(x - 0.1, y) + \varphi(x - 0.2, y),$$
   $$- \varphi(x, y + 0.2) + 8\varphi(x, y + 0.1)$$
   $$- 8\varphi(x, y - 0.1) + \varphi(x, y - 0.2)).$$

This version of the central difference converges $O(\Delta h^4)$. It is often used for situations like this. In Part 2, we develop the extended differences.

c. Draw the vector field using the Arrow function in *Mathematica*.

d. If the vectors are too short or too long, the viewer will have difficulty identifying the flow. In this case, uniformly shorten or lengthen the arrows to ameliorate the problem.

2.2. Temperature Distribution along a Rod, the Derivation and Solution of the 1D Heat Equation

The 1D heat equation provides an opportunity to see the process described in the introduction realized in an actual setting. In this section, we will see the development of the mathematical model, derivation of the solution to the PDE for restricted cases, implementation of a numerical method (FDM) to resolve a broader class of cases. We save stability analysis for the next section and convergence for Chapter 7.

The 1D heat equation resolves the temperature distribution along a narrow rod. For particular boundary values, it has a closed form solution and so needs no numerical technique. However, this particular setting will allow us to introduce FDM on the one hand while on the other it will give us the opportunity to compare the FDM generated output to the actual data. As there are several FDM variations, we use the comparison process to compare output for one variation against another. In the following section, we will see how to extend the ideas developed here to the two-dimensional case. Later, we will see that this equation is equivalent to the Black–Scholes equation for pricing market futures. Any work we do here will prove useful later when we consider some basic questions of financial mathematics.

In this section, there are three distinct events, the heat equation derivation, the derivation of the closed form solution, the development of the FDM solution (explicit, implicit, and Crank–Nicolson).

We begin with the derivation of the one-dimensional Heat Equation. First, recall that heat energy drives the small scale motion of micro particles in a mass (Brownian motion). The macroscopic real-ization is temperature. Suppose we have a rod of small circular cross-section. Suppose further that the rod is insulated along its length so that heat is lost or gained only at the end points.

Let $u(t, x)$ represent the temperature at time equal to t and loca-tion x. Further let A denote the area of the circular cross-section. Further, we identify a location on the rod as x_0 and a narrow seg-ment at x_0 with width Δx. With this notation, we make the following two assumptions.

(A) The change in thermal energy, denoted ΔH, across a thin slice is proportional to the mass of the slice times the change in temperature. In particular,

$$\Delta H = c(\rho(A\Delta x))\Delta u,$$

where the mass is density, ρ, times volume $A\Delta x$, and c, the constant, is called the specific heat of the object.

(B) The rate of change of thermal energy across either face of the slice is proportional to the area of the face times the rate of change of temperature at the face. For the face at $x = x_0$,

$$\frac{\Delta H}{\Delta t} = kA\frac{\partial u}{\partial x}\big|_{x_0},$$

where k is the heat conductivity constant.

Dividing the equation in (A) by Δt yields a single equation with-out reference to ΔH. The second term on the right reflects the fact that the small volume in (A) has two faces.

$$c\rho(A\Delta x)\frac{\Delta u}{\Delta t} = kA\frac{\partial u}{\partial x}\big|_{x_0} - kA\frac{\partial u}{\partial x}\big|_{x_0+\Delta x}.$$

Setting $\alpha = k/c\rho$ and dividing through by $A\Delta x$ yields

$$\frac{\Delta u}{\Delta t} = \alpha\frac{1}{\Delta x}\left(\frac{\partial u}{\partial x}\big|_{x_0} - \frac{\partial u}{\partial x}\big|_{x_0+\Delta x}\right).$$

Taking the limit as $\Delta t \to 0$ and $\Delta x \to 0$ gives the usual 1D heat equation,

$$\frac{\partial u}{\partial t} = \alpha\frac{\partial^2 u}{\partial x^2}. \tag{2.2.1}$$

We now demonstrate the solution to (2.2.1). For this purpose, we make a digression into Fourier transforms. We state several results without proof. A complete development of Fourier transforms can be found in Rudin (1986). There is a less complete but also less demanding development in Ablowitz *et al.* (1993).

To provide justification for the Fourier transform, we consider the following. Suppose that f is a complex valued function defined on the vertical stripe, $[a, b] \times \mathbb{R}$. Further suppose that f is periodic of period $2L = b - a$. In particular, f is periodic of period $2L$ provided $f(t, x) = f(t, x + 2L)$ for any t in $[a, b]$. Since the stripe can be thought of as a subset of $\mathbb{C}$, then we may suppose that f is a function of a complex variable. We add that f is also differentiable (analytic).

Recall that the complex exponential function is periodic. In particular, for each integer n, the function $f_n(t, x) = e^{nt} e^{in x \pi / L} = e^{nt} [\cos(\pi n x / L) + \mathbf{i} \sin(\pi n x / L)]$ is a periodic function of period $2L$ defined on $[a, b] \times \mathbb{R}$ with values in $\mathbb{C}$. Note that $f_1(t, x)$ maps the stripe domain $[a, b] \times \mathbb{R}$ onto the annulus $\{z \in \mathbb{C} : e^a < |z| < e^b\}$. Hence, if we define H on the annulus with $H \circ f_1 = f$, then H is analytic. Additionally, using results of complex analysis (Ahlfors, 1979), H has a representation as a convergent Laurent series. Hence, we have the following representation for f:

$$f(t, x) = \sum_{-\infty}^{\infty} a_n e^{nt} e^{\left(in x \frac{\pi}{L} \right)} = \sum_{n=-\infty}^{\infty} a_n (e^t e^{\left(\mathbf{i} x \frac{\pi}{L} \right)})^n, \qquad (2.2.2)$$

where convergence is uniform on the interior of the stripe and the coefficients are given by

$$a_n = \frac{1}{2\pi \mathbf{i}} \int_0^{2\pi} \zeta^{-n+1} H(\zeta) d\zeta = \frac{1}{2\pi \mathbf{i}} \int_{-L}^{L} f_1^{-n+1} f(\eta) \frac{\pi \mathbf{i}}{L} f_1(\eta) d\eta$$

$$= \frac{1}{2L} \int_{-L}^{L} f_1^{-n} f(\eta) d\eta = \frac{1}{2L} \int_{-L}^{L} f(\eta) (e^t e^{\left(\mathbf{i} x \frac{\pi}{L} \right)})^n d\eta, \qquad (2.2.3)$$

using the variable change $\zeta = f_1(\eta)$. The a_n are called the Fourier coefficients of the periodic function f. We consider a_n as a function $n \in \mathbb{Z}$ and taking values in $\mathbb{C}$ via (2.2.3) and then if we think of a as extended to the real line, we define an analogous function,

$$\hat{F}(k) = \int_{-\infty}^{\infty} f(x)e^{-ikx}dx. \tag{2.2.4}$$

This is the *Fourier transform*. More precisely, if f is absolutely summable, then $\int_{-\infty}^{\infty} |f|dx$ is convergent and the function $\hat{F}$ is called the *Fourier transform* of f.

We now state without proof some of the properties of the Fourier transform. First, we set the context for (2.2.4). The integral is the real Lebesgue integral extended to complex valued functions by dividing the integrand into the real and complex parts. The function f belongs to the space L^1, of all absolutely integrable functions on $\mathbb{R}$. Moreover, if f is also square summable, then the Fourier transform determines a Hilbert space isomorphism of the subspace of absolutely summable elements of L^2. This is important, but not used in the following. However, the next three properties of the Fourier transform are critical to our purposes.

The Fourier transform if f, $\hat{F}$ is also in L^1, and h, the inverse transform is defined via

$$h(x) = \frac{1}{2\pi} \int_{-\infty}^{\infty} \hat{F}(k)e^{ikx}dk. \tag{2.2.5}$$

In particular, the inverse transform h is equal to f almost everywhere. If g is any other function in L^1 with Fourier transform $\hat{G}$, then

$$\hat{F}(k)\hat{G}(k) = \int_{-\infty}^{\infty} f * g(x)e^{-ikx}dx, \tag{2.2.6}$$

where $f * g(x) = \int_{-\infty}^{\infty} f(x') * g(x - x')dx' = \int_{-\infty}^{\infty} f(x - x') * g(x')dx'$ is called the *convoluted product* of f and g. The statement that the integral in (2.2.6) converges requires Fubini's theorem (Marsden and Tromba, 2003).

Moving on, if f has n derivatives, then $\hat{F}_n$, the transform of $f^{(n)}$, the nth derivative of f, is related to the transform of f by

$$\hat{F}_n(k) = \int_{-\infty}^{\infty} f^{(n)}(x)e^{ikx}dx = (ik)^n \hat{F}(k). \qquad (2.2.7)$$

We now return to the 1D heat equation (2.2.1). Looking at u as a function of the spatial variable alone, we apply the Fourier transform to the differential equation. Now, there is no reason to suppose that u, the temperature function, should have a well-defined Fourier transform. However, if we ignore the issue for the moment and proceed, we arrive at an expression that indeed satisfies (2.2.1). That is the goal. So we begin.

Setting $\hat{U}$ to be the Fourier transform of u with respect to spatial variable and applying (2.2.7) to (2.2.1), we get

$$\frac{\partial}{\partial t}\hat{U}(t,k) = -\alpha k^2 \hat{U}(t,k). \qquad (2.2.8)$$

With (2.2.8), we can use results of Calculus 1 to write down $\hat{U}$.

$$\hat{U}(t,k) = \hat{U}(0,k)e^{-\alpha t k^2}, \qquad (2.2.9)$$

where $\hat{U}(0,k)$ denotes the initial state, $t = 0$ or the initial condition (IC). Taking $\hat{U}(0,k) = 1$, we compute the inverse transform for $e^{-\alpha t k^2}$ by (2.2.5). For this purpose, we use

$$-t\left(k\sqrt{\alpha} - i\frac{x}{2t\sqrt{\alpha}}\right)^2 = -t\left(\alpha k^2 - \frac{x^2}{4\alpha t^2} - 2i\frac{kx\sqrt{\alpha}}{2t\sqrt{\alpha}}\right)$$

$$= -\alpha t k^2 + ikx + \frac{x^2}{4\alpha t}$$

and $\int_{-\infty}^{\infty} e^{u^2} du = \sqrt{\pi}$ from standard complex variables. This yields

$$g(x) = \frac{1}{2\pi}\int_{-\infty}^{\infty} e^{-\alpha t k^2} e^{ikx} dk = \frac{1}{2\pi}\int_{-\infty}^{\infty} e^{\frac{-x^2}{4\alpha t^2}} e^{-t\left(k\sqrt{\alpha} - i\frac{x}{2t\sqrt{\alpha}}\right)^2} dk$$

$$= e^{\frac{-x^2}{4\alpha t^2}}\frac{1}{2\pi}\int_{-\infty}^{\infty} e^{-\alpha t\left(k - i\frac{x}{2\alpha t}\right)^2} dk = \frac{1}{2t\sqrt{\pi\alpha}}e^{\frac{-x^2}{4\alpha t^2}}.$$

We recognize g as the density function of a normal variable of mean 0 and variance $2\alpha t$. Indeed, it is easy to verify that the density function of a normal variable satisfies the heat equations. We mention in passing that this is not the only occasion that (2.2.1) is connected to settings far from physics. Indeed, we have already mentioned that there is a relationship to the Black–Scholes equation for pricing market futures.

More generally, if we set the (IC) for (2.2.1) as $h(x) = u(0, x)$, then $\hat{U}(t, k)$ is the transform of the convolution of h and g and $u(t, x)$ is the inverse transform of that. Hence, u is the convoluted product of h and g.

$$u(t, x) = \frac{1}{2t\sqrt{\pi\alpha}} \int\limits_{-\infty}^{\infty} h(\zeta)g(\zeta - x)d\zeta = \frac{1}{2t\sqrt{\pi\alpha}} \int\limits_{-\infty}^{\infty} h(\zeta)e^{\frac{-(\zeta-x)^2}{4\alpha t^2}} d\zeta.$$

$$(2.2.10)$$

We have made no stipulation about u at the boundary (end points) of the rod. The solution for (2.2.1) given by (2.2.10) supposes a rod extending indefinitely in both directions. In addition, if we look at $|u|$ for large t and fixed x, then the exponential term is effectively 1. Since $|\int_{-\infty}^{\infty} hd\zeta|$ is bounded, then $|u|$ is dominated by $1/t$, which converges to zero. Hence, whatever conditions are being implied for $|x|$ large, they must be zero for large t. Thus, the solution we have found can only represent heat that is dissipating in time. Additionally, there are solutions similar to (2.2.10) under the assumption that the rod has a fixed, finite boundary on one end and a stipulated boundary value for u at that location (see Simmons and Robertson (1991)).

The following image was generated using Plot3D. It shows u for the case $\alpha = 1$ and (IC): $h(x) = 20$ for $x \in [-5, 5]$ and zero outside. The time interval is $[0.1, 40]$. Note that Mathematica has rendered the surface to help us resolve the 3D surface. In general, we understand 3D surfaces by interpreting how light reflects from them. The idea is that the intensity of the reflected light is a function of the angle between the surface normal and the light ray. Hence, when we see how light reflects from the surface, we intuitively know the surface normals. In turn, we know the surface geometry. Looking at this image, we see that the simulated light source is over the viewer's right shoulder. In practice, the color intensity is set proportional to the cosine of the angle between the normal and light source. We note that Mathematica does give the user access to a light source parameter to change the direction of the source.

Next, we turn to the FDM rendering of (2.2.1). We start with uniform partitions of the t and x axes, $40 = t_1 < t_2 \cdots < t_N = T$ and $a = x_1 < x_2 \cdots < x_m = b$. We set $k = t_{n+1} - t_n$ and $h = x_{i+1} - x_i$. We agree to write the FDM calculated value of $u(t_n, x_i)$ as u_i^n and the nth state vector as u^n. Rendering the time derivative via the forward difference and the spatial derivative via the second central difference, we have the *explicit FDM formulation*,

$$\frac{u_i^{n+1} - u_i^n}{k} = \alpha \frac{u_{i+1}^n - 2u_i^n + u_{i-1}^n}{h^2}. \tag{2.2.11}$$

Solving for u_i^{n+1} and setting $\alpha k / h^2 = \lambda$, we arrive at the usual forward time, central space (FTCS) FDM rendering

$$u_i^{n+1} = \lambda u_{i+1}^n + (1 - 2\lambda) u_i^n + \lambda u_{i-1}^n.$$

In terms of state vectors, $u^{n+1} = A u^n$, where A is the usual tri-diagonal symmetric matrix with entries $A_{i,i} = 1 - 2\lambda$ and $A_{i,i-1} = A_{i,i+1} = \lambda$. We implement boundary values as Dirichlet values. For instance, if the boundary values are time independent, then we set row 1 of A to the m-tuple $(1, 0, 0, \ldots, 0)$, row m of A to $(0, 0, 0, \ldots, 1)$ and u_1^n to the left-hand value, u_1^m to the right-hand value.

Changing the forward difference on the left-hand side of (2.2.11) to the backward difference yields the *implicit FDM (BTCS) formulation.*

$$\frac{u_i^{n+1} - u_i^n}{k} = \alpha \frac{u_{i+1}^{n+1} - 2u_i^{n+1} + u_{i-1}^{n+1}}{h^2}. \tag{2.2.12}$$

The resulting iterative formulation is

$$u_i^n = -\lambda u_{i+1}^{n+1} + (1 + 2\lambda) u_i^{n+1} - \lambda u_{i-1}^{n+1},$$

$$u^{n+1} = B^{-1} u^n,$$

where B is the tri-diagonal matrix with entries $B_{i,i} = 1 + 2\lambda$ and $B_{i,i-1} = B_{i,i+1} = -\lambda$.

Next, we derive the Crank–Nicolson formulation. Begin by adding (2.2.11) to (2.2.12) and dividing by 2 to get

$$\frac{u_i^{n+1} - u_i^n}{k} = \alpha \frac{u_{i+1}^{n+1} - 2u_i^{n+1} + u_{i-1}^{n+1}}{2h^2} + \alpha \frac{u_{i+1}^n - 2u_i^n + u_{i-1}^n}{2h^2}.$$

Add and subtract a fictitious state $u_i^{n+(1/2)}$ on the left-hand side.

$$\frac{u_i^{n+1} - u_i^{n+(1/2)}}{k} + \frac{u_i^{n+(1/2)} - u_i^n}{k}$$

$$= \alpha \frac{u_{i+1}^{n+1} - 2u_i^{n+1} + u_{i-1}^{n+1}}{2h^2} + \alpha \frac{u_{i+1}^n - 2u_i^n + u_{i-1}^n}{2h^2}.$$

Separate this into two separate processes,

$$\frac{u_i^{n+(1/2)} - u_i^n}{k} = \alpha \frac{u_{i+1}^n - 2u_i^n + u_{i-1}^n}{2h^2},$$

$$\frac{u_i^{n+1} - u_i^{n+(1/2)}}{k} = \alpha \frac{u_{i+1}^{n+1} - 2u_i^{n+1} + u_{i-1}^{n+1}}{2h^2},$$

to yield a two step time process. The first step from u^n to $u^{n+(1/2)}$ is explicit while the second from $u^{n+(1/2)}$ to u^{n+1} is implicit. Together the two processes are called *Crank–Nicolson*. The two matrices are tri-diagonal and symmetric. The entries are now one half the value computed for explicit and implicit.

In Exercise 2, we ask the reader to compute values for (2.2.1) by all three FDM procedures. The output shows that Crank–Nicolson outperforms the other two. For most numerical analysts, it is procedure of choice.

Before closing this section, we mention that *Mathematica* provides its own variant of FDM via the function NDSolve. In the exercises, we explore all three solutions to (6.2.1): the solution given by (6.2.10), the FDM described above and NDSolve.

Recall that when we setup the FDM, we required that we had a uniform partition of the spatial and temporal variables. In practice, this is a severe restriction. Often, we want to concentrate our observation on particular areas of interest, areas where there is the most change. And we want to do this without significantly increasing the computational cost measured as the number of spatial intervals. Researchers have developed FDM techniques that allow for non-uniform partitions. The procedures do work well in some circumstances. These are the extended differences. They are alternative versions of the usual first- and second-order differences that have higher-order convergence. On the other hand, they tend to be less stable than the standard difference formula. We consider stability in the next section and extended differences in Chapter 7.

Exercises

1. Consider the solution to the 1D heat equation as shown in Figure 2.2.1. Plot cross-sections for the solution at $t = 0.5, 1, 2,$ and 3.
2. Consider the heat problem from Exercise 1. Set $k = 0.1$ and $h = 0.5$. Use implicit, explicit and Crank–Nicolson to compute values for the 1D heat equation at $t = 0.5, 1, 2,$ and 3. For each time, use a single axis to display the actual values along with the three FDM realization.

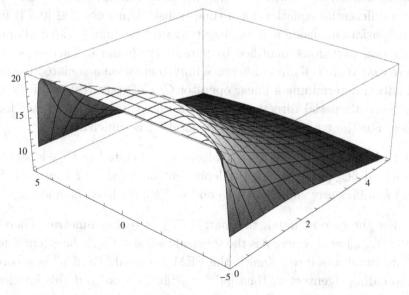

Figure 2.2.1.   The solution to the heat equation, $\alpha = 2$, $h(\zeta) - 20$, $x \in [-5, 5]$ and $t \in [0.1, 5]$.

3. Use Crank–Nicolson FDM to solve the 1D heat equation on the spatial interval $[-5, 5]$ and time interval $[0, 20]$ with the following assumptions:

   (IC) $u(0, x) = 0$ for every x,
   (BC) $u(t, -5) = 0$, $u(t, 5) = 40$ for all $t > 0$,
   Set $k = 0.0001$, $h = 0.2$ and $\alpha = 2$.

4. Execute FDM described in Exercise 3 using NDSolve. Compare these results against your data for $t = 0.01$ and $t = 0.1$.

2.3. The Heat Equation and FDM Stability

The purpose of this section is to develop Neumann stability analysis. In the three cases of FDM, we have seen there is a linear operator $C_{h,k}$, depending on the partition, such that $u^{n+1} = C_{h,k} u^n$ for every state vector u^n. Further, since $C_{h,k}$ is a linear transformation of a finite dimensional vector space, then $C_{h,k}$ is bounded and $\|u^{n+1}\| = \|C_{h,k}^{n+1} u^0\| \leq \|C_{h,k}^{n+1}\| \|u^0\|$. This relation leads to the following definition. First, we develop the notation. Let $\partial u / \partial t = Lu - f$ be a differential equation for a time transient process ($t \in [0, T]$) in one spatial dimension and suppose that we have uniform spatial and temporal partitions identified by parameters h and k. Further, suppose that there is a finite difference implementation associated to the partition determining a linear operator $C_{h,k}$ of state vectors. If there are several spatial directions, then we would need a parameter for each. For the moment, the current notation is sufficient.

Definition 2.3.1. The FDM realization associated to $C_{h,k}$ is *stable* provided $\|C_{h,k}^n\|$ is bounded independent of n ($nk \leq T$) for each h and k sufficiently small. ($h \leq h_0$ and $k \leq k_0$, for fixed h_0 and k_0.)

For the case at hand, the matrix $C_{h,k}$ is real symmetric. Therefore, $\|C_{h,k}\| = \rho^n$ where ρ is the spectral radius of $C_{h,k}$, the eigenvalue of maximal absolute value. If the FEM is unstable and u^0 is a corresponding eigenvector, then $\|u^n\| = \rho^n \|u^0\| \to \infty$. But this implies that the coordinates of the state vectors are growing in absolute value without bound. As this conclusion is independent of the application, then there is a serious problem. Therefore, for an FDM realization to be meaningful, it must be stable.

Alternatively, we have the following definition which coincides with Definition 2.3.1 for the case when $C_{h,k}$ is symmetric (see Exercise 2). More generally, stability implies the time state inequality of the following definition. We prove this in Section 7.4.

Definition 2.3.2. An FDM realization is *Neumann stable* provided time states satisfy $\|u^{n+1}\| \leq \|u^n\|$, for every $n \geq n_0$.

To develop Neumann stability analysis, we start with the discrete Fourier transform or interpolation. Corollary 2.3.1 connects Neumann stability to the discrete Fourier transform. We take a complex valued function f defined on a uniform partition of an interval. With a simple variable transformation, we may suppose that the interval is $[-\pi, \pi]$. In particular, we have values $f(x_i)$ for $-\pi = x_0 < x_1 < \cdots < x_{N-1} < \pi$. It is immediate that the set of all functions defined on the partition forms a linear space. In turn, we define a form σ on the space,

$$\sigma(f, g) = \sum_{i=0}^{N-1} f(x_i)\overline{g(x_i)}. \tag{2.3.1}$$

It is not difficult to show that this is indeed a positive semi-definite Hermitian form defined on the space of functions with values at $x_0 \cdot x_1, \ldots, x_{N_1}$ (see Exercise 4a). Note that we do not include x_N. Since Δx is 2π, then this is required because we are not assuming that $f(x_0) = f(x_N)$. In standard terminology, we do not assume that we have the function space of periodic functions. The following lemma states that the functions e^{ijx} and e^{ikx} are σ-orthogonal for j not equal to k.

Lemma 2.3.1. *For any integers j and k,*

$$\sigma(e^{ijx}, e^{ikx}) = N,$$

if $(j - k)/N$ is an integer and

$$\sigma(e^{ijx}, e^{ikx}) = 0,$$

otherwise.

Proof. In any case, setting $h = 2\pi/N$,

$$\sigma(e^{ijx}, e^{ikx}) = \sum_{i=1}^{N-1} e^{ijx_i} e^{ikx_i} = \sum_{i=0}^{N-1} e^{i(j-k)ih} = \sum_{i=0}^{N-1} (e^{i(j-k)h})^i.$$

If $(j-k)/N$ is an integer, then since $h = 2\pi/N$, we have

$$e^{i(j-k)h} = e^{i(j-k)2\pi/N} = (e^{i(j-k)})^{2\pi/N} = 1$$

and

$$\sum_{i=0}^{N-1} (e^{i(j-k)h})^i = N.$$

On the other hand, if $(j-k)/N$ is not an integer, then $e^{i(j-k)h} = (e^{i2\pi})^{(j-k)/N} = \zeta$, an Nth root of unity. We apply the usual expression for the partial geometric sum to get

$$\sum_{i=0}^{N-1} (e^{i(j-k)h})^i = \sum_{i=0}^{N-1} \zeta^i = \frac{1-\zeta^N}{1-\zeta} = 0. \qquad \square$$

We now define

Definition 2.3.3. Let f be a complex valued function defined on the set $\{x_i : x_i = ih, i = 0, 1, \ldots, N-1$ and $h = (2\pi)/N\}$. The *discrete Fourier interpolation* of f is given by

$$\sum_{k=-M_0}^{M_1-1} c_k e^{ikx}, \qquad (2.3.2)$$

where each $c_k = \sigma(f, e^{ikx})/N$, and $M_0 = M_1 = N/2$ for N even, $M_0 = M_1 = (N-1)/2$ for N odd. The c_k are called the (*discrete*) *Fourier coefficients* for f.

The expression in (2.3.2) is an interpolation of f as a trigonometric polynomial. If we were to take f and form the Fourier transform and then the reverse transform, then with the proper assumptions, the result would be a function equal to f almost everywhere. In Exercise 3, we see that the discrete Fourier interpolation is the inverse transform provided we replace the integration by the numerical integral using the trapezoid form. The following theorem states the basic properties of the discrete Fourier interpolation.

Theorem 2.3.1. *If* $f(x_j) = \sum_{k=-M_0}^{M_1-1} d_k e^{ikx_j}$ *for each* $j = 0, 1, \ldots, N-1$, *then each* $d_k = c_k$ *as given in Definition 11.1.1. Conversely, for each* j, $\sum_{k=-M_0}^{M_1-1} c_k e^{ikx_j} = f(x_j)$.

Proof. See Exercises 4c and 4d. $\square$

Our next result justifies Neumann stability analysis. The result identifies the norm of f as an N-tuple and the norm of f defined by σ.

Theorem 2.3.2. *(Perceval) For* f *defined on the given partition,* $f = (f(x_0), \ldots, f(x_{N-1}))$, *then* $\|f\| = \|f\|_\sigma$, *where* $\|f\|_\sigma$ *is the* σ-*norm,* $(\sigma(f, f))^{1/2}$.

Proof. As a complex N-tuple, the norm squared of f is

$$\|f\|^2 = \sum_{i=0}^{N-1} |f(x_i)|^2 = \sum_{i=0}^{N-1} f(x_i)\overline{f(x_i)} = \sigma(f, f). \qquad \square$$

Next suppose that u^n is a state vector of an FDM and suppose that u^n has discrete Fourier coefficients c_i^n, then Theorem 2.3.2 restated in terms of stability yields.

Corollary 2.3.1. *An FDM realization is Neumann stable provided for each* n *and* i, $|c_i^n| \geq |c_i^{n+1}|$.

Proof. From the basic properties of the discrete Fourier interpolation and Theorem 2.3.2,

$$\|u^{n+1}\|^2 = \|u^{n+1}\|_\sigma^2 = \sigma\left(\sum_{i=-M_0}^{M_1-1} c_i^{n+1} e^{iix}, \sum_{j=-M_0}^{M_1-1} c_j^{n+1} e^{ijx}\right)$$

$$= \sum_{i,j=-M_0}^{M_1-1} c_i^{n+1} \overline{c_j^{n+1}} \sigma(e^{iix}, e^{ijx}) = N \sum_{i=-M_0}^{M_1-1} |c_i^{n+1}|^2$$

since $o(e^{iir}, e^{ijr}) = 0$ if $i \neq j$ and $o(e^{ijr}, e^{ijr}) = N$. By hypothesis,

$$N \sum_{i=-M_0}^{M_1-1} |c_i^{n+1}|^2 \leq N \sum_{i=-M_0}^{M_1-1} |c_i^n|^2 = \|u^n\|_\sigma^2 = \|u^n\|^2.$$

The result follows.   □

We now return to the 1D heat equation and the explicit formulation as derived from (2.2.11).

$$u_i^{n+1} = \lambda u_{i+1}^n + (1 - 2\lambda)u_i^n + \lambda u_{i-1}^n.$$

If we write this equation as the discrete Fourier interpolation, we get

$$\sum_{k=-M_0}^{M_1-1} c_k^{n+1} e^{ikx_i}$$

$$= \lambda \sum_{k=-M_0}^{M_1-1} c_k^n e^{ikx_{i+1}} + (1 - 2\lambda) \sum_{k=-M_0}^{M_1-1} c_k^n e^{ikx_i} + \lambda \sum_{k=-M_0}^{M_1-1} c_k^n e^{ikx_{i-1}}$$

$$= \sum_{k=-M_0}^{M_1-1} (\lambda c_k^n e^{ik\Delta x} + (1 - 2\lambda)c_k^n + \lambda c_k^n e^{-ik\Delta x})e^{iki\Delta x}$$

or

$$\sum_{k=-M_0}^{M_1-1} c_k^{n+1} e^{iki\Delta x} = \sum_{k=-M_0}^{M_1-1} (\lambda c_k^n e^{ik\Delta x} + (1 - 2\lambda)c_k^n + \lambda c_k^n e^{-ik\Delta x})e^{iki\Delta x}.$$

By equating corresponding Fourier coefficients (Theorem 2.3.1),

$$c_j^{n+1} = \lambda c_j^n e^{ij\Delta x} + (1 - 2\lambda)c_j^n + \lambda c_j^n e^{-ij\Delta x}.$$

Now divide through by c_j^n and take the absolute value,

$$\frac{|c_j^{n+1}|}{|c_j^n|} = |\lambda e^{ij\Delta x} + (1 - 2\lambda) + \lambda e^{-ij\Delta x}|.$$

Expressing the right-hand side in terms of cos and sin yields,

$$\frac{|c_j^{n+1}|}{|c_j^n|} = |\lambda(\cos(j\Delta x) + \mathbf{i}\sin(j\Delta x))$$

$$+ (1 - 2\lambda) + \lambda(\cos(j\Delta x) - \mathbf{i}\sin(j\Delta x))|$$

or

$$\frac{|c_j^{n+1}|}{|c_j^n|} = |2\lambda(\cos(j\Delta x) + (1 - 2\lambda)| = |1 + 2\lambda(\cos(j\Delta x) - 1)|.$$

Now, stability will occur when the absolute value of the left-hand side is less than or equal to 1, or $-1 \leq 1 - 2\lambda(1 - \cos(j\Delta x)) \leq 1$. Hence, $-2 \leq -2\lambda(1 - \cos(j\Delta x)) \leq 0$. Therefore, Neumann stability is satisfied provided $1 \geq \lambda(1 - \cos(j\Delta x))$. But the maximal value of $1 - \cos(\varphi)$ is 2. Hence, stability requires that $\lambda \leq 1/2$. It turns out that for this equation implicit and Crank–Nicolson are unconditionally stable (see Exercise 1).

We end this presentation of the 1D heat equation by looking ahead with a preliminary discussion of convergence and consistency. As we see convergence and stability are closely related. We will provide formal definitions for convergence and consistency in Chapter 7. Since FDM is a discrete process, it generates error. At the nth state, the error is

$$e^n = (u(t_n, x_1), \ldots, u(t_n, x_m)) - C_{h,k} u^{n-1}$$

$$= (u(t_n, x_1), \ldots, u(t_n, x_m)) - C_{h,k}(u(t_{n-1} x_1), \ldots, u(t_{n-1}, x_m))$$

$$+ C_{h,k}(u(t_{n-1}, x_1), \ldots, u(t_{n-1}, x_m)) - C_{h,k} u^{n-1}$$

$$= (u(t_n, x_1), \ldots, u(t_n, x_m)) - C_{h,k}(u(t_{n-1}, x_1), \ldots, u(t_{n-1}, x_m))$$

$$+ C_{h,k}((u(t_{n-1}, x_1), \ldots, u(t_{n-1}, x_m)) - u^{n-1}).$$

Setting $(u(t_n, x_1), \ldots, u(t_n, x_m) - C_{h,k}(u(t_{n-1}, x_1), \ldots, u(t_{n-1}, x_m))) = \tau^n$, we have $e^n = \tau^n + C_{h,k} e^{n-1}$. Continuing the process,

$$e^n = \tau^n + C_{h,k}\tau^{n-1} + C_{h,k}^2\tau^{n-2} + \cdots + C_{h,k}^{n-1}\tau^1, \tag{2.3.3}$$

since $e^0 = (u(t_0, x_1), \ldots, u(t_0, x_m)) - u^0 = 0$. The term τ^n is called the nth truncation error. In Section 7.1, we define consistency for the

FDM realization. We then see that for a consistent FDM realization, τ^n goes to zero. On the other hand, convergence means that the error $e^n \to 0$ as the partition size vanishes. If we take the norm of both sides of (2.3.3), we see that consistency and stability will imply convergence. Indeed, the Lax equivalence theorem states that with consistency, stability and convergence are equivalent.

Now, we turn to the 2D heat equation. This equation is sometimes referred to as the *diffusion equation*. We let $u = u(t, x, y)$ be a function of time and two spatial variables satisfying

$$\frac{\partial u}{\partial t} = \alpha \left(\frac{\partial^2 u}{\partial x^2} + \frac{\partial^2 u}{\partial y^2} \right) = \alpha \nabla^2 u. \tag{2.3.4}$$

As in the 1D case, we are given a uniform partition of the time variable. Now, we also have uniform and independent partitions in the x and y directions. We begin by rendering the time variable with forward difference while using the second central difference in the two spatial directions.

$$u_{i,j}^{n+1} - u_{i,j}^n = \lambda_1(u_{i-1,j}^n - 2u_{i,j}^n + u_{i+1,j}^n) + \lambda_2(u_{i,j-1}^n - 2u_{i,j}^n + u_{i,j+1}^n),$$

where $\lambda_1 = \alpha \Delta t / \Delta x^2$ and $\lambda_2 = \alpha \Delta t / \Delta y^2$. This yields the following relation between time states:

$$u_{i,j}^{n+1} = \lambda_1 u_{i-1,j}^n + \lambda_2 u_{i,j-1}^n + (1 - 2(\lambda_2 + \lambda_1))u_{i,j}^n$$
$$+ \lambda_1 u_{i+1,j}^n + \lambda_2 u_{i,j+1}^n. \tag{2.3.5}$$

We can write (2.3.5) in the form $u^{n+1} = Au^n$ provided we write the states as an $m_x \times m_y$-tuple, where m_x is the number of elements in the x-axis partition, and m_y is defined similarly. In this case, the entries of u^n associate to spatial positions as follows: $(x_0, y_0), (x_0, y_1), (x_0, y_2), \dots, (x_0, y_{m_y}), \dots, (x_{m_x}, y_{m_y})$.

The corresponding implicit rendering of the diffusion equation is

$$u_{i,j}^n = \lambda_1 u_{i-1,j}^{n+1} + \lambda_2 u_{i,j-1}^{n+1} + (1 - 2(\lambda_2 + \lambda_1))u_{i,j}^{n+1}$$
$$+ \lambda_1 u_{i+1,j}^{n+1} + \lambda_2 u_{i,j+1}^{n+1}. \tag{2.3.6}$$

The immediate question is what conditions are needed to ensure that FDM realization for (2.3.4) is stable. For this purpose, we need to extend the discrete Fourier interpolation to two variables. We begin by identifying two inner product spaces and then forming their tensor product. In particular, we use (2.3.1) for f and g defined on $\{x_0, x_1, x_2, \ldots, x_{m_x-1}\}$ or f and g defined on $\{y_0, y_1, y_2, \ldots, y_{m_y-1}\}$ and we write

$$\sigma_x(f,g) = \sum_{i=0}^{m_x-1} f(x_i)\overline{g(x_i)}, \quad \sigma_y(f,g) = \sum_{i=0}^{m_y-1} f(y_i)\overline{g(y_i)}.$$

In particular, from the one variable case, the set of linear combinations of the e^{iix}, $0 \le i < m_x$ denoted V_x is an inner product space with orthogonal basis e^{iix}. The same is true for the linear span V_y of e^{ijy}. Hence, $V_x \otimes V_y$ is an inner product space with Hermitian form $\sigma_x \otimes \sigma_y$ and orthogonal basis $e^{iix} \otimes e^{ijy}$. Furthermore, the elements of $V_x \otimes V_y$ define functions of $\{x_0, x_1, x_2, \ldots, x_{m_x-1}\} \times \{y_0, y_1, y_2, \ldots, y_{m_y-1}\}$ via

$$\left[\sum_{k=M_{x,0}}^{M_x-1} \sum_{l=M_{y,0}}^{M_{y,1}-1} c_{k,l} e^{ikx} \otimes e^{ily} \right](x_i, y_j) = \sum_{k=M_{x,0}}^{M_{x,1}-1} \sum_{l=M_{y,0}}^{M_{y,1}-1} c_{k,l} e^{ikx_i} e^{ily_j}.$$

Hence, by a dimension argument, $V_x \otimes V_y$ is isomorphic, as a linear space, to the space $\mathcal{V}$ of functions on $\{x_0, \ldots, x_{m_x-1}\} \times \{y_0, \ldots, y_{m_y-1}\}$. Furthermore, for any f in $\mathcal{V}$, the coefficient $c_{k,l}$ is equal to $\sigma_x \otimes \sigma_y(f, e^{ikx} \otimes e^{ily})$ and that the norm of f as an $m_x \times m_y$-tuple is equal to $\|f\|_{\sigma_x \otimes \sigma_y}$. Therefore, we may extend our earlier development to the two variable case. In particular, the FDM realization in (2.3.6) is Neumann stable provided the state coefficients $c_{k,l}^n$ satisfy $|c_{k,l}^{n+1}| \le |c_{k,l}^n|$.

Analogous to the one variable case, we begin with (2.3.5), introduce the discrete Fourier interpolation and simplify to get

$$\sum_{k=M_{x,0}}^{M_{x,1}-1} \sum_{l=M_{y,0}}^{M_{y,1}-1} c_{k,l}^{n+1} e^{iki\Delta x} \otimes e^{ilj\Delta y}$$

$$= \sum_{k=M_{x,0}}^{M_{x,1}-1} \sum_{l=M_{y,0}}^{M_{y,1}-1} [\lambda_2 c_{k,l}^n 1 \otimes e^{-il\Delta y} + \lambda_1 c_{k,l}^n e^{-ik\Delta x} \otimes 1]$$

$$+ [1 - 2(\lambda_2 + \lambda_1)]c_{k,l}^n + \lambda_1 c_{k,l}^n e^{ik\Delta x} \otimes 1$$

$$+ \lambda_2 c_{k,l}^n 1 \otimes e^{il\Delta y}]e^{iki\Delta x} \otimes e^{ilj\Delta y}.$$

Taking the inner product with $e^{i+si\Delta x} \otimes e^{i+tj\Delta y}$ and dividing by $c_{s,t}^n$, we get

$$\frac{c_{s,t}^{n+1}}{c_{s,t}^n} = \lambda_2 1 \otimes e^{-it\Delta y} + \lambda_1 e^{-is\Delta x} \otimes 1 + (1 - 2)(\lambda_2 + \lambda_1)$$

$$+ \lambda_1 e^{is\Delta x} \otimes 1 + \lambda_2 t \otimes e^{it\Delta y}.$$

Taking absolute values and resolving expressions $f \otimes 1$ as f, we have

$$\left|\frac{c_{s,t}^{n+1}}{c_{s,t}^n}\right| = |1 + 2\lambda_1(\cos(s\Delta x) - 1) + 2\lambda_2(\cos(t\Delta y) - 1)|.$$

Next, we use standard multivariate calculus techniques to determine the maximal values of the expression on the right-hand side. The result will be an expression in terms of λ_1 and λ_2. Setting the maximum to 1 yields the desired stability result. In particular, (2.3.5) is stable provided $\lambda_1 + \lambda_2 \leq 1/2$.

Additionally, the implicit and Crank–Nicolson forms are unconditionally stable. Note that for the preceding schemes, the FDM matrix has five non-zero diagonal stripes. The following modified Crank–Nicolson technique has two matrices with only three non-zero entries in each row and column. In addition, the scheme separates the x-axis variation from the y-axis variation. Hence, the stability issue is resolved by appealing directly to the 1D case. The process is based on the central difference for the time variable. Recall that the central difference is second-order convergence while the forward and backward differences are only first-order convergent. This FDM schema based on the central time difference are often referred to as a *leap frog* technique. This particular procedure is called *fractional step method* by Chung (2002) and *Peaceman–Rachford* by Thomas (1995).

$$u_{i,j}^{n+1/2} - u_{i,j}^n = \frac{\lambda_1}{2}\left(u_{i-1,j}^{n+1/2} - 2u_{i,j}^{n+1/2} + u_{i+1,j}^{n+1/2}\right)$$

$$+ \frac{\lambda_2}{2}\left(u_{i,j-1}^n - 2u_{i,j}^n + u_{i,j+1}^n\right). \tag{2.3.7}$$

$$u_{i,j}^{n+1} - u_{i,j}^{n+1/2} = \frac{\lambda_1}{2}\left(u_{i-1,j}^{n+1/2} - 2u_{i,j}^{n+1/2} + u_{i+1,j}^{n+1/2}\right)$$

$$+ \frac{\lambda_2}{2}\left(u_{i,j-1}^{n+1} - 2u_{i,j}^{n+1} + u_{i,j+1}^{n+1}\right). \quad (2.3.8)$$

Equation (2.3.7) is derived from the diffusion equation by using FTCS in the y-direction and backward time plus central space in the x-direction.

$$u_{i,j}^{n+1/2} - u_{i,j}^{n} = \lambda_2\left(u_{i,j-1}^{n} - 2u_{i,j}^{n} + u_{i,j+1}^{n}\right),$$

$$u_{i,j}^{n+1/2} - u_{i,j}^{n} = \lambda_1\left(u_{i-1,j}^{n+1/2} - 2u_{i,j}^{n+1/2} + u_{i+1,j}^{n+1/2}\right).$$

Then, we add the equations and divide by 2. We may identify this as a central difference scheme as it averages the forward and backward differences. The derivation of (2.3.8) is similar.

To demonstrate stability, we begin by rearranging each equation so that terms relating to different time steps are separated.

$$-\frac{\lambda_1}{2}u_{i-1,j}^{n+1/2} + \left(1 + 2\frac{\lambda_1}{2}\right)u_{i,j}^{n+1/2} - \frac{\lambda_1}{2}u_{i+1,j}^{n+1/2}$$

$$= \frac{\lambda_2}{2}u_{i,j-1}^{n} + \left(1 - 2\frac{\lambda_2}{2}\right)u_{i,j}^{n} + -\frac{\lambda_2}{2}u_{i,j+1}^{n}, \quad (2.3.9)$$

$$-\frac{\lambda_2}{2}u_{i,j-1}^{n+1} + \left(1 + 2\frac{\lambda_2}{2}\right)u_{i,j}^{n+1} - \frac{\lambda_2}{2}u_{i,j+1}^{n+1}$$

$$= \frac{\lambda_1}{2}u_{i-1,j}^{n+1/2} + \left(1 - 2\frac{\lambda_1}{2}\right)u_{i,j}^{n+1/2} + -\frac{\lambda_1}{2}u_{i+1,j}^{n+1/2}. \quad (2.3.10)$$

The right-hand side of (2.3.9) is known, so if we replace it by a vector, then the equation is exactly in the form of implicit FDM in one spatial dimension. This step is unconditionally stable. Substituting into (2.3.10), then for exactly the same reason, the time step described in (2.3.10) is unconditionally stable.

The idea underlying this technique is commonly used when applying the FDM to cases with more than one spatial dimension (see Chung (2002)). In Section 7.1, we introduce consistency. Consistency for this technique is proved in Thomas (1995).

We summarize the preceding work with Neumann stability analysis in the following theorem.

Theorem 2.3.3. *Neumann stability and stability are equivalent for the diffusion equation. For the 1D diffusion equation,*

a. *Explicit FDM is stable* $\lambda \leq 1/2$.
b. *Implicit FDM is unconditionally stable.*
c. *Crank–Nicolson FDM is unconditionally sable.*

For the 2D diffusion equation,

a. *Explicit FDM is stable* $\lambda_1 + \lambda_2 \leq 1/2$.
b. *Implicit FDM is unconditionally stable.*
c. *Crank–Nicolson FDM is unconditionally stable.*
d. *Leap-frog FDM is unconditionally stable.*

Exercises

1. Resolve stability for implicit and Crank–Nicolson methods for the 1D heat equation.

   a. For implicit FDM, derive the expression

   $$\left| \frac{c_j^{n+1}}{c_j^n} \right| = \frac{1}{|1 + 2\lambda(1 - \cos(j\Delta x))|}.$$

   b. For Crank–Nicolson, derive the analogous relation

   $$\left| \frac{c_j^{n+1}}{c_j^n} \right| = \frac{|1 - 2\lambda(1 - \cos(j\Delta x))|}{|1 + 2\lambda(1 - \cos(j\Delta x))|}.$$

   c. Prove that both techniques are unconditionally Neumann stable.

2. Determine the relationship between Neumann stability and stability for symmetric matrices. Prove that if $C_{h,k}$ is symmetric, then the following conditions are equivalent.

   i. $\|u^{n+1}\| \leq \|u^n\|$, for any u^0,
   ii. $\|C_{h,k}\| \leq 1$,
   iii. $\|C_{h,k}^m\|$ is bounded for all m.

3. The Fourier coefficients for a periodic function f of period 2π are given by

$$c_k = \frac{1}{2\pi} \int\limits_0^{2\pi} f(x)e^{ikx}dx.$$

Set $h = 2\pi/N$ and $x_i = ih$. Prove that if you approximate the integral by a trapezoid scheme at $0 = x_0 < \cdots < x_N = 2\pi$, then

$$c_k = \frac{1}{N} \sum_j f(x_j)e^{ikx_j}.$$

Conclude that the discrete Fourier interpolation is the trapezoid form of the inverse Fourier Transform.

4. This exercise leads you through the development of the discrete Fourier interpolation. We begin with an interval $[a, b]$ with a uniform partition $a = x_0 < \cdots < x_{N-1} < b$, where each $x_i = x_{i-1} + \Delta x = x_0 + i\Delta x$. Taking the usual variable change, we may suppose that $a = 0$, $b = 2\pi = N\Delta x$ and $x_i = i\Delta x$. For two complex valued functions f and g defined on $\{x_0, x_1, \ldots, x_{N-1}\}$, we set $\sigma(f, g) = \sum_{i=0}^{N-1} f(x_i)\overline{g(x_i)}$.

   a. Prove that σ defines a positive definite Hermitian form on the space of all complex valued functions with domain $\{x_0, \ldots, x_{N-1}\}$. And σ is positive semi-definite on the space of functions with domain $[a, b]$.

   Using $\Delta x = 2\pi/N$, prove each of the following statements.

   b. Verify that the expression

$$\sum_{k=-M_0}^{M_1-1} c_k e^{ikx},$$

   may be recast as

$$\sum_{s=0}^{N-1} c_{s-M_0} e^{i(s-M_0)x}.$$

   c. Suppose that for each j, $f(x_j) = \sum_{k=-M_0}^{M_1-1} c_k e^{ikj\Delta x}$, then each $c_k = \sigma(f, e^{ik})/N$. (*Hint*: compute $\sigma(f, e^{ilx})$ for $-M_0 \le l \le M_1 - 1$.)

d. Prove the converse to part b, namely prove that if $g(x_j) = \sum_{k=-M_0}^{M_1-1} c_k e^{ikj\Delta x}$, where the coefficients are given as in part b, then for each j, $g(x_j) = f(x_j)$. (*Hint*: Expand

$$\sum_{k=-M_0}^{M_1-1} c_k e^{ikj\Delta x} = \frac{1}{N} \sum_{k=-M_0}^{M_1-1} \sigma(f, e^{ikx}) e^{ikj\Delta x}$$

$$= \frac{1}{N} \sum_{k=-M_0}^{M_1-1} \left(\sum_{l=0}^{N-1} f(x_l) e^{-ikx_l} \right) e^{ikj\Delta x}$$

and keep in mind that $e^{ikj\Delta x} = e^{ikx_j} = e^{ijx_k}$.)

5. Prove that implicit FDM for the 1D heat equation is unconditionally stable.

6. Prove that Crank–Nicolson FDM for the 1D heat equation is unconditionally stable.

7. Complete the proof of stability for the 2D diffusion equation.

   a. For the explicit scheme, complete the max/min and conclude that it is stable for $\lambda_1 + \lambda_2 \leq 1/2$.
   b. Prove that the implicit scheme is unconditionally stable.
   c. Prove that Crank–Nicolson process is unconditionally stable.

8. An $n \times n$ real matrix $A = [\alpha_{i,j}]$ is called tridiagonal if the only non-zero entries of A occur for $\alpha_{i,j}$ with $j = i-1, i$ and $i+1$ and there are non-zero reals a, b and c so that for each i, $a = \alpha_{i,i}$, $b = \alpha_{i,i+1}, c = \alpha_{i,i-1}$.

   a. Determine that implicit and explicit FDM for the single variable heat equation give rise to tridiagonal matrices.
   b. Verify that the eigenvalues of a tridiagonal matrix are given by $\lambda_i = a + 2b(\sqrt{c/b})\cos(i\pi/(n+1))$ with corresponding eigenvector v_i with jth entry given by $2(c/b)^{j/2}\sin(ij\pi/(n+1))$.
   c. Determine the spectral radius of the an explicit FDM matrix for the 1D heat equation.
   d. Repeat c for implicit FDM.
   e. Using Definition 2.3.1, determine when implicit and explicit FDM are stable.

9. Use FTCS FDM to solve the 2D diffusion equation on the spatail domain $[0, 100] \times [0, 200]$ for the time iterval $[0, 5]$ and the following assumptions.

   (IC) $u(0, x, y) = 0$ for every (x, y),

   (BC-1) $u(t, x, 0) = 20$ for each x and $t \geq 0$, (BC-2) $u(t, x, y) = 0$ for each $x > 0$, every y and $t \geq 0$,

   Select Δx, Δy and Δt to ensure stability. Show the output for $t = 0.5$, 2 and 5 by displaying level curves for u.

10. In Exercise 9 replace BC-1 by a periodic boundary condition $u(y, x, 0) = 10(1 - \sin(t))$. Select the time interval so that this periodic condition goes through several cycles during the execution.

11. Redo the 10 using the fractional step method.

Chapter 3

Numerical Solutions
to Steady-State Problems

Introduction

In this chapter, we use two steady-state problems to introduce the finite element method (FEM). This technique is generally considered the 'gold standard' of numerical solution procedures for PDE. It was developed initially in the late 1930s by mechanical engineers (Clough, 1980). They used it to calculate the joint deflections on a frame when the object is fully loaded. The particular setting was the frame of an aircraft or automobile or bridge. In the 1950s, the engineering community extended the technique to be a general purpose PDE solver. At this time, it was proposed as an alternative to FDM. One primary appeal was that they were able to get good results without requiring a uniform spatial partition. Beginning in the late 1940s, mathematicians and physicists published papers using FEM. They used locally defined or piecewise polynomials to define a globally continuous function. It was not until 1959 that a paper appeared which pulled all of this together and coined the term FEM. Beginning in the mid 1960s, the mathematical community extended the known theory and proved that the computed FEM data did approximate the actual solution for a large class of linear elliptical PDE. Currently, engineers use the FEM well beyond the mathematically supported cases. In addition, biologists and economists are applying it to PDEs that arise in their respective fields.

FEM in its standard context is based on a domain partition. The basic idea is to approximate the solution to the PDE by means of piecewise defined polynomials. In particular, the approximate solution is a polynomial on each partition element. In 2D, the partition is usually rectangular or triangular. We see both cases in this chapter. The idea behind the use of piecewise polynomials lies with the small degree polynomials. Usually, it is more efficient to deal with many small degree polynomials than a few large degree polynomials. However, with massively parallel computing, this is not necessarily the case. We revisit this issue in subsequent chapters. One additional advantage to the FEM over the FDM is that we can set the boundary values to model the forces at the boundary. They are called Neumann or natural boundary conditions. It is possible to implement Neumann boundary conditions for FDM, however, they arise more naturally in this context.

The FEM is a complex procedure involving several steps. The first time through it seems daunting. Nevertheless, each step is straightforward and mechanical. This process is easily automated.

In this chapter, we introduce this technique with two examples. Both cases lie within the area that is mathematically supported. We do not attempt to describe the method in its entirety but rather to initiate the study. Later in Part 1, we will encounter FEM applied to a transient process. In Part 2, we will broaden the description of the procedure and give the mathematics that supports it. Our present goal is to start the reader toward FEM by means of two very simple examples. In the first section, we highlight the steps in the process. We hold the formal definition of the FEM to the second section.

The first example derives from the previous work with the Joukowski airfoil. But this time we work directly with the Laplace equation. The second example looks at the Helmholtz equation. In this case, we see a variant of the FEM where the final solution step solves for an eigenvector.

3.1. Ideal Fluid Flow, the Laplace Equation via FEM and a Rectangular Partition

In this section, we simulate ideal fluid flow through a partially obstructed channel. For this purpose, we introduce Galerkin finite element method, commonly referred to as FEM or GFEM. FEM is an involved multi-step solution procedure. Even so, many researchers consider it the best numerical technique to solve a PDE. As an introduction, we restrict our attention to one case, the 4-nodal, rectangular element, quadratic basis function version. In the literature, this is referred to as Q_1^4. This is sufficient for the case at hand. In the next section, we look at a variant procedure. There is a more complete development of FEM that is given in Part 2, Chapters 8 and 9. We develop additional examples in Chapter 6.

There is a fundamental difference in point of view between the FDM and the FEM. Suppose you have a differential equation with unknown solution u. The FDM returns approximate values for u on the mesh. On the other hand, the FEM actually returns a function u_h, which approximates u. We can use u_h to estimate u. But we can also use u_h to estimate functions related to u. For instance, in this section, we use the FEM solution to estimate the gradient. Hence, the FEM provides more information than the FDM. On the other hand, it is more difficult to implement and the mathematical foundations are more complicated.

As we go through the example, we identify each step. At the end, we summarize.

We now turn to the problem at hand. Consider the following diagram. It represents a partially obstructed channel with left to right flow. The top and bottom edges represent the boundaries for the channel. The left and right edges are the inflow and outflow edges, respectively. We refer to the interior of the diagram as U, and the interior together with the edges (boundary) as D. The boundary itself is Γ.

As with the airfoil, we will suppose that the flow is continuous, inviscid, irrotational and incompressible, laminar and steady.

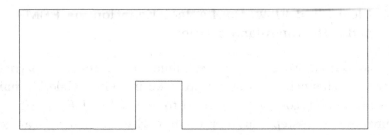

Figure 3.1.1. The partially obstructed channel.

Hence, we may represent the flow by a continuous vector field $u(x,y) = (u_1(x,y), u_2(x,y))$ defined on D. Exactly, as in the case of the airfoil, we know that there is an analytic function $F(z) = \varphi(x,y) - \mathbf{i}\psi(x,y)$ with $z = x + \mathbf{i}y$ and $u(x,y) = \nabla\varphi$.

At this point, there is a serious problem. As simple as the geometry of the problem may seem, there is no known analytic function F. Indeed, the theory of analytic functions assures us that F exists. The problem is that we do not know what it is. Hence, we do not and cannot write F. This is not really a problem as FEM provides means to estimate F well enough to compute $\nabla\varphi$. Recall that the function φ must satisfy the Laplace equation. In the presentation of the flow about an airfoil, this was an incidental fact. Now, it is the central idea behind the implementation of the FEM.

For a mathematically supported application of the FEM, it is essential that we know the existence of a solution to the PDE. For the case in hand, this is not a problem. We know that F exists. Hence, φ exists.

Step 1. Set the geometric model, define the elements and nodes

The first step is to divide the D into rectangles. The rectangles are called the elements and the points of the grid are called nodes (see Figure 3.1.2). Note that a given node is the vertex for several rectangles. Also, the rectangles do not have to be of the same size. Thus, we have more freedom than with the FDM where we were required to have a uniform partition for each of the spatial and the temporal

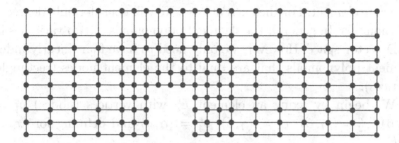

Figure 3.1.2. The elements and nodes.

variables. The only restriction we have is that the vertices of each rectangle must coincide with the vertices of the adjacent rectangles. Otherwise stated, the vertex of one rectangle may not lie on the interior of the edge of another rectangle. Next, we identify the nodes by their coordinates and the elements by their nodes. The standard procedure is to list the vertices and elements from bottom to top and left to right. For reasons that will be apparent later, it is important that we have the nodes (vertices) of each element listed in counterclockwise order.

The following diagram shows the nodes and the elements. There are 188 nodes and 155 elements. Note that the elements are smaller in the vicinity of the obstruction. This is the area of primary interest. We want to see how the flow is deflected by the obstruction.

We are heading toward a 188 by 188 linear system of equations called the *global linear system*. In the terminology of FEM, the size of the linear system is called the degrees of freedom. For this case, there is a degree of freedom for each node or 188 degrees of freedom. The solution to the system will provide an approximation, φ_h, of φ. Note the subscript h. This is called the *mesh parameter*. It identifies the partition via the largest diagonal of any rectangle in Figure 3.1.2.

Step 2. Determine the Polynomial Model

The FEM solution φ_h will be a continuous, piecewise polynomial function on D. In particular, it is a sum of functions φ_e, where each φ_e is a polynomial on an element E_e with support contained within the same element. For the current case, we will consider only degree 2 polynomials. In general, the degree 2 polynomials in two variables is a 6D vector space. However, we only need to have four linearly independent polynomials that are dual to the element nodes (rectangle vertices).

We begin by fixing an element E_e with vertices (nodes) $v_1^e = (\alpha, \beta)$, $v_2^e = (\gamma, \beta)$, $v_3^e = (\gamma, \delta)$, $v_4^e = (\alpha, \delta)$ and defining for (x, y) in E_e

$$N_1^e(x,y) = \frac{(\gamma - x)\,(\delta - y)}{(\gamma - \alpha)\,(\delta - \beta)}; \quad N_2^e(x,y) = \frac{(x - \alpha)\,(\delta - y)}{(\gamma - \alpha)\,(\delta - \beta)};$$

$$N_3^e(x,y) = \frac{(x - \alpha)\,(y - \beta)}{(\gamma - \alpha)\,(\delta - \beta)}; \quad N_4^e(x,y) = \frac{(\gamma - x)\,(y - \beta)}{(\gamma - \alpha)\,(\delta - \beta)}.$$

Further, it is convenient to consider each function N_i^e as a function of D. We do this by setting $N_i^e(x,y) = 0$ if (x,y) is not in E_e. It is easy to check that $N_i^e(v_j^e) = \delta_{i,j}$, the Kronecker delta. Figure 3.1.3

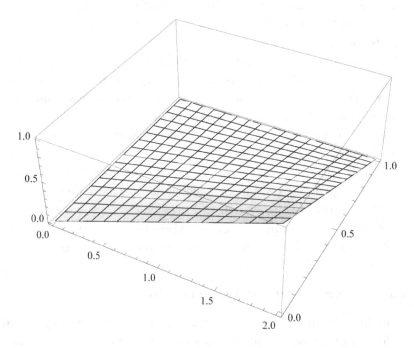

Figure 3.1.3. $f(x,y) = 0.5(x - xy)$.

shows the plot of N_2^e where E_e is the rectangle with vertices $(0,0), (2,0), (2,1)$ and $(0,1)$. Note that N_2^e takes the value of 1 at one vertex and 0 at the other three.

The polynomials defined in this manner are called *Lagrange polynomials*. These four polynomials are the two variable, first-order, degree 2 Lagrange polynomials. It is easy to see that the four functions N_i^e form a linearly independent subset of $L^2(U)$, the vector space of square Lebesgue integrable functions on U. In fact, they span the subset of all polynomials $f^e(x,y) = a + bx + cy + dxy$, (x,y) in E_e and $f^e(x,y) = 0$ for (x,y) not in E_e. We will refer to this space as V_e^h. Once again, we use h to represent the mesh.

Step 3. Determine the element linear systems

Our FEM solution φ_h is a piecewise polynomial $\varphi_h = \sum_e \varphi_e$, where φ_e belongs to V_e^h. We write V^h for the space of piecewise polynomials. In terms of the four Lagrange polynomials defined on E_e,

$$\varphi_h = \sum_e \sum_{i=1}^4 \alpha_i^e N_i^e. \qquad (3.1.1)$$

Furthermore, it must satisfy the weak form of the PDE. In particular, the *FEM solution* to the Laplace equation φ_h is given by the relation,

$$\int_D (\nabla^2 \varphi_h)\psi = 0, \qquad (3.1.2)$$

for any ψ in V^h. The form of the Laplace equation given in (3.1.2) is called the weak form of the PDE. We can state the goal of FEM to identify the 4×155 coefficients α_i^e so that the resulting function satisfies the weak form of the PDE.

Moving forward, it is important to keep in mind that we have an established partition of D. In this context, piecewise polynomial means polynomial on the elements of that partition. Recall that we have defined the N_i^e as functions of D, and in that context, they are not continuous. Hence, we will need to take care that φ_h is continuous. To achieve continuity, we must satisfy the following. If E_e and $E_{e'}$ are adjacent elements, then φ_e must equal $\varphi_{e'}$ on the common boundary. Since the Lagrange polynomials are dual to the element vertices, if vertex P_i^e of E_e is the same node as vertex $P_i^{e'}$ of $E_{e'}$, then $\alpha_i^e = \varphi_h(P_i^e) = \varphi_h(P_i^{e'}) = \alpha_i^{e'}$. Therefore, instead of having 4×155 coefficients to determine, we actually have one coefficient for each node, or 188 coefficients to determine.

Let V^h denote the space of all continuous piecewise polynomial functions on D associated to the partition. This is a finite dimensional subspace of the Hilbert space L^2, which contains φ_h. As usual, L^2 has the inner product $\sigma(f,g) = \int_D fg$. Since finite dimensional subspaces are closed, then V^h has an orthogonal complement W^h. In particular, $L^2 = V^h \oplus W^h$. Up to this point, we have not stated which element of V^h we are looking for in order to define φ_h. We want to take φ_h so that $\nabla^2 \varphi_h$ lies in W^h. However, we have not required that φ_h be differentiable, only continuous. Therefore, we must explain what we mean by $\nabla^2 \varphi_h$. On any element E_e, φ_h is a polynomial, hence, there is no problem inside the elements. If Γ_e denotes the boundary of E_e, then the issue lies in $\cup_e \Gamma_e$. But the union of boundaries of the elements forms a set of measure zero in L^2 and the elements of L^2 are only known uniquely up to sets of measure zero, then φ_h may be considered differentiable as an element of L^2. Later, we will introduce the idea of the weak derivative. At that time, we will have a more formal means to define $\nabla^2 \varphi_h$.

Next, substituting the expression for φ_h in terms of functions defined on the elements, we have

$$0 = \int_D \left[\nabla^2 \sum_e \varphi_e \right] \psi = \sum_e \int_{E_e} [\nabla^2 \varphi_e] \psi_e,$$

where ψ_e denotes ψ restricted to E_e. Note that the summation on the right is over all E_e that support ψ. For now, we will focus on the summands of this expression, $\int_{E_e} [\nabla^2 \varphi_e] \psi_e$.

Before proceeding, we modify the last expression. First, as ψ lies in V^h, then ψ_e lies in V_e^h. In particular, our expression states that $\nabla^2 \varphi_e$ is orthogonal to V_e^h. But then it must be orthogonal to any basis of V_e^h. Hence, we may rewrite $\int_{E_e} (\nabla^2 \varphi_e) \psi_e$ as $\int_{E_e} (\nabla^2 \varphi_e) N_i^e$, $i = 1, 2, 3, 4$. In particular, the weak form gives rise to four linear equations with unknown φ_e.

The next step is to apply the divergence theorem.

$$\int_{E_e} (\nabla^2 \varphi_e) N_i^e = \int_{E_e} (\nabla \cdot \nabla \varphi_e) N_i^e = \int_{E_e} \nabla \cdot (\nabla \varphi_e) N_i^e$$

$$= \int_{E_e} \nabla \cdot (\nabla \varphi_e N_i^e) - \int_{E_e} \nabla \varphi_e \cdot \nabla N_i^e$$

$$= \int_{\Gamma_e} (\nabla \varphi_e) \cdot \mathbf{n} N_i^e - \int_{E_e} \nabla \varphi_e \cdot \nabla N_i^e,$$

where $\mathbf{n}$ is an outward pointing normal. Finally, writing $\varphi_e = \sum_{j=0}^{4} \alpha_{e,j} N_j^e$, we get a linear system of equations

$$\sum_{j=0}^{4} \left[\int_{E_e} \nabla N_j^e \cdot \nabla N_i^e \right] \alpha_{e,j} = \int_{\Gamma_e} \left(\frac{\partial \varphi_e}{\partial \mathbf{n}} \right) N_i^e,$$

where the coefficients of φ_e are the unknowns. Alternatively, we write

$$K^e \alpha^e = R^e, \tag{3.1.3}$$

where K^e is the matrix $K_{i,j}^e = \int_{E_e} \nabla N_i^e . \nabla N_j^e$, α^e has jth entry $\alpha_{e,j}$ and R^e has ith entry $\int_{\Gamma_e} (\partial \varphi_e / \partial \mathbf{n}) N_i^e$. We have $\varphi_h = \sum_e \sum_{j=1}^{4} \alpha_{e,j} N_j^e$. This system is called *the element linear system*.

Step 4. Determine the global linear system

Solving (3.1.3) for each element would produce a piecewise polynomial function. But the result would not necessarily be continuous. To force continuity, we must assemble the element systems into a single global linear system. The requirement behind the assembly process that if a node is shared by two elements, then the global system must reflect this fact. Now, a node that belongs to two elements is dual to a basis function in each of the polynomial spaces. And each basis function associates to a coefficient α_i^e. Hence, each of the element linear systems contains constraints for the node, and we force simultaneity for the node by adding the corresponding constraints. To accomplish this, we use the node list that is associated to each element.

The element in the lower left corner in Figure 3.1.2 has (in counterclockwise order) nodes 1, 11, 12, and 2. Looking at the list of nodes for the first element, we have $\{1, 11, 12, 2\}$ and for the second, we have $\{2, 12, 13, 3\}$. With the notation of (3.1.3), $K_{4,3}^1$ is accumulated to position $(2, 12)$ of the global coefficient matrix and $K_{1,2}^2$ is also added to position $(2, 12)$ of the global matrix K. In general, each element has 4 nodes that are listed $\{n_1, n_2, n_3, n_4\}$, where $1 \leq n_i \leq 188$ for each i. The coefficient computed for $K_{i,j}^e$ is accumulated to position (n_j, n_i) of the global matrix. This procedure is conceptually awkward but easy to program. Looking back at Step 3, we see that this accumulation process has merely recreated the linear system that we decomposed into the local linear systems.

Provided we accumulate the entries of R^e into a 188-tuple R, the result of this step is the global linear system,

$$K\alpha = R. \tag{3.1.4}$$

We look more closely at the entries of R. They are accumulated from expressions of the form,

$$R_i^e = \int_{\Gamma_e} \frac{\partial \varphi_e}{\partial \mathbf{n}} N_i^e,$$

where the integral is executed in counterclockwise fashion. Also, as Γ_e is the boundary of a rectangle, we can divide it into 4 edges, $\Gamma_{e,i}$, $i = 1, 2, 3, 4$. If an edge is interior, then that edge appears as the boundary of two adjacent rectangles. In addition, the edge is traversed in opposite direction. For instance, the edge connecting $Node - 2$ and $Node - 12$ is common to the first and second elements, $\{1, 11, 12, 2\}$ and $\{2, 12, 13, 3\}$. However, in element 1, the edge appears as $Node - 12 \rightarrow Node - 2$. For element 2, the edge appears as $Node - 2 \rightarrow Node - 12$.

Next, we claim that the integrand $(\partial \varphi_e / \partial n) N_i^e$ must be the same for common edges. Indeed, we suppose that φ_h is continuous. Hence, if elements E_e and $E_{e'}$ are adjacent along an edge, then $\varphi_e = \varphi_{e'}$ along the same edge. It remains to consider the basis functions N_i^e. Only two of the basis functions are non-zero along this edge. One evaluates as a line connecting the first end point at height 1 to the second at height 0. The second is the reverse. And the same is true for both elements.

Therefore, the integrands are the same, but the integration for one element is the reverse direction of the integration for the second.

The result is that we accumulate zero. Hence, the entry for R at any interior node is zero. The remaining entries of R are associated to boundary values.

Step 5. Set Boundary values

The purpose of boundary conditions is to introduce the specific context. The same differential equation may arise in completely unrelated settings. At this point, we have rendered the continuous PDE in a discrete form. It is now possible to implement the discrete boundary values.

There are three common types of boundary conditions. We will only consider two here: Dirichlet conditions and Neumann conditions. In the first case, we designate a value of φ or equivalently φ_e at a boundary location. In the second case, we set a value for the normal derivative at a location. We have already seen Dirichlet boundary values in the context of FDM. For FEM, we implement Dirichlet boundary values exactly as with FDM. We will therefore concentrate on Neumann conditions.

We can divide the boundary into the channel edges at the top and bottom and the inflow/outflow edges at the left and right. We will suppose that the fluid does not penetrate the channel boundary. We implement this decision by setting $\partial \varphi_e / \partial \mathbf{n} = 0$ at the top and bottom. Therefore, $R_i = 0$ for all nodes along those edges. In order to generate flow, there must be a force that pushes the fluid through the channel. We can model this by setting $\partial \varphi_e / \partial \mathbf{n} = -1$ along the inflow edge and $\partial \varphi_e / \partial \mathbf{n} = 1$ at the outflow edges. We use a negative value at the inflow as the normal is outward pointing.

Suppose that the kth node lies on the inflow edge. Further, suppose that this node belongs to E_e and that N_i^e equals 1 at the node. With the notation set, we have

$$R_i^e = \int\limits_{\Gamma_e} \frac{\partial \varphi_e}{\partial \mathbf{n}} N_i^e = -\int\limits_{\Gamma_e} N_i^e,$$

which is easily calculated from the definition of N_i^e. This is the value that is subsequently accumulated to R to yield (3.1.4).

In order for the linear system to have a unique solution, we must set one Dirichlet boundary point. For this purpose, we impose $\varphi_h = 0$ at the node located at the lower front corner of the obstruction, node 61. Consequently, we have the corresponding row of the coefficient matrix set to the standard unit vector e_{61}. The result is a non-singular coefficient matrix.

In the next section, we will say more about boundary values.

Step 6. Solve the linear system

Apply the linear solve to the system in (3.1.4) for the vector α. For any (x, y) in U, we calculate $\varphi_h(x, y)$ as follows. Identify an element that contains (x, y). Suppose it is E_e with nodes $\{n_1, n_2, n_3, n_4\}$. Then, $\varphi_h(x, y) = (\alpha_{n_1} N_1^e + \alpha_{n_2} N_2^e + \alpha_{n_3} N_3^e + \alpha_{n_4} N_4^e)(x, y)$.

Step 7. Do post-processing

Post-processing depends on the particular setting. In our case, we want to show the vector field and the streamlines. To show the flow field, we will show the flow vector at each element centroid. If element E_e has nodes $\{n_1, n_2, n_3, n_4\}$ and centroid (x_e, y_e), then we get the flow vector by calculating $\nabla \varphi_h(x_e, y_e)$ (see Figure 3.1.4). The length of the velocity vectors shows the speed. We should expect the longest vectors to be at the top of the obstruction. They are near the upper corners of the obstruction. This is because of the singularity in the flow caused by the sharp corners.

Next, we infer the streamlines by applying forward Euler to the vector field. In particular, if (x, y) is a point on a streamline, then the next point will be at $(x, y) + \nabla \varphi_h(x, y)(\Delta x, \Delta y)$.

Finally, we state the FEM steps:

(1) Set the geometric model, define the elements and nodes.
(2) Determine the polynomial model.

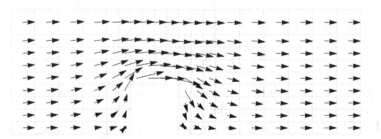

Figure 3.1.4.   The flow vectors at the element centroids.

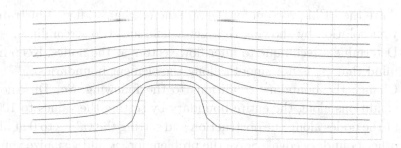

Figure 3.1.5. The streamlines calculated using forward Euler.

(3) Determine the element linear systems.
(4) Determine the global linear system.
(5) Set Boundary values.
(6) Solve the linear system.
(7) Do post-processing.

Exercises

1. Consider the following diagram.

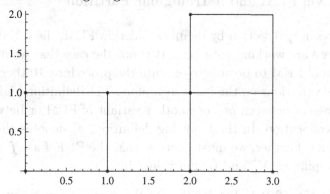

a. Number to nodes and elements.
b. Identify the node numbers for the vertices of each element.
c. Suppose that the element matrices are all equal to

$$K^e = \begin{pmatrix} 2 & 1 & 0 & 1 \\ 1 & 2 & 1 & 0 \\ 0 & 1 & 2 & 0 \\ 1 & 0 & 0 & 2 \end{pmatrix}.$$

Compute the global matrix K.

2. Solve the FEM problem for the obstructed channel flow. Illustrate your solution by showing the vector field and two streamlines.

3. Determine a technique to infer flow vectors at the nodes. Keep in mind that φ_h is not smooth along the element boundaries.

4. Change the boundary conditions to the following. Set Dirichlet conditions along the entire boundary by setting the values to 100 at the nodes along the obstruction and setting the values to 0 at all other boundary edges. Solve the problem for φ_h and visualize your solution by drawing level curves for φ_h. Recall that the Laplace equation will also represent a steady-state heat problem. With these boundary conditions, you have setup a heat problem. In particular, think about a flat rectangular plate with a notch at one edge. There is a heat source at the notch. The heat flows from the notch, across the plate and escapes at the other edges. The function φ_h now represents the temperature. The level curves you have drawn are the isotherms.

3.2. Waves in an Enclosed Pool, Helmholtz Equation via FEM and a Triangular Partition

We begin this section by defining Galerkin FEM, the variant of FEM that we are working with here. It is not the case that we need a definition of FEM to be able to execute the procedure. Rather, there are many variations on the basic technique. The definition highlights the differences between one or another variant of FEM or between FEM and collocation. In the following definition, σ_2 denotes the L^2 inner product. Further, we must suppose that the PDE $Lu - f$ determines a mapping on V^h and taking values in L^2.

Definition 3.2.1. Let V^h denote the space of continuous piecewise polynomial functions associated to a finite element partition of a domain D and let $Lu - f$ be a PDE defined on D. Then $\varphi_h \in V^h$ is the *FEM solution* to the PDE provided $\sigma_2(L\varphi_h - f, \psi) = 0$ for every ψ in V^h.

Our goal in this section is to implement FEM for a triangular partition. Triangular partitions are able to approximate a wide variety of shapes, many more than is possible with rectangular partitions. On the other hand, the mathematics is more complex. To provide a context for the implementation, we look at the Helmholtz equation. This equation is associated to the second-order wave equation. In this context, it arises from several different and interesting settings. These include acoustic vibrations, electromagnetic waves and chemotaxis. For our demonstration, we will consider a standing wave in a shallow pool.

To begin, we look at the order 2 forced wave equation.

$$\nabla^2 u - \frac{1}{c^2}\frac{\partial^2 u}{\partial t^2} = F(x,t),$$

where $c > 0$ is the speed of propagation of the unforced wave. If the external forcing function F represents a periodic vibration with frequency ω, then we may write it as $F(x,t) = f(x)e^{i\omega t}$. Factoring u as $u(x,t) = \varphi(x)e^{i\omega t}$, then upon substitution, we arrive at

$$\nabla^2 \varphi + \frac{\omega^2}{c^2}\varphi = f(x).$$

And in the unforced case, we have the Helmholtz equation.

$$\nabla^2 \varphi + \frac{\omega^2}{c^2}\varphi = 0. \tag{3.2.1}$$

Note that equation (3.2.1) is an eigenvalue problem for the linear operator ∇^2 defined on an appropriate function space. In its discrete form, (3.2.1) becomes an eigenvalue problem for a linear operator on a finite dimensional vector space. The eigenvalue is often referred to as the wave number. In this context, it may be written as $2\pi/gT$ where g is gravitational acceleration and T is the wave frequency. In the context of an electron, (3.2.1) is the steady state from the Schrodinger equation and the eigenvalue is $\sqrt{2mE}/\hbar$ where $\hbar$ is Planck's constant, m is the particle mass and E is kinetic energy.

We are interested in modeling the following situation. Suppose you are looking at a shallow pool. The water is still. The shape of the pool is rectangular and enclosed by smooth concrete walls. You stand at one edge and drop pebbles into the placid water. This action will generate a series of concentric circular ripples emanating from where the pebble fell. As the ripples travel across the pool, they strike the walls and reflect back to intermingle with the following waves as well as other waves reflecting from the other sides. In time, the waves will combine to form an overall choppy surface. The once distinguishable wave patterns are lost in an apparently chaotic intermingling of all the waves. This is the state we want to model.

We start with a rectangular pool. The problem is to determine the depth of the water (measured as the offset from quiescence) at any location. Hence, we have a 2D problem and will use the 2D version of (3.2.1). Let $\varphi = \varphi(x, y)$ be the dependent variable representing the wave height above quiescence at location (x, y). Additionally, we suppose that the depth at quiescence is constant and equal to one unit. Now, we traingulate the region to yield a finite element partition (see Figure 3.2.1).

With the geometry set, the next step is to define the polynomial spaces V_e^h. Given triangular elements, the usual means for setting V_e^h is through a *reference element* technique. We begin with a triangle R with vertices $(0, 0)$, $(1, 0)$ and $(0, 1)$. R is called the reference triangle. Using the technique developed in Section 1.3, we define the three degree 1 polynomials on R dual to the vertices. If P_j denotes the jth vertex and φ_i is the ith polynomial, then we want $\varphi_i(P_j) = \delta_{i,j}$.

Suppose the E_e is an element with vertices $(\alpha, \beta), (\gamma, \delta)$, and (μ, ν) and define the *affine mapping* $A = A(T, (\alpha, \beta))$ where

$$T = \begin{pmatrix} a & c \\ b & d \end{pmatrix},$$

$a = \gamma - \alpha$, $b = \delta - \beta$, $c = \mu - \alpha$, $d = \nu - \beta$. In particular, $A(T, (\alpha, \beta))(x, y) = T(x, y) + (\alpha, \beta)$. Now, it is easy to verify that A maps R to E_e. Moreover, $A^{-1} = A(T^{-1}, -T^{-1}(\alpha, \beta))$ maps the element E_e to the reference R. Setting $\varphi_i^e = \varphi_i \circ A^{-1}$, $i = 1, 2, 3$, defines polynomial functions on the element. And since the φ_i are dual to the vertices of R, then the three polynomials φ_i^e are dual to the vertices of E_e. Hence, we use these three polynomials to define the function space V_e^h. Next, we define V^h exactly as in the prior section. In particular, V^h is contained in the continuous, piecewise polynomial functions associated to the partition in Figure 3.2.1.

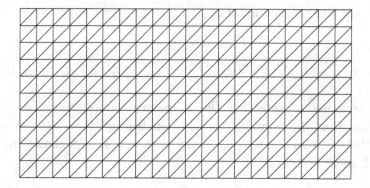

Figure 3.2.1. Rectangular domain with a triangular partition.

The next thing to do is to write the weak form of the PDE and use it to build the element linear system. For the equation in (3.2.1), the weak form is

$$0 = \int_D (\nabla^2 \varphi_h + \lambda \varphi_h)\psi = \sum_e \int_{E_e} (\nabla^2 \varphi_e + \lambda \varphi_e)\psi_e. \qquad (3.2.2)$$

Note that we have carried forward the notation from the prior section. In particular, φ_e and ψ_e are elements of V_e^h. In addition, the summation is over all elements that support ψ. Applying the divergence theorem to the diffusion term yields

$$\int_{E_e} (\nabla^2 \varphi_e + \lambda \varphi_e)\psi_e = \int_{E_e} (\nabla^2 \varphi_e)\psi_e + \lambda \int_{E_e} \varphi_e \psi_e$$

$$= \int_{E_e} \nabla \cdot (\nabla \varphi_e)\psi_e - \int_{E_e} \nabla \varphi_e \cdot \nabla \psi_e + \lambda \int_{E_e} \varphi_e \psi_e$$

$$= \int_{\Gamma_e} \left(\frac{\partial \varphi_e}{\partial \mathbf{n}}\right)\psi_e - \int_{E_e} \nabla \varphi_e \cdot \nabla \psi_e + \lambda \int_{E_e} \varphi_e \psi_e.$$

If we replace φ_e by $\sum_{j=1}^3 \alpha_{e,j}\varphi_j^e$ and ψ with φ_i^e, we have the element system,

$$\sum_{j=1}^3 \alpha_{e,j} \left(\int_{E_e} \nabla \varphi_j^e \cdot \nabla \varphi_i^e - \lambda \int_{E_e} \varphi_j^e \varphi_i^e \right) = \int_{\Gamma_e} \left(\frac{\partial \varphi_e}{\partial \mathbf{n}}\right)\varphi_i^e. \qquad (3.2.3)$$

In matrix and vector formats,

$$(K^e - \lambda M^e)\alpha = R^e,$$

where $K^e_{i,j} = \int_{E_e} \nabla\varphi^e_j \cdot \nabla\varphi^e_i$, $M^e_{i,j} = \int_{E_e} \varphi^e_j\varphi^e_i$ and $R^e_i = \int_{\Gamma_e} (\partial\varphi_e/\partial\mathbf{n})\varphi^e_i$.

It remains to establish the boundary values. Since we have assumed that the boundary is non-penetrating, then we may suppose that $\partial\varphi_e/\partial\mathbf{n} = 0$ for every boundary edge. Hence, the assembled global linear system is

$$(K - \lambda M)\alpha = 0. \tag{3.2.4}$$

As in the case of the Laplace equation, the solution to (3.2.3) is not uniquely determined. As the equation has been reduced to a finite size linear system, the problem is that the coefficient matrix is singular. We resolve this by setting a Dirichlet boundary value. Usually, this is done so that $\lambda = 1$ is an eigenvalue. We do this by setting the first row of each matrix equal to $e_1 = (1, 0, 0, \ldots)$. The result is that e_1 is an eigenvector of eigenvalue 1 for a row equivalent linear system.

Before proceeding, note that equation (3.2.3) is the discrete form of the Helmholtz (3.2.1). When the eigenvalue problem is setup and solved, then each eigenvalue yields an eigenvector which determines the fluid displacement at each node. In addition, the size of the linear system is equal to the number of nodes. There will be as many eigenvalues as there are nodes. The larger ones are usually not useful as the resulting wave pattern is too irregular or choppy. The $\lambda = 1$ solution corresponds to the one known solution, $\varphi(x) = e^{-\|x\|}$. One solution is shown in Figure 3.2.2.

Before ending this section, we have a few comments. Some authors consider the triangulation displayed in Figure 3.2.1 as biased. Note for instance, the definite lower left to upper right pattern of wave in Figure 3.2.2. This is certainly reminiscent of the pattern of triangles in Figure 3.2.1. One means around this problem is to use the *Union Jack pattern* shown in Figure 3.2.3 (see Heubner *et al.* (2001)). Another means to triangulate a rectangular region uses Delaunay triangulation. Consider the following procedure.

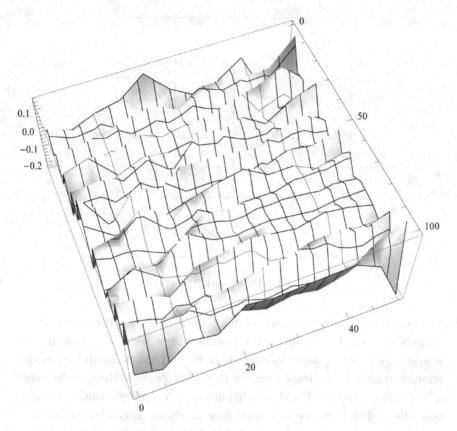

Figure 3.2.2. The solution for $\lambda = 3.01906$.

Step 1. Determine a nearly regular partition of the rectangle boundary. Use the Mathematica provided Delaunay triangulation procedure to triangulate the rectangle based on the list of boundary points.

Step 2. Determine the midpoint of each triangle edge, add these points to the boundary point list and execute the Delaunay triangulation procedure again.

Finally, we comment on boundary values. First of all, boundary values in the numerical setting are discrete nodal values. Nevertheless, we normally use the notation already developed for the continuous model. In the standard setting, Γ is divided into disjoint subsets

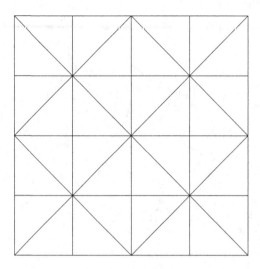

Figure 3.2.3.   Union Jack Triangulation.

Γ_D and Γ_N. Γ_D includes all nodes where Dirichlet boundary conditions are set. Γ_N includes the points where Neumann conditions are set. No node should belong to both and every boundary node should belong to at least one. In the continuous setting, it is usually obvious that Dirichlet conditions and Neumann conditions are equivalent. For instance in a fluid flow problem, Dirichlet conditions would often represent the velocity field and Neumann values would then represent the external force that is causing the flow. And of course, velocity and force are functionally equivalent. However, in the numerical setting, these interpretations are not necessarily valid. Hence, the equivalence is no longer obvious. In Part 2, Chapter 8, we revisit this question and derive a functional relationship between them. That is, every Neumann value has a corresponding Dirichlet value and vice versa. As with the continuous case, it is usually necessary to set a Dirichlet value in order to ensure that the coefficient matrix for the linear system is non-singular. In the flow setting, this point represents a stagnation point, a point of zero flow.

In addition to the usual implementation of Dirichlet and Neumann boundary values just described, it is possible to set both Neumann and Dirichlet values at some boundary nodes while setting no values at other nodes. This is called setting a *Robin value*. This situation naturally arises in the Black–Scholes equation. There are even settings where the researcher wants to set values at non-boundary points. For instance, there is the interface between heterogeneous materials in a solid and the edge of a crack that has occurred within a solid. These cases may be called *Nitsche values*.

Exercises

1. Solve the 2D Helmholtz equation on a triangular partition for a pool 100' by 50'. Partition the pool uniformly into 200 triangular elements. Display the result for three values of λ. Select a large, mid range and small value for λ. How do the different eigenvalues affect the image?
2. Repeat Exercise 1 using a rectangular model.
3. The configuration of triangles shown in Figure 2.4 is considered to be biased. An alternative called the Union Jack pattern introduced at the end of the section. Repeat Exercise 1 using a Union Jack triangulation.
4. Implement the Delauney-based triangulation procedure described in this section.
5. In Section 12.3, we develop several other FEM models that may be used for a triangular element configuration. Redo 2 using 6 nodal triangles with quadratic polynomials. How is the output changed?

Chapter 4

Population Models

Introduction

In this chapter, we introduce some basic population models. For each case, we are concerned with whether or not the model can produce reasonable results, something that one might observe in nature. On the one hand, we look at predator/prey interactions. As the dependent variable is population by time, then we are dealing with a set of interlocking ODE. We use these examples as an occasion to first introduce stability analysis and second to develop the Runge–Kutta method as the preferred means to simulate families of ODE (Gerald and Wheatley, 2003). On the other hand, population dynamics between competing groups is an area of current interest (see for instance, Wodarz and Komarova (2015) for applications to turmor development).

We also look at a mathematical model for herding. In this case, the Fokker–Planck equation and natural selection are used to guide the researcher to a model for directed movement. The analysis uses the birth/death function to determine the direction. In other terms, the direction of motion is governed by optimizing survival potential in the local environment. Simulating this model illustrates the formation of herds. In a later chapter, we will consider chemotaxis. In that case, the motion is induced by the presence of a chemical attractant, a signaling substance that initializes cell motility. As motion is involved here, the dependent variable is population density by location and time. Hence, the differential equation is a PDE. In our treatment of this problem, we will allow random perturbations in the direction variable. This provides an opportunity to introduce the stochastic collocation method.

In general, mathematical models for biological systems are much simpler than the event being modeled. As such, they only capture a small fraction of the actual process. The examples presented in this chapter are no exception. Nevertheless, these models do provide useful information. This is especially the case when used to simulate unobservable processes. Furthermore, as mathematical modeling of biological processes matures, the models capture more and more of the actual processes.

We have just mentioned stochastic collocation. This technique is a recent development (Xiu, 2010). PDE with random coefficients attempt to better address the actual complexity of most physical and biological processes. In particular, it is not reasonable to suppose that we can know the coefficients of a PDE exactly. Rather, they should only be known to us as random variables with mean and variance and perhaps known distribution.

In conjunction with Runge–Kutta, we introduce a technique to estimate the error inherent in any numerical solution to a differential equation. This theorem supposes that there are two techniques that estimate the solution at different orders of convergence. The error is then estimated by the difference between the two simulations.

The chapter is organized as follows. In Section 4.1, we introduce stability for a pair of ODE. For this purpose, we use three versions of the predator/prey model. Continuing this development, in Section 4.2, we introduce the Runge–Kutta numerical procedure as the means to simulate the predator/prey relationships. We end the chapter in Section 4.3 with stochastic collocation applied to a population model that considers group formation.

4.1. Predator/Prey Models, Stability Analysis

Predator/prey population models go back at least the early part of the last century. They are largely identified with the work of the Italian mathematician Volterra. In this section, we consider three separate models. We begin with a simplified version of the classical Volterra model. In this case, there is an analytic solution. Next, we turn to the classical Volterra Predator/Prey model. This model is not solvable. However, it is possible to infer information about a solution. Indeed, enough information is available in this manner to indicate that the model is better behaved than the one with known solution. Finally, we consider an extension of the basic Volterra model. In this case, the prey population is limited not only by the existence of the predator but also by the presence of a food resource. Again, we determine that the model does produce reasonable outcomes.

We propose that a successful population is one that neither goes extinct nor expands in an unsustainable manner. In other words, the population should tend towards a non-zero constant or orbit about a fixed value. In other terms, the model is stable. Our focus is on population models and stability. We use the examples as a means to define the concepts and motivate the analysis.

To begin, we consider two populations as a function of time $x = x(t)$ and $y = y(t)$ and suppose they are linked via the following pair of ODE, $x' = \alpha x - \beta y$ and $y' = \gamma x - \delta y$, $\alpha, \beta, \gamma, \delta > 0$. We see these as a pair of birth/death equations where y is the predator and x is the prey. Here, y benefits from the presence of x and is diminished by competition within its own group. On the other hand, x is diminished by the presence of the predator and enhanced by its own population level.

We may write this equation pair as a smooth curve in the real plane $t \to (x(t), y(t))$. The two ODE now yield the following linear relation:

$$\begin{pmatrix} x' \\ y' \end{pmatrix} = \begin{pmatrix} \alpha & \beta \\ \gamma & \delta \end{pmatrix} \begin{pmatrix} x \\ y \end{pmatrix}. \tag{4.1.1}$$

Recall that for any real $n \times n$ matrix, $\exp[tA] = \sum_{n=0}^{\infty} \frac{t^n}{n!} A^n$ converges to a non-singular matrix (see for instance, Loustau and Dillon (1993)). In addition, the function $t \to Exp[tA]$ represents a smooth arc in $\mathbb{R}^4$ with tangent vector $d\exp[tA]/dt = A\exp[tA]$. We set

$$\exp[tA] = \begin{pmatrix} \varphi_1(t) & \psi_1(t) \\ \varphi_2(t) & \psi_2(t) \end{pmatrix}$$

and multiply by a vector to get a smooth curve in $\mathbb{R}^2$.

$$\exp[tA] \begin{pmatrix} a \\ b \end{pmatrix} = \begin{pmatrix} a\varphi_1(t) + b\psi_1(t) \\ a\varphi_2(t) + b\psi_2(t) \end{pmatrix}.$$

It is easy to verify that $x(t) = a\varphi_1(t) + b\psi_1(t)$, $y(t) = a\varphi_2(t) + b\psi_2(t)$ satisfies (4.1.1) with initial conditions $x(0) = a$ and $y(0) = b$.

To better understand the solution, we consider the Jordan canonical form of A. For a 2×2 matrix, there are two possible Jordan forms:

$$J = \begin{pmatrix} \xi & 0 \\ 0 & \eta \end{pmatrix}; \quad J = \begin{pmatrix} \zeta & 1 \\ 0 & \zeta \end{pmatrix}$$

and take non-singular P with $P^{-1}JP = A$. In this context, there are three distinct cases depending on whether ξ, η are real or complex and conjugate or ζ is a single eigenvalue of multiplicity 2. Thinking of P as a change of basis transformation, or simply as a variable change, we rewrite (4.1.1)

$$\begin{pmatrix} \overline{x} \\ \overline{y} \end{pmatrix} = P^{-1} \begin{pmatrix} x \\ y \end{pmatrix}.$$

$$\begin{pmatrix} \overline{x'} \\ \overline{y'} \end{pmatrix} = P^{-1} \begin{pmatrix} x' \\ y' \end{pmatrix} = P^{-1}A \begin{pmatrix} x \\ y \end{pmatrix} = J \begin{pmatrix} \overline{x'} \\ \overline{y'} \end{pmatrix}.$$

We first consider the case, ξ, η real and compute

$$\exp[tJ] = \begin{pmatrix} e^{t\xi} & 0 \\ 0 & e^{t\eta} \end{pmatrix}.$$

This will result in the following solution:

$$\begin{aligned} x(t) &= C_1 e^{t\xi} + C_2 e^{t\eta}; \quad y(t) = D_1 e^{t\xi} + D_2 e^{t\eta}, \\ a &= C_1 + C_2, \quad b = D_1 + D_2. \end{aligned} \tag{4.1.2}$$

If ξ and η are complex conjugates, then $\exp[tJ]$ will have real and imaginary parts. Each part will yield a real solution in the format given in (4.1.2).

For the third case, we write $J = \zeta I + U$. Since $IU = UI$, it follows that $\exp[tJ] = \exp[t\zeta I]\exp[tU] = e^{t\zeta}(I + tU)$. Therefore, $\exp[tA] = e^{t\zeta}(I + tP^{-1}UP)$. The resulting solution will be

$$x(t) = ae^{t\zeta} + Cte^{t\zeta}, \quad y(t) = be^{t\zeta} + Dte^{t\zeta}.$$

To proceed from this point, we need to have a precise definition for stability. In particular, we consider a pair of order 1 ODE $x' = F(x, y)$, $y' = G(x, y)$. As before, we recast this in terms of the curve $(x(t), y(t))$ with tangent (x', y'). An isolated critical point, $(x(t_0), y(t_0))$ occurs when t_0 determines an isolated zero for the tangent (x', y'). If we consider the curve as a path, then there is no velocity at an isolated critical point. Hence, no object moving along the path can pass through such a point. For this reason, we refer to the location as a stagnation or equilibrium point.

Definition 4.1.1. The ODE pair is *stable* at the equilibrium point $\nu = (x_0, y_0)$ provided there is a neighborhood N of v and a point $(x(t_0), t(t_0)) \in N$, so that $(x(t), y(t))$ lies in N for all $t \geq t_0$. Additionally, the pair is *asymptotically stable* at v if there exists a neighborhood N of v so that if $(x(t), y(t)) \in N$, then $(x(t), y(t)) \rightarrow v$ as $t \rightarrow \infty$. Finally, if it is stable but not asympotically stable, then the equilibrium is called a *center*.

We return now to the prey/predator model. Since the function $D \rightarrow P^{-1}DP$ is a homeomorphism, then we may suppose that A is in Jordan form. Next, if A is singular, then one population is constant and the pair are not interacting. However, our basic problem involves interacting populations. Hence, we reject this possibility. If A is nonsingular, then there is a unique isolated equilibrium point at $(0, 0)$.

Continuing, for a positive eigenvalue, the corresponding solution increases without bound as $t \rightarrow \infty$ and converges to zero for negative eigenvalue. Hence, (4.1.1) is asymptotically stable provided the eigenvalue is negative and unstable if positive. Since the stagnation point is $(0, 0)$, then stability here means that the population is asymptotically converging to extinction. Alternatively, it grows out of control. Hence, even though the equation pair (4.1.1) is fully solvable, the result is not satisfying. This model does not represent an interesting biological setting.

In the next example, the equilibrium is a center.

The classical Volterra model is given by the equation pair,

$$x' = \alpha x - \beta xy; \quad y' = -\gamma y + \delta xy, \qquad (4.1.3)$$

where $\alpha, \beta, \gamma, \delta > 0$. The equation pair is well known to have no solution that can be written in terms of standard functions. However, we can get a great deal of information from the pair as is.

To begin, we look for a equilibrium point other than $(0, 0)$. Since the x and y are necessarily non-negative, then such a stagnation point should be interesting. If we set $x' = y' = 0$ and multiply the second equation by β/δ and add the pair, we get $x = (\beta\gamma/\alpha\delta)y$. Substituting into the first equation,

$$0 = \frac{\beta\gamma}{\delta}y - \frac{\beta^2\gamma}{\alpha\delta}y^2 = \left(\frac{\beta\gamma}{\delta} - \frac{\beta^2\gamma}{\alpha\delta}y\right)y.$$

Hence, $y = 0$ or $y = \alpha/\beta$. In the latter case, $x = \gamma/\delta$. Therefore, there is an isolated non-zero critical or equilibrium point. As noted, the curve cannot pass through the point $(\gamma/\delta, \alpha/\beta)$. We claim that the curve is stable here. We see shortly that the point is also a center.

Starting with (4.1.3) as an expression for dy/dx, we write

$$\frac{\alpha - \beta y}{y}dy = \frac{y - \delta x}{x}dx$$

and integrate. Next, raise the result of the integration over e to yield the following equation:

$$Kx^{-\gamma}e^{\delta x} = w = z = y^\alpha e^{-\beta y}, \tag{4.1.4}$$

where K depends on $(x(t_0), y(t_0))$ and where the equality holds for points $(x(t), y(t))$.

Both w and z have a single critical point at the equilibrium $(\gamma/\delta, \alpha/\beta)$. In each case, it is a minimum (see Exercise 3). Moreover, the minimum for z occurs at $\overline{y}$, which identifies two curve points $(\overline{x}_1, \overline{y})$ and $(\overline{x}_2, \overline{y})$. Similarly, the minimum for w identifies two more curve points $(\overline{x}, \overline{y}_1)$ and $(\overline{x}, \overline{y}_2)$. Since $w(\overline{x}_1) = w(\overline{x}_2)$, then we may suppose that $\overline{x}_1 \leq \overline{x}_2$ and conclude that for any t, $\overline{x}_1 \leq x(t) \leq \overline{x}_2$. Similarly, $\overline{y}_1 \leq y(t) \leq \overline{y}_2$. Therefore, the curve is bounded by the rectangle with side midpoints $(\overline{x}_1, \overline{y})$, $(\overline{x}_2, \overline{y})$, $(\overline{x}, \overline{y}_1)$, and $(\overline{x}, \overline{y}_2)$. Since the points on the curve are identified by values of $w = z$, then the curve is necessarily periodic. Hence, the point is a center. In the following section, we introduce numerical techniques to plot the interacting populations.

Next, we turn to yet another Volterra type predator/prey model. This time the prey is not only limited by interactions with the predator but also by a food source or resource (Previte and Hoffman, 2013).

$$x' = \alpha x - \kappa x^2 - \beta xy; \quad y' = -\gamma y + \delta xy, \quad (4.1.5)$$

where $\alpha, \beta, \gamma, \delta, \kappa > 0$. Note that there is net growth independent of predation exactly when $x < \alpha/\kappa$. This quotient is called the *carrying capacity*. Before proceeding, we simplify the equation slightly. First, replace t with τ/α so that $dx/d\tau = (1/\alpha)(dx/dt)$. Hence, (4.1.5) becomes

$$x' = x - \frac{\kappa}{\alpha}x^2 - \frac{\beta}{\alpha}xy; \quad y' = -\frac{\gamma}{\alpha}y + \frac{\delta}{\alpha}xy. \quad (4.1.6)$$

Next, set $X = (\delta/\alpha)x$ and $Y = (\beta/\alpha)y$ so that

$$\frac{dX}{d\tau} = \frac{\delta}{\alpha}\frac{dx}{d\tau} = X - \frac{\kappa}{\delta}X^2 - XY,$$

$$\frac{dY}{d\tau} = \frac{\beta}{\alpha}\frac{dy}{d\tau} = -\frac{\gamma}{\alpha} + XY.$$

Finally, we simplify the notation by setting X to x, Y to y, τ to t, κ/δ to κ and γ/α to γ. Hence, we are now working with (4.1.5) in the equivalent form,

$$x' = x - \kappa x^2 - xy; \quad y' = -\gamma y + xy. \quad (4.1.7)$$

To identify equilibria, we set $x' = y' = 0$ and suppose that neither population will be zero. Hence, the second equation implies that $x = \gamma$. We insert this value for x in the first equation to get $0 = \gamma - \kappa\gamma^2 - \gamma y$ or $1 - \kappa\gamma = y$. In addition, with the carrying capacity equal to $1/\kappa$, then we have $1 > \kappa\gamma$.

To continue the analysis, we introduce Liapunov functions. For this purpose, we generalise the basic setting. Suppose we have an ODE system $dx/dt = G(x, y)$, $dy/dt = H(x, y)$ with isolated critical point (a, b). Additionally, we set D an open subset of the plane that contains all paths (x, y) for any initial condition. By applying a translation, we may suppose that the critical point is at the origin.

Definition 4.1.2. Given an ODE pair, a real valued function F defined on D is called a *Liapunov function* provided

i. F is C^1 on D,
ii. $F(0,0) = 0$ and $F(x,y) > 0$ otherwise,
iii. for the ODE pair, $\nabla F(x(t), y(t)) \cdot (x'(t), y'(t)) \leq 0$ provided $t > 0$.

We now prove

Theorem 4.1.1. *Let F be a Liapunov function for an ODE pair, then (x, y) is stable in a neighborhood of the equilibrium point. Further, if $\nabla F(x(t), y(t)) \cdot (x'(t), y'(t)) < 0$, then (x, y) is asymptotically stable.*

Proof. Let C be a circle about the origin. Since C is compact and F is continuous, then F takes a minimal value $M > 0$ on C. Since $F(0,0) = 0$, then there is a circle C_ρ of radius ρ about the origin with $F(x,y) < M$ on and inside C_ρ. Otherwise, there would be a sequence of points (x_i, y_i) that converges to the origin with $F(x_i, y_i) \geq M$ for every i. In this case, $F(0,0) > 0$. Now, take a curve and a t_0 with $(x(t_0), y(t_0))$ inside C_ρ. Since F is non-increasing along the path, then for all $t > t_0$, $F(x(t), y(t)) < M$. It now follows that the path must remain inside C. Hence, the first assertion follows.

For the second assertion, we want to prove that $(x(t), y(t)) \rightarrow (0,0)$ as $t \rightarrow \infty$. By Definition 4.1.2ii, it suffices to prove

$$\lim_{t \to \infty} F((x(t), y(t)) = 0. \qquad (4.1.8)$$

By hypothesis, $F((x(t), y(t))$ is strictly decreasing and non-negative on and inside C_ρ. Therefore, we may suppose that the limit (4.1.7) exists and is non-negative. If the limit is positive, then take r, $0 < r < M$ so that the curve $((x(t), y(t))$ remains inside the annulus between C_ρ and C_r for all t greater than a fixed t_0. Since F is C^1, we may write

$$F((x(t), y(t)) = F((x(t_0), y(t_0)) + \int_{t_0}^{t} \nabla F(x(\tau), y(\tau)) . (x'(\tau), y'(\tau)) d\tau.$$

But the right-hand side is bounded above by $F((x(t_0), y(t_0)) - \kappa(t - t_0)$ where $-\kappa$ is the maximal value of the derivative of F along any curve contained in the closed annulus. Since $t-t_0$ is unbounded, it follows that there are negative values of $F((x(t), y(t)))$. This contradicts the definition of a Liapunov function. The result now follows. $\quad\square$

Consider a function in the form $F(x, y) = ax + b\ln(x) + cy + d\ln(y)$. The basic form is suggested from the same considerations that gave (4.1.4). Computing the gradient $\nabla F = (a + b/x, c + d/y)$, we have $\nabla F(\gamma, 1 - \kappa\gamma) = (0, 0)$. Since any multiple of a Liapunov function also satisfies Definition 4.1.2, then we may take $\alpha = 1$. Now we have $1 + b/\gamma = 0$, $c + d/(1 - \kappa\gamma) = 0$. Hence, $b = -\gamma$ and $d = -c(1 - \kappa\gamma)$. Next, we set $c = 1$ and compute $\nabla F(x(t), y(t)).(x'(t), y'(t)) = -\kappa(x - \gamma)^2 < 0$. Therefore, condition iii of Definition 4.1.2 is satisfied and the equilibrium point is asymptotically stable. Using a result of Poincarè (Simmons and Robertson, 1991), we may verify that paths near the equilibrium spiral. We will demonstrate this in the next section.

Exercises

1. Consider the ODE pair given by (4.1.1), prove that if A is singular, then one population is constant.
2. Prove that the functions in equation (4.1.4) have a single extrema at the critical point.
3. Complete the derivation of (4.1.4). Prove that w and z both have minima at the equilibrium.

4.2. Visualizing Predator/Prey Population Models with the Runge–Kutta Method

In this section, we develop the often used *Runge–Kutta* method (Gerald and Wheatley, 2003; Hildebrand, 1974). This technique has a built-in corrector. When the estimator/corrector version is used, then it may be referred to as *Runge–Kutta–Fehlberg*. The idea behind Runge–Kutta is that it approximates the solution of a first-order ODE by developing a polynomial form for the function from values, which are conveniently available in the ODE. In addition, the first several terms of the Runge–Kutta polynomial is identical to the Taylor expansion. And the more Runge–Kutta terms that are computed, the more the expression coincides with the Taylor expansion. Hence, convergence for Runge–Kutta rests very solidly on the convergence for the Taylor series and the rightmost terms of the Runge–Kutta expansion.

We begin with an order 1 ODE, $u' = f(x, u(x))$ and set $\Delta x = h$. $u(x) = y$. As this procedure is based on the Taylor expansion, then we will need high order derivatives. In particular, we suppose that f and u are sufficiently smooth to support all necessary derivatives.

The basic Runge–Kutta method predicts y_{n+1} from x_n and y_n via

$$y_{n+1} = y_n + h(\alpha_0 f(x_n, y_n) + \alpha_1 f(x_n + \mu_1 h, y_n + b_1 h)$$

$$+ \alpha_2 f(x_n + \mu_2 h, y_n + b_2 h) + \cdots + \alpha_p f(x_n + \mu_p h, y_n + b_p h)),$$

$$(4.2.1)$$

where the α_i, μ_i, and b_i are to be determined.

For fixed values of p and α_i, it is easy to see that Runge–Kutta generalizes some of the standard elementary procedures for approximating the solution to an ODE. For instance, if $p = 0$ and $\alpha_0 = 1$ in (4.2.1), then we have the first-order truncated Taylor expansion or the forward Euler method. If $p = 1$, $\alpha_0 = \alpha_1 = 0.5$, and $\mu_1 = b_1 = 1$, then we have the corrected Euler method. And if $p = 1$, $\alpha_0 = 0, \alpha_1 = 1, \mu_1 = 0.5$ and $b_1 = (x_n, y_n)/2$, $\mu_1 = 0.5$, and $b_1 = f(x_n, y_n)/2$, then we have the midpoint method. To develop Runge–Kutta,

1. We develop (4.2.1) using Taylor expansions for each of the terms $f(x_n + \mu_i h, y_n + \lambda_i h)$.
2. Then collect up the coefficients of the powers of h.
3. Solve for each the α_i, μ_i, b_i so that the resulting expression corresponds to the Taylor expansion of u up to a given number of terms.

If we have r terms of the Taylor series for u, then the calculated value will converge to the actual $O(h^{r+1})$ provided the Runge–Kutta remainder is well behaved. Hence, we have built in a means to evaluate convergence.

In particular, we rewrite (4.2.1) as follows:

$$y_{n_1} = y_n + \alpha_0 k_0 + \alpha_1 k_1 + \alpha_2 k_2 + \cdots + \alpha_p k_p. \qquad (4.2.2)$$

so that the k_i are expressions in $b_i h$, the prior k_j and parameters set so the expression (4.2.2) has maximal overlap with the Taylor expansion. In particular,

$$k_0 = hf(x_n, y_n), \quad k_1 = hf(x_n + \nu_1 h, y_n + \lambda_{10}k_0),$$
$$k_2 = hf(x_n + \mu_2 h, y_n + \lambda_{20}k_0 + \lambda_{21}k_1)$$

and so forth.

In order to get a better idea of this, we look at the case $p = 1$,

$$y_{n+1} = y_n + \alpha_0 k_0 + \alpha_1 k_1$$

with $k_0 = hf(x_n, y_n)$, and $k_1 = hf(x_n + \mu_1 h, y_n + \lambda_{10}k_0)$. To simplify the notation, we write

$$f = f(x_n, y_n), \quad f_x = \frac{d}{dx}f(x_n, y_n), \quad f_{xx} = \frac{d^2}{dx^2}f(x_n, y_n)$$

and then express $f(x_n \mu_1 h, y_n + b_1 h)$ terms of the second degree Taylor expansion about (x_n, y_n). The result is

$$k_1 = h \left[f + (\mu_1 h f_x + \lambda_{10} h f f_y) + \frac{1}{2}(\mu_1^2 h^2 f_{xx} \right.$$

$$\left. + 2\mu_1 \lambda_{10} h^2 f f_{xx} + \lambda_{10} h^2 f f_{xy} + \lambda_{10}^2 h^2 f^2 f_{yy}) \right] + O(h^4)$$

$$= hf + h^2(\mu_1 f_x + \lambda_{10} f f_y) + \frac{h^3}{2}(\mu_1^2 f_{xx} + 2\mu_1 \lambda_{10} f f_{xy} + \lambda_{10}^2 f^2 f_{yy})$$

$$+ O(h^4).$$

Hence, in powers of h, we have

$$y_{n+1} = y_n + h(\alpha_0 + \alpha_1)f + h^2 \alpha_1(\mu_1 f_x + \lambda_{10} f f_y)$$

$$+ \frac{h^3}{2}\alpha_1(\mu_1^2 f_{xx} + 2\mu_1 \lambda_{10} f f_{xy} + \lambda_{10}^2 f^2 f_{yy}) + O(h^4). \quad (4.2.3)$$

On the other hand, we can express y_{n+1} as a Taylor polynomial about y_n. By comparing coefficients, we are able to assign values to the parameters so as to maximize the overlap between (4.2.3) and the Taylor polynomial. In particular, the Taylor expansion is

$$y_{n+1} = y_n + hf + \frac{h^2}{2}(f_x + ff_y)$$

$$+ \frac{h^3}{6}(f_{xx} + 2ff_{xy} + f^2 f_{yy} + f_y(f_x + ff_y)) + O(h^4)$$

and $\alpha_0 + \alpha_1 = 1$, $\alpha_1 \mu_1 = 1/2$, $\alpha_1 \lambda_{10} = 1/2$. Setting $\alpha_1 = c$, we get $\alpha_0 = 1 - c$, $\mu_1 = 1/2c = \lambda_{10}$. Substituting into (4.2.3) yields,

$$y_{n+1} = y_n + hf + \frac{h^2}{2}(f_x + ff_y) + \frac{h^3}{8c^2}(f_{xx} + 2ff_{xy} + f^2 f_{yy})$$

$$+ O(h^4), \tag{4.2.4}$$

$$k_0 = hf(x_n, y_n), \quad k_1 = hf\left(x_n + \frac{1}{2c}h, y_n + \frac{1}{2c}k_0\right).$$

The first three terms on the right-hand side of (4.2.4) are exactly the first three terms of the Taylor expansion. The fourth term of the Taylor expansion is

$$\frac{h^3}{6}(f_{xx} + f_x f_y + ff_{xy} + f_{xy} + f_y f_y + ff_{yy}).$$

Hence, subtracting the order 3 Taylor expansion from (4.2.4) yields the following expression for the order 3 Runge–Kutta method truncation error.

$$\frac{h^3}{24c^2}[(3 - 4c^2)f_{xx} + (6 - 4c^2)ff_{xy} + (3 - 4c^2 f)f_{yy}$$

$$- 4c^2 f_x f_y - 4c^2 f_y f_y] + O(h^4). \tag{4.2.5}$$

We remark that as the order of the Runge–Kutta expression increases, so does the number of free variables. In practical terms, there is a great deal of flexibility in selecting the Runge–Kutta coefficients. At order 0, there were no free variables. At order 1 there was one. At order p, there will be p.

It is usual to take $c = 1/2$, so that for $p = 1$,

$$y_{n+1} = y_n + \frac{1}{2}(k_0 + k_1) + O(h^3),$$

$$k_0 = hf(x_n, y_n), \quad k_1 = hf(x_n + h, y_n + k_0).$$

The usual implementations for $p = 2$ and 3 are

$$y_{n+1} = y_n + \frac{1}{4}(k_0 + 3k_2) + O(h^4),$$

$$k_0 = hf(x_n, y_n), \quad k_1 = hf\left(x_n + \frac{1}{3}h, y_n + \frac{1}{3}k_0\right), \qquad (4.2.6)$$

$$k_2 = hf\left(x_n + \frac{2}{3}h, y_n + \frac{2}{3}k_1\right)$$

and

$$y_{n+1} = y_n + \frac{1}{6}(k_0 + 2k_1 + 2k_2 + k_3) + O(h^5),$$

$$k_0 = hf(x_n, y_n), \quad k_1 = hf\left(x_n + \frac{1}{2}h, y_n + \frac{1}{2}k_0\right), \qquad (4.2.7)$$

$$k_2 = hf\left(x_n + \frac{1}{2}h, y_n + \frac{1}{2}k_1\right), \quad k_3 = hf(x_n + h, y_n + k_2).$$

Since the rate of convergence for the Runge–Kutta method is faster when p is larger than the special cases for $p = 3$ or 2 provides an occasion to use the following theorem. In general, when presented with two numerical methods, which converge at differing rates, then the difference between the results may be used to estimate the error (Zienkiewicz *et al.*, 2005).

Theorem 4.2.1. *Suppose $\overline{y}$ estimates y order $O(h^m)$ and $\hat{y}$ estimates y order $O(h^{m+1})$. If we set $e = |y - \overline{y}|$ and $\overline{e} = |\hat{y} - \overline{y}|$, then $e/\overline{e} \to 1$ as $h \to 0$.*

Proof. We set $\hat{e} = |y - \hat{y}|$ and compute

$$e = |y - \overline{y}| = |y - \hat{y} + \hat{y} - \overline{y}| \leq \overline{e} + \hat{e}.$$

Similarly,

$$\overline{e} - \hat{e} \leq |\overline{e} - \hat{e}| \leq e.$$

Therefore,

$$\overline{e} - \hat{e} \leq e \leq \overline{e} + \hat{e}.$$

Dividing this inequality through by $\overline{e}$, we get

$$1 - \frac{\hat{e}}{\overline{e}} \leq \frac{e}{\overline{e}} \leq 1 + \frac{\hat{e}}{\overline{e}}. \qquad (4.2.8)$$

The quotient $\hat{e}/\bar{e}$ is the ratio of two terms that both converge to zero. However, as the denominator converges to 0 order h^m and the numerator order h^{m+1}, then their ratio converges to 0 order h. In particular, both the left-hand side and the right-hand side of (4.2.8) converge to 1 order h. The result now follows $e/\bar{e} \to 1$ order h. □

Hence, we may conclude that $\bar{e}$, the difference between the two estimates, is a reasonable estimator for the error. This idea is the core of the Runge–Kutta–Fehlberg method.

Finally, we simulate the population models developed in Section 4.1. Since there are two dependent variables, then $f = f(t,x,y)$ and $g = g(t,x,y)$. The corresponding revised versions of (4.2.6) and (4.2.7) follow. Note that the expression for m_i involves both m_j and k_j with $j < i$. The same is true of k_i.

$$x_{n+1} = x_n + \frac{1}{6}(m_0 + 2m_1 + 2m_2 + m_3) + O(h^5),$$

$$y_{n+1} = y_n + \frac{1}{6}(k_0 + 2k_1 + 2k_2 + k_3) + O(h^5),$$

$$m_0 = hf(t_n, x_n, y_n),$$

$$m_1 = hf\left(t_n + \frac{1}{2}h, x_n + \frac{1}{2}m_0, y_n + \frac{1}{2}k_0\right),$$

$$m_2 = hf\left(t_n + \frac{1}{2}h, x_n + \frac{1}{2}m_1, y_n + \frac{1}{2}k_1\right),$$

$$m_3 = hf\left(t_n + \frac{1}{2}h, x_n + m_2, y_n + k_2\right),$$

$$k_0 = hg(t_n, x_n, y_n),$$

$$k_1 = hg\left(t_n + \frac{1}{2}h, x_n + \frac{1}{2}m_0, y_n + \frac{1}{2}k_0\right),$$

$$k_2 = hg\left(t_n + \frac{1}{2}h, x_n + \frac{1}{2}m_1, y_n + \frac{1}{2}k_1\right),$$

$$k_3 = hg(t_n + h, x_n + m_2, y_n + k_2).$$

We want to model an instance of the ODE pair (4.1.3). In particular, we take

$$x' = 0.9x - 0.1xy, \quad y' = -0.1y + 0.5xy.$$

The equilibrium point is at (0.2, 9.0). For this demonstration, we take the initial state as (0.198, 9.0) and $h = 0.001$. The following plots show x, prey and y, predator through the first 20144 iterations of the two dependent variable Runge–Kutta method with $p = 3$.

The results are remarkably correct. There is no apparent error in the two plots. At iteration 20,144, $x = 0.198$ correct to 12 decimal places and $y = 9.0$ correct to 7 places. Next, we compare output for $p = 2$ and $p = 3$. At iteration 20,144, the corresponding values for y vary only at the 11th decimal place. The absolute error is approximately $3 * 10^{-11}$. By Theorem 4.2.1, this is an estimator for the actual error. Next, we look at the third population model for the

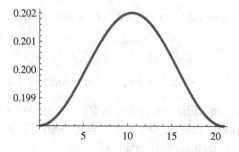

Figure 4.2.1.   The prey for 125 iterations.

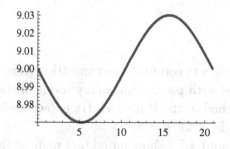

Figure 4.2.2.   The predator for 125 iterations.

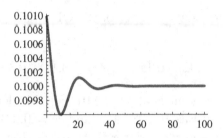

Figure 4.2.3.   The prey for 500 iterations.

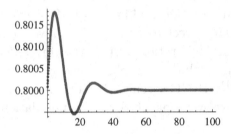

Figure 4.2.4.   The predator for 500 iterations.

following case:

$$x' = x - 2x^2 - xy; \quad y' = -0.1y + xy.$$

This time the equilibrium point is $(0.1, 0.8)$. We take $(0.101, 0.8)$ for the initial point and again set $h = 0.0001$. The following plots show x and y through 500 iterations.

As anticipated, it is asymptotically stable. The following plot shows (x, y). We see that it spirals to the equilibrium.

Exercises

1. Use Mathematica to compute the truncation terms for the Runge–Kutta method with $p = 2$. How many terms of the Taylor expansion are matched in the Runge–Kutta expansion?
2. Repeat Exercise 1 for $p = 3$.
3. Figures 4.1.1 and 4.1.2 show individual plots of the predator and prey for the basic Volterra model (4.1.3).

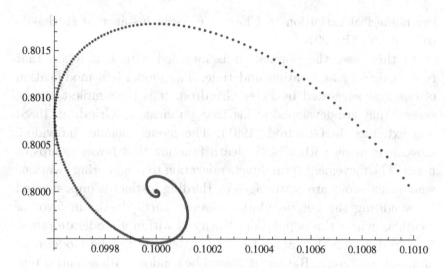

Figure 4.2.5. The model 4.2.5 is asymptotically spiral.

4. Execute the Runge–Kutta method ($p = 3$) for the differential equation $N''(t) = 0.09N(t) - B(t)N(t)^{1.7}$, where $B(t) = t^2$ and $N(0) = 100$ and $\Delta t = 0.01$ and $t \leq 5$. Plot the values of N.

5. Consider the equation pair,

$$\frac{dx}{dt} = r_x x \frac{1 - \alpha_{yx} x}{\kappa}, \quad \frac{dy}{dt} = r_y y \frac{1 - \alpha_{xy} y}{\kappa},$$

which represents two competing populations that share the same resource. This sort of model arises with cancer research where the single reseource is the blood supply available to the two cell types Wodarz and Komarova (2015). Consider the following specific setting, $\kappa = 0.5$, $r_x = 0.1$, $r_y = 0.5$, $\alpha_{yx} = 0.9$ and $\alpha_{yx} = 0.5$ and investigate the stability properties of the equation pair. In particular, select sevaral initial values for x and y and use Runge–Kutta to develop the particular instance.

4.3. Population Model with Herding Instinct Using Stochastic Collocation Method

In this section, we begin to develop a population model that reports density at location and time. We setup the problem here and reserve the numerical execution to Chapter 6 as an instance of stochastic collocation (Xiu, 2010).

In this case, the population is modeled with a function that reports density at location and time. The model is a modification of one first suggested by Peter Grindrod. It is the simplest among several that are developed in his 1988 publication (Grindrod, 1988) and extended in Grindrod (1991). The model includes individual movement along with a birth/death function that favors groups or herds. The movement term directs migration to neighboring locations where conditions are advantageous. Herding instinct is implemented by weighting the net population change (birth/death) in favor of locations where the population density is within a moderate range. Grindrod argues that animal movement should not be modeled as an intelligent event. Rather, it should be random with a central tendency. In particular, he refers to molecular motion described by the Fokker–Planck equation. In that case, the velocity vector is described by a distribution. In our case, the central tendency should be associated to a biological principle such as natural selection. In mathematical terms, the favored direction is parallel to the local gradient of the birth/death function. Hence, the direction is determined by local, not global, advantage.

We begin with a domain D and a time interval $[0, T]$. We will suppose that D is a compact subset of $\mathbb{R}^n$, $n = 1$ or 2. We let $u(t, x)$ denote the population density at the location x and time t for an animal that we refer to as A. We then model the change in density by time in terms of a directed flux term and a reaction term

$$\frac{\partial u}{\partial t} = -\nabla \cdot (vu) + f(t, x, u). \qquad (4.3.1)$$

In particular, f models the arrival of individuals into the system and v is the direction of motion. Since f denotes the net population change, then we may write $f = uE$, where $E(t, x, u) = $ *birth-rate minus death-rate*. In particular, we are supposing that no individual arrives or departs across the region boundary. Hence, (4.3.1) becomes

$$\frac{\partial u}{\partial t} = -\nabla \cdot (vu) + uE. \qquad (4.3.2)$$

A direction vector parallel to the gradient ∇E must be considered as optimal in terms of the local environment. In particular, it is advantageous for A to move toward a location where E is larger, an adjacent location where the chance of survival is greater. We designate this direction as w. Additionally, if we suppose that E is a function of u and independent (directly) from location and time, then $\nabla E = \left(\frac{dE}{du}\right)\nabla u$ and ∇u is parallel to w. Additionally, we take z to be a unit vector orthogonal to w and set $v = -\delta(\nabla u)/u) + \beta z + w$ where δ and β are small random parameters. In particular, β introduces small variability in direction and δ does the same for speed. If we suppose that δ is independent of location, then we derive the following from (4.3.2):

$$\frac{\partial u}{\partial t} = -\nabla \cdot (vu) + uE = -\nabla \cdot \left[\left(-\delta\frac{\nabla u}{u} + \beta z + w\right)u\right] + uE$$

$$= -\nabla \cdot [(-\delta\nabla u + (\beta z + w)u] + uE = \delta\nabla^2 u$$

$$- \nabla \cdot [(\beta z + w)u] + uE.$$

This yields a PDE model with two random coefficients,

$$\frac{\partial u}{\partial t} = \delta\nabla^2 u - \nabla \cdot (\beta z + w)u + uE. \qquad (4.3.3)$$

In order to make progress from here, we need to pin down E. For this purpose, we set $E(u) = (1 - u)(u - \alpha)$. Figure 4.3.1 shows a

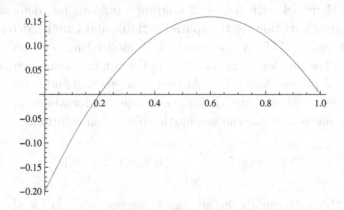

Figure 4.3.1.   Plot of E for $\alpha = 0.2$.

plot of E for $\alpha = 0.2$. Note that this function provides reproductive advantage to organisms that cluster into groups of moderate density. In addition, isolated organisms cannot reproduce, and hence, they are not persistent. We use this choice of E for our implementation, Eq. (4.3.4).

$$\frac{\partial u}{\partial t} = \delta \nabla^2 u - \nabla \cdot [(\beta z + w)u] + u(1 - u)(u - \alpha). \qquad (4.3.4)$$

As we proceed, we make a simplifying assumption. We suppose that δ is constant leaving β as the sole random coefficient.

The technique we want to introduce for representing the solution to a PDE with random coefficients is called stochastic collocation. It has aspects of Monte Carlo method. We begin by designating N locations in $P_1, \ldots, P_N$ in D and writing $\xi = (\xi_1, \ldots, \xi_N)$, where each $\xi_i = \xi_{P_i}$ is the $(0, 0.1)$ normal variable at P_i, then write $\beta = \beta(x, y; \xi)$ and $u = (t, x, y; \xi)$. Note that we designate the solution to the PDE as a function of space, time and the random variable. We see below that the quality of the solution will depend on the quality of the sampling locations, $P_1, \ldots, P_N$.

The next step is to define a sample set or stochastic collocation set $\Theta_N^Q = \{(\hat{\xi}_1, \ldots, \hat{\xi}_N) : \hat{\xi}_i \text{ is a sample value of } \xi_i, \ j = 1, \ldots, Q\}$. For each j, take $\beta = \beta(x, y; \xi^{(j)})$ so that (4.3.4) is now a usual PDE. Hence, we may resolve $u(t, x, y; \xi^{(j)})$ by any numerical procedure of our choice. For instance, as D is rectangular, we may choose to use the FEM model with degree 2 Lagrange polynomials defined on a rectangular partition for the spatial variable and Crank–Nicolson for time stepping. This is the same FEM model that we used in the channel flow problem of Section 3.1. In Section 6.4, we implement this numerical process for (4.3.4). At that time, we will find it convenient to use the element centroids for the designated locations, P_i.

It remains to write the stochastic collocation solution.

$$u(t, x, y; \xi) = \sum_{j=1}^{Q} (u(t, x, y; \xi^{(j)}) \Phi_j(\xi), \qquad (4.3.5)$$

where the polynomials Φ_j are the basis polynomials for the space of degree 1 polynomials in N variables. There are $N + 1$ such polynomials, therefore we must take $Q = N + 1$. Note that (4.3.5) represents the discrete solution to the PDE as piecewise polynomial interpolation in the spatial direction, spectral polynomial interpolation in the random coefficient and an FDM time stepping procedure.

Grindrod goes on to consider the two competing populations. In this case, we have two population density functions, u_1, u_2. If we set the corresponding $E_i = a_i - b_i u_i - c_j u_j$, $i \neq j$, and substitute into (4.3.4), then we have a pair of PDE representing the interaction of the two populations with spatial dispersion included. Note that if $b_i = 0$ and diffusion is ignored, then we have the Volterra equations discussed in Section 1, $u'_i = u_i E_i$. In general, if $a_2 < 0$, $b_i, c_i, a_1 > 0$, then the relationship is prey/predator. When all six parameters are positive, then two populations are competing and spatially exclusive. In this case, $b_1 c_2 < b_2 c_1$ indicates that interspecies competition dominates over intra-species competition. For δ large, diffusion dominates over net birth/death. For δ near zero, the opposite is true.

Exercises

1. For the two species population model discussed in this section,

   a. Derive the two equation model analogous to Eq. (4.3.4),
   b. Explain why a_2 negative and each of b_i, c_i, and a_1 positive determines a prey/predator relationship,
   c. Explain why diffusion is contradictory in a prey/predator setting,
   d. Explain the two species situation for the case $b_i, c_i, a_i > 0$ and $b_1 c_2 < b_2 c_1$.

Chapter 5

Transient Problems
in One Spatial Dimension

Introduction

The world is not a static place. Arguably, everything that we observe is in transition. Even if it is in effect static, it arrived at its current state via some transient process. There are researchers (Gresho and Sani, 1998) who write that all fluid flows have a life cycle. When simulating a flow, we should initiate the simulation at zero (no flow) and build up to the current state as the end state of a startup process. Alternatively, for the transient heat problem in Chapter 2, we saw the transition from an initial state to an intermediate state or even the asymptotic transition to a limit state. In that case, we used FDM to simulate the process. It was then when we first mentioned that all transient process time stepping is usually FDM. The population models in Sections 4.1 and 4.2 are transient processes but have no spatial dimension. However, the problem presented in Section 4.3 is a transient problem with a spatial dimension. In that case, there was also a stochastic element. We reduced the problem to one that would be solved using techniques developed here or in the next chapter.

Most often numerical techniques to model transient processes are hybrid techniques involving some method such as FEM applied to the spatial variables and FDM applied to the temporal. Besides FEM

115

or FDM, there are several other alternatives. In this chapter, we introduce one of these, collocation method.

We present two variations of collocation method: spectral collocation based on polynomial interpolation and Gaussian collocation based on an FEM style partition. In this case, the result is a piecewise polynomial function. In any case, collocation differs from FEM as it works with the strong form of the PDE. Moreover, no integrals are computed. Hence, it is faster to execute as it requires less computer power. On the other hand, collocation is often not as accurate or as broadly applicable as FEM. Alternatively, we can approximate the integrals that arise in FEM. However, this alternative introduces error into the process. And there is the approach introduced in Section 1.4. This approach is accurate and fast but not always applicable. We will return to the question of collocation verses FEM in later chapters.

Evaluating a transient process via a hybrid technique in effect amounts to a separation of variables. For instance, if we were to use FEM for the spatial variables and FDM for time, then the process would develop as follows. First, the FEM spatial rendering is executed under the assumption that $d\varphi/dt = 0$. Then, the resulting matrices are used to setup the FDM. In particular, we would write the piecewise polynomial FEM solution as $\varphi_h(t, x, y) = \sum_i a_i(t) B_i(x, y)$, where the B_i are continuous piecewise polynomial functions associated to the finite element partition and where the coefficients are time-dependent. Specifically, for the two-dimensional transient heat equation, we develop the weak form of the equation to yield the following linear relation, $M a_t = K a(t) + R$. Then, we apply FDM to the time derivative a_t. For instance, forward Euler is realized as a relation between time states, $M a^{n+1} = (M + (\Delta t)K)a^n + \Delta t R$. As expected, stability is an issue.

As in previous chapters, we present an array of different applications. We begin with the Black–Scholes equation for options pricing. In Topper (2005), financial PDEs are approached using the techniques developed here. Indeed, this author goes so far as to say that Gaussian collocation is the technique of choice for economists. This example returns us to the 1D heat equation in Section 5.2. Then, we look at the transport–diffusion equation. The specific context for this example is the problem of tracking the progress of a chemical spill. Finally, we look at a nonlinear PDE. In particular, we consider a model for traffic congestion based on vehicle density, system throughput and mean system vehicle velocity.

5.1. Options Pricing, the Black–Scholes Equation, and Collocation

Unlike the previous applications, this one is not part of everyone's daily experience. This time we must begin by explaining what a stock option is and why someone might want to buy one. As this section is introductory, we will restrict our attention to European single stock call and put options.

A *call option* contract is an agreement you make with the seller of a stock. When you purchase the call option, you are buying the right to purchase the stock in the future at an agreed price. The future date is called the contract expiration date or *execution date*. The agreed price is called the *strike price*. When you make the contract with the seller, you pay a fee, the value or *price of the option*. For instance, you contract with a dealer to buy 100 shares of Stock-A at 35 per share six months from now. The broker charges you (the price) 0.15 per share. The price is low because right now Stock-A is selling for only 30. The current price is called the *spot price*. The option is *out of the money* since the spot is less than the strike. So you pay 15 dollars and take your contract. Six months down the road, the stock is selling for 28. You let the contract expire and the option seller keeps the 15 dollars. But if the stock is selling for 37, you give the seller 3,500 dollars and he gives you the shares, which you immediately sell and pocket the difference. In this case, you pocket 200 dollars minus the 15 dollars you paid upfront and minus the interest you would have made on the 15 dollars if you had kept it in your pocket. In effect, you have made nearly 200 dollars on an investment of only 15 dollars plus interest.

A *put option* is the opposite case. For a put option, you agree to sell the stock. The option we have described is called *European* since there is a fixed expiration date. An option which can be executed at any time is called an *American option*. With Internet trading, truly American options are possible. Even so American options do have expiration dates. An option with certain fixed execution dates and a fixed expiration date is called a *Bermuda option*.

Why would someone want to buy a put option? Suppose you own a hundred shares of Stock-A. This gives your current holding a value of 3,000. So you buy a put option at 25 dollars per share paying 15 dollars for the option. Then, disaster strikes. At the execution date, the stock is only worth 22 per share. Now, your holding is only worth 2,200 giving you a net loss of 800 dollars. But, you exercise your option and sell your shares for 2,500 to the dealer making your loss 500 plus the 15 dollars you paid for the option and the interest you lost on the 15 dollars. In this case, the option buyer is hedging his investment. Holders of high risk investments often hedge in this manner. When their investments pay, then they return high profits to their investors. When they fail, the options set a cap on the loss. Funds often write call options (out of the money) against their current positions. Assuming that some items in their portfolio are always rising, then they have a steady cash flow.

Selling options is a lucrative business. Most options are sold against blocks of tens of thousands of shares. Should disaster strike across a broad spectrum of the market, then even a large reserve will not keep the broker afloat. Clearly, selling options is a risky business. You can sustain heavy losses very quickly. Think of the dealers who sold put options on Enron when it was worth 30 dollars a share and then in a matter of months sunk to 0.05 per share. The only protection that the seller has is his knowledge of the market. Hence, his ability to charge a fair price for a host of options on a broad range of stocks will depend on this knowledge. We will see that this in large measure amounts to estimating volatility. But even then there is no protection against market free-fall which occurred in October 2008.

Now, we need some notation so we can render this setting mathematically. The question here is how to price an option contract. Since the contract seller prices the contract, then we work from the seller's point of view. Let S denote the spot price, V the price of the option contract and K the strike price. As usual t is the time variable and T is the time to contract expiration. We let X denote the product price variable. This is a random variable with mean μ and standard deviation σ. In the financial world, μ is called the drift and σ is called the volatility. The interest rate r is the rate for a risk free loan, for instance, a US Treasury bill.

We now develop the Black–Scholes equation for pricing market options. There are several derivations of this equation in the literature (Espinosa and Jorgenson, 2003; Topper, 2005). The one we use is based on Ito's lemma. It requires levels of differentiability that may not be reflected by observation, hence many authors argue against it. However, it is efficient and arrives at the same result. It will do for our purposes.

First, if y is a real value lying in an interval, then Δy denotes the difference of two values of y. If $|\Delta y|$ is close to zero, then we write δy.

We make the following assumptions:

(1) S and X satisfy the following differential equation (Ito)

$$\delta S = \mu S \delta t + \sigma S \delta X. \qquad (5.1.1)$$

(2) The interest rate, r, is constant.
(3) The stock does not pay dividends or interest during the life of the option.
(4) Trading in the product is continuous. There are no transaction costs.
(5) There are no arbitrage opportunities.
(6) Short selling is allowed.

The usual terminology is that S satisfying (5.1.1) is an *Ito process*. An Ito process is a special case of a Wiener process or Brownian motion. Assumptions (2), (3), and (6) may be relaxed but will result in a more complicated model. The first part of item (4) states that t is a continuous variable. It is implicit in (5.1.1) and essential for this approach. As with (2), (3), and (6), the second item may be relaxed. Again, the resulting model will be somewhat more complicated.

By short selling, we mean that the seller of shares is allowed to sell shares that he has borrowed. The idea is that you expect the price to fall before you have to return the loan.

Item (5) is essential. Generally, an arbitrage is an investment without cost or risk. For instance, an investor could hold a position in a stock S, sell a call V_c and buy a put V_p both for the same strike K and expiration date T. Then, at expiration, one of the options would be executed leaving him $V_c - V_p + K - S$. Hence, in this case, the no arbitrage requirement is

$$K(e^{-rT}) = S - V_c + V_p,$$

indicating that the cost of the investment must be the interest on the expected return. In the literature, this case is referred to as *put/call parity is not violated.*

Ito's lemma states that a (sufficiently smooth) function of S (for instance $V = V(S,t)$, where S is also a function of t) will satisfy

$$\delta V = \left(\frac{\partial V}{\partial S}\mu S + \frac{\partial V}{\partial t} + \frac{1}{2}\sigma^2 S^2 \frac{\partial^2 V}{\partial S^2}\right)\delta t + \frac{\partial V}{\partial S}\sigma S\delta X, \qquad (5.1.2)$$

where t is the time to completion. In particular, $t = 0$ will give the price at completion. Now, if we solve (5.1.1) for $\sigma S\delta X$ and apply that to (5.1.2) to get

$$\delta V = \left(\frac{\partial V}{\partial t} + \frac{1}{2}\sigma^2 S^2 \frac{\partial^2 V}{\partial S^2}\right)\delta t + \frac{\partial V}{\partial S}\delta S. \qquad (5.1.3)$$

Equation (5.1.3) is precisely the reason for the general acceptance of the Black–Scholes approach. We have eliminated the random variable X. It is the least understood of all the variables associated to options pricing. Moreover, as a random variable, we cannot assume that X is much more than measurable. On the other hand, it is reasonable to suppose that S is a differentiable function of t and (as we have already done) suppose that V is a differentiable function of t and S.

Next, consider a portfolio of value Π consisting of $\partial V/\partial S$ shares of a stock at spot price S and a call option of price V, $\Pi = (\partial V/\partial S)S - V$. Over a short time interval, we may suppose that

$\partial V / \partial S$ is unchanged. Hence, for two very close time values, we have $\Pi_1 = (\partial V / \partial S)S_1 - V_1$ and $\Pi_2 = (\partial V / \partial S)S_2 - V_2$. Now, if we subtract the two equations,

$$\Pi_1 - \Pi_2 = \left(\frac{\partial V}{\partial S}\right)(S_1 - S_2) - (V_1 - V_2)$$

$$\sim \frac{V_1 - V_2}{S_1 - S_2}(S_1 - S_2) - (V_1 - V_2) = 0.$$

Hence, this is an essentially risk-less portfolio for at least a short interval of time. Using (5.1.3), we have

$$\delta\Pi = \frac{\partial V}{\partial S}\delta S - \delta V = -\left(\frac{\partial V}{\partial t} + \frac{1}{2}\sigma^2 S^2 \frac{\partial^2 V}{\partial S^2}\right)\delta t. \qquad (5.1.4)$$

Alternatively, to avoid arbitrage, the return of this risk-less portfolio over the short time interval is equal to the interest earned,

$$\delta\Pi = r\left(\left(\frac{\partial V}{\partial S}\right)S - V\right)\delta t. \qquad (5.1.5)$$

Equating (5.1.4) and (5.1.5), dividing by δt and rearranging terms yields the Black–Scholes equation:

$$\frac{\partial V}{\partial t} = -\frac{1}{2}\sigma^2 S^2 \frac{\partial^2 V}{\partial S^2} - rS\frac{\partial V}{\partial S} + rV. \qquad (5.1.6)$$

This is a second-order PDE that is often called a transport–diffusion equation. As such, it is in the same format as the equation encountered in Section 5.3. In the next section, we will see that it is equivalent to the 1D heat equation we saw in Section 2.2.

There are several means to generate numeric solutions to (5.1.6). In this section, we introduce one that has not yet appeared here. It is called orthogonal spline collocation (OSC) or Gaussian collocation. In subsequent sections, we will see another form of collocation method.

Gaussian collocation is based on a finite element partition. Hence, it uses the same piecewise polynomial model and has the same assembly process. However, it uses the strong form of the PDE. Therefore, there is no integration. As a result, it is easier to program and faster to execute.

We begin with a definition of collocation method. As usual, we have a PDE $Lu - f$, defined on a spatial domain D. But now there is a finite set Z of points of D. For (5.1.6), the spatial part is the right-hand side. The linear operator is

$$L = -\frac{\sigma^2 S^2}{2} \frac{\partial^2}{\partial S^2} - rS \frac{\partial}{\partial S} + rI,$$

where I is the identity operator and $f = 0$. For the next few paragraphs we focus on the method in general. After describing the general method, we return the specific case, the Black–Scholes equation.

If V is a space of piecewise polynomial functions on D, then we may denote by $\mathcal{P}_Z$ the piecewise polynomial interpolation operator, $\mathcal{P}_Z : C^0[D] \to V$. In particular, for v in $C^0[D]$, $\mathcal{P}_Z(v)(z) = v(z)$ for every z in Z.

Definition 5.1.1. Let Z be a finite subset of D. The *collocation solution* for the PDE $Lu - f$ on D is a piecewise polynomial function u_C that satisfies $\mathcal{P}_Z L u_C = \mathcal{P}_Z f$ where $\mathcal{P}_Z$ is a piecewise polynomial interpolation operator associated to Z and the partition. In this case, the points of Z are called *collocation points*.

Note that u_C is the collocation solution provided, $Lu_C(z) = f(z)$ for every z in Z. We will use this second formulation to define a system of linear equations. The solution of this system will yield u_C.

We now turn to the case at hand. With only one spatial dimension, then D is a closed interval. The elements are subintervals and the nodes are the interval end points. We begin by setting the notation. For element E_e with end points (nodes), P_{e_1} and P_{e_2}, we define the local polynomial space V_e^h as the degree 3 polynomials. In prior cases, the dimension of V_e^h was equal to the number of element nodes and we used that quantity to define a basis for V_e^h that is dual to the nodes. Now, there are two nodes and the polynomial space is dimension 4. In order to recover the idea of duality, we introduce degrees of freedom. Note that we revisit the idea of FEM partition and degrees of freedom in Chapter 8. At that time, we will expand on the ideas available here.

Definition 5.1.2. Suppose we are given an element E_e in an FEM partition with nodes denoted as P_j and with basis B for the local polynomial space V_e^h. Then, the local degrees of freedom are elements of $\text{Hom}(V_e^h, \mathbb{R})$ dual to the elements of B. Similarly, the global degrees of freedom are elements of a basis for $\text{Hom}(V^h, \mathbb{R})$ dual to the given basis for V^h.

For instance, if we select basis polynomials $H_{i,j}^e$, $j = 0, 1$, $k = 0, 1$ so that $f_{\alpha,\beta}(H_{j,k}^e) = \partial^\beta H_{j,k}^e P_\alpha = \delta_{j,\alpha}\delta_{k,\beta}$, then the $f_{\alpha,\beta}$ form a basis for $\text{Hom}(V_e^h, \mathbb{R})$ dual to the basis for V_e^h. The following four polynomials form a basis for the degree 3 polynomials on $[\alpha, \beta]$ and satisfy the relation $f_{\alpha,\beta}(H_{j,k}^e) = \delta_{j,\alpha}\delta_{k,\beta}$:

$$H_{00} = 3\left(\frac{\beta - x}{\beta - \alpha}\right)^2 - 2\left(\frac{\beta - x}{\beta - \alpha}\right)^3,$$

$$H_{01} = 3\left(\frac{x - \alpha}{\beta - \alpha}\right)^2 - 2\left(\frac{x - \alpha}{\beta - \alpha}\right)^3,$$

$$H_{10} = \frac{(\beta - x)^2}{\beta - \alpha} - \frac{(\beta - x)^3}{(\beta - \alpha)^2},$$

$$H_{11} = -\frac{(x - \alpha)^2}{\beta - \alpha} + \frac{(x - \alpha)^3}{(\beta - \alpha)^2}.$$

These polynomials are called the *Hermite polynomials*. Visually, the functions for the interval $[0, 3.5]$ are shown in Figure 5.1.1. Note that there are two degrees of freedom at each node.

Analogous to Section 3.2, we set a reference element to $[0, 1]$ and use affine transformations to determine the polynomials on each element. In particular, for element E_e with end points P_{e-1} and P_e, the map $A_e^h(x) = (x - P_{e-1})/(P_e - P_{e-1})$ sends $[P_{e-1}, P_e]$ to $[0, 1]$. For a given partition with mesh parameter h and element E_e, we write $\varphi_{e,i,j}^h = H_{ij}A_e^h$. Hence, the collocation solution $u_C = u_C(t, x)$ will take the form,

$$u_C(t, x) = \sum_e \sum_{i,j=0}^{1} a_{e,i,j}^h(t)\varphi_{e,i,j}^h(x). \tag{5.1.7}$$

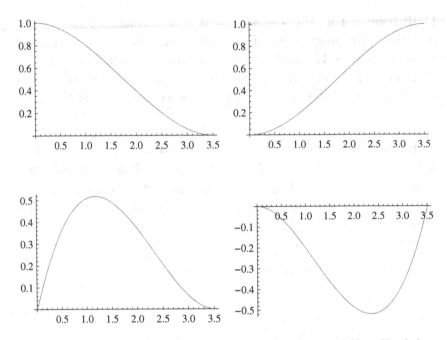

Figure 5.1.1.   The 4 Hermite polynomials $H_{0,0}$, $H_{0,1}$ above and $H_{1,0}$, $H_{1,1}$ below.

Note that we have separated x and t so that the polynomial model carries the spatial variable while the coefficients are functions of time. It may seem strange that we are treating time separately from space. It may seem more natural to just increase the number of polynomial variables. There are indeed reasons to do it as presented in (5.1.7). We will return later to this question.

The summation in (5.1.7) is over all elements. However, as in Chapter 3, there will be an assembly process. As a result, the actual number of distinct coefficients $\alpha_{e,i,j}^{h}(t)$ for any t is twice the number of nodes. This number is the number of degrees of freedom over all of the nodes. We accomplish this through the assembly by requiring $\alpha_{e,0,1}^{h}(t) = \alpha_{e+1,0,0}^{h}(t)$ and $\alpha_{e,1,1}^{h}(t) = \alpha_{e+1,1,0}^{h}(t)$. When this is done, then u_C will be C^1. Note that the common values associate to a nodal degree of freedom. For node P_e, we can denote the values as $B_{e,0}(t)$ and $B_{e,1}(t)$.

The next step in Gaussian collocation is to define the collocation points. First, we take ζ_0, ζ_1 to be the abscissa used for 2-point Gaussian quadrature. Next, with B_e^h the inverse of A_e^h, we set $z_{e,i} = B_e^h \zeta_i$ and $Z = \bigcup_e \{z_{e,0}, z_{e,1}\}$ In this case, we write the differential equation as $u_t = Lu - f$. Using (5.1.7) and Definition 5.1.1, u_C is given by the $2N$ equations,

$$\sum_{i,j=0}^{1} \frac{\partial}{\partial t} \alpha_{e,i,j}^h(t) \varphi_{e,i,j}^h(z_{e',k})$$

$$= \sum_e \sum_{i,j=0}^{1} \alpha_{e,i,j}^h(t) L\varphi_{e,i,j}^h(z_{e',k}) - f(z_{e',k}), \qquad (5.1.8)$$

where $z_{e',k}$ is in Z and N is equal to the number of intervals in the partition. Further, since the support of $\varphi_{e,i,j}^h$ is contained in E_e, then (5.1.8) reduces to

$$\sum_{i,j=0}^{1} \frac{\partial}{\partial t} \alpha_{e,i,j}^h(t) \varphi_{e,i,j}^h(z_{e,k}) = \sum_{i,j=0}^{1} \alpha_{e,i,j}^h(t) L\varphi_{e,i,j}^h(z_{e,k}) - f(z_{e,k})$$

$$(5.1.9)$$

for each e and k. These equations will assemble into a linear system of the form $M\alpha_t = K\alpha + R$. Each collocation point will determine an equation and each nodal degree of freedom will determine an entry of α. So, we expect $2N$ equations in $2N + 2$ unknowns. By setting two boundary values, we get a square system.

First, we load the matrices. To do this, we need only to identify each row and column with a particular collocation point and a particular basis polynomial. The collocation points are naturally ordered by the element number and then 0 or 1, $Z = \{z_{1,0}, z_{1,1}, z_{2,0}, z_{2,1}, \ldots\}$. We can then associate each collocation point to a matrix row by using the order in Z. For the columns, we sequentially list the nodal degrees of freedom, node number and 0 or 1. Hence, the four polynomials for a given element associate to columns as follows:

$$\varphi_{e,0,0}^h \to e, 0; \quad \varphi_{e,1,0}^h \to e, 1; \quad \varphi_{e,0,1}^h \to e+1, 0; \quad \varphi_{e,1,1}^h \to e+1, 1.$$

$$(5.1.10)$$

For each element, we compute (5.1.9) at the two collocation points in that element and the four basis polynomials load values to the matrices using the associations listed in (5.1.10). It is not difficult to devise a look up list, so that the assembly process becomes mechanical. For instance, the first two intervals would associate to rows (collocation points) 1, 2 and 3, 4 respectively and columns (degrees of freedom) $\{1, 2, 3, 4\}$ and $\{3, 4, 5, 6\}$. Alternatively, devise combinatorial formulae to a given element collocation point, element polynomial to a row and column. But, the mechanical option results in the more robust program.

The spatial domain of a numerical process is always a closed and bounded set, compact in $\mathbb{R}^n$. For the case at hand, it is an interval $[a, b]$. However, the domain for the Black–Scholes equation is the set of allowable spot prices, $(0, \infty)$. Hence, we need to take an interval with a close enough to zero and b large enough so that for practical purposes the corresponding interval will function like the half line $(0, \infty)$. The issue is the strike price K. We must have $a \ll K$ and $b \gg K$. For our example we will use 40 for the strike price and $a = 0.01$ and $b = 100$.

To set boundary values we have $u_C(t, 0.01) = 0$ and $u_C(t, 100) = 100 - Ke^{-r(T-t)}$, where $T - t$ is the time to maturity and r is the interest rate. We implement the boundary values by adding two rows to the M and K matrices and the R vector so that during time stepping u_C will take the required values. For instance, if we are implementing forward Euler time stepping, then we have $M\alpha_t = K\alpha + R$ and we compute.

$$\frac{1}{\Delta t} M(\alpha^{n+1} - \alpha^n) = K\alpha^n + R,$$

$$\alpha^{n+1} = \alpha^n + \Delta t M^{-1} K \alpha^n + \Delta t M^{-1} R.$$

Now u_C at the left-hand end point and at time t_n is α_1^n and at the right, it is α_{N+1}^n. So, we add two rows to M and K. We add a row equal to the vector e_1 and another equal to e_{N+1}. To accommodate these changes, we add two entries to R. One in position 1 to ensure that the first entry of $\Delta t M^{-1} K \alpha^n + \Delta t M^{-1} R$ is zero and a second in position $N+1$ to ensure that α_{N+1}^{n+1} is $100 - Ke^{-r(T-t_{n+1})}$. We now have a square system. However, R is now time sensitive.

For our implementation of the Black–Scholes equation (5.1.6), we set the following parameter values:

$$\text{Spot} - \text{Max} = 100, \quad \text{Spot} - \text{Min} = 0.01,$$
$$K = 40, \quad r = 0.1, \quad \sigma = 0.2,$$
$$T = 0.5 \ (= 6 \text{ months}), \quad \Delta t = 0.0005.$$

We know u_C at maturity, $u_C(T, x) = 0$ for $x < 40$ and $x - 40$ for $x \geq 40$ (see Figure 5.1.2). Our goal is to compute u_C at time 0, the price of the contract now. Hence, we revise time step from maturity $t = T$ to $t = 0$. Figure 5.1.3 shows the price of the contract using implicit time stepping.

Before ending the section, we comment on the degree of the polynomial model for collocation method. When using FEM, we are working from the weak form of the PDE. Inside the integral, there is the opportunity to decrease the order of integration using integration by parts. If the result only uses order 1 terms, then we can use degree 1 piecewise polynomial interpolation. Indeed, we did exactly that in Section 2.2. On the other hand, collocation does not arise from the weak form of the PDE. Hence, there is no opportunity to use

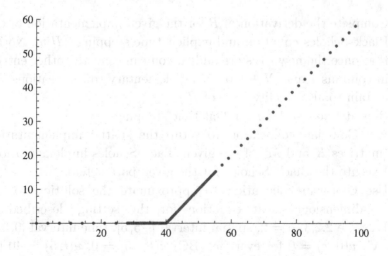

Figure 5.1.2.   The initial state or state at maturity.

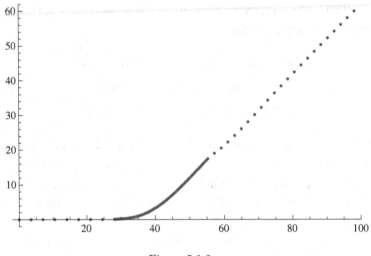

Figure 5.1.3.

integration by parts and so no opportunity to decrease the order of differentiation. Therefore, we must use at least a degree 2 model when using collocation.

Exercises

1. Complete the derivation of R for the given implementation of the Black–Scholes equation and explicit time stepping. (*Hint*: Notice that once the new rows are added, you can clear the other entries in columns 1 and $N+1$ by using elementary row operations. Be certain to also modify rows of R.)
2. Repeat Exercise 1 for implicit time stepping.
3. Use Gaussian collocation to setup the spatial implementation (matrices K and M) of the given Black–Scholes implementation.
4. Execute the Black–Scholes for the given parameters.
5. Use Gaussian collocation to approximate the solution to the one-dimensional heat equation for the setting described in Exercise 2.2.3, $\alpha = 2$, spatial interval $[-5, 5]$, time interval $[0, 20]$; (IC) $u(0, x) = 0$ for every x, (BC) $u(t, -5) = 0$, $u(t, 5) = 40$ for all $t > 0$. $\Delta t = 0.0001$.

5.2. Options Pricing, the Black–Scholes Equation, and the 1D Heat Equation

It is a remarkable fact that with an appropriate variable change, we can derive the Black–Scholes equation (5.1.6) from the one-dimensional heat equation (2.2.1) of Chapter 2. This means we can use the techniques developed in Chapter 2 and the experience gained there when dealing with options pricing issues. Indeed, if we were to start with an option pricing question as laid out in Exercise 1.1, then first we partition the spot price interval $[1, 100]$. Next, we use the variable transformation to recast this partition into the spatial axis of the heat equation. We may then solve the pricing issues via an FEM solution of the heat equation. Alternatively, we can start with a partition of the spatial variable in the heat equation, solve it via FDM or NDSolve and infer prices for a corresponding market option. In any event, at the end of this section, we will investigate these alternatives.

We start by considering (2.2.1) for the case of $\alpha = 1/2$. For our current purpose, it is convenient to use τ to denote time for the heat equation and t time for the Black–Scholes equation:

$$\frac{\partial u}{\partial \tau} = \frac{1}{2}\frac{\partial^2 u}{\partial x^2}. \tag{5.2.1}$$

Now, we set the following variable changes:

$$x = \log S + rt - \frac{\sigma^2 t}{2} \quad \tau = \sigma^2 t \tag{5.2.2}$$

or equivalently,

$$\mathrm{Exp}\left(x - rt + \frac{\sigma^2 t}{2}\right) = S, \quad t = \frac{\tau}{\sigma^2}. \tag{5.2.3}$$

In particular, S is a function of x and τ. Hence, using the chain rule and keeping in mind that r and σ are constants, we have

$$\frac{\partial}{\partial S} = \frac{\partial x}{\partial S}\frac{\partial}{\partial x} + \frac{\partial \tau}{\partial S}\frac{\partial}{\partial \tau} = \frac{1}{S}\frac{\partial}{\partial x} + 0,$$

since τ (time) is independent of S. Thus,

$$S\frac{\partial}{\partial S} = \frac{\partial}{\partial x}. \tag{5.2.4}$$

Again, using the chain rule along with (5.2.4), we have

$$\frac{\partial}{\partial t} = \frac{\partial x}{\partial t}\frac{\partial}{\partial x} + \frac{\partial \tau}{\partial t} = \left(r - \frac{\sigma^2}{2}\right)\left(S\frac{\partial}{\partial S}\right) + \sigma^2\frac{\partial}{\partial \tau}.$$

Hence,

$$\frac{\partial}{\partial \tau} = \frac{1}{\sigma^2}\left[\frac{\partial}{\partial t} - \left(r - \frac{\sigma^2}{2}\right)\left(S\frac{\partial}{\partial S}\right)\right]. \tag{5.2.5}$$

Next, from (5.2.4) and the product formula,

$$\frac{\partial^2}{\partial x^2} = S\frac{\partial}{\partial S}\left(S\frac{\partial}{\partial S}\right)$$

$$= \left(S\frac{\partial S}{\partial S}\right)\frac{\partial}{\partial S} + S\left(S\frac{\partial^2}{\partial S^2}\right) = S\frac{\partial}{\partial S} + S^2\frac{\partial^2}{\partial S^2}.$$

Hence,

$$\frac{\partial^2}{\partial x^2} = S\frac{\partial}{\partial S} + S^2\frac{\partial^2}{\partial S^2}. \tag{5.2.6}$$

Now, substitute (5.2.6) and (5.2.5) into (5.2.1),

$$\frac{1}{\sigma^2}\left[\frac{\partial u}{\partial t} - \left(r - \frac{\sigma^2}{2}S\frac{\partial u}{\partial S}\right)\right] = \frac{1}{2}\left(S\frac{\partial u}{\partial S} + S^2\frac{\partial^2 u}{\partial S^2}\right).$$

Expanding the left-hand side and multiplying through by σ^2, we have

$$\frac{\partial u}{\partial t} - rS\frac{\partial u}{\partial S} + \frac{\sigma^2}{2}\left(S\frac{\partial u}{\partial S} + S^2\frac{\partial^2 u}{\partial S^2}\right)$$

or

$$\frac{\partial u}{\partial t} - rS\frac{\partial u}{\partial S} - \frac{\sigma^2}{2}S^2\frac{\partial^2 u}{\partial S^2} = 0. \tag{5.2.7}$$

Setting $V = e^{-rt}u$, then

$$\frac{\partial V}{\partial t} = e^{-rt}\frac{\partial u}{\partial t} - rV, \quad \frac{\partial V}{\partial S} = e^{-rt}\frac{\partial u}{\partial S}, \quad \frac{\partial^2 V}{\partial S^2} = e^{-rt}\frac{\partial^2 u}{\partial S^2}.$$

Now, multiply (5.2.7) by e^{-rt} to get

$$\begin{aligned} 0 &= e^{-rt}\frac{\partial u}{\partial t} - rSe^{-rt}\frac{\partial u}{\partial S} - \frac{\sigma^2}{2}S^2 e^{-rt}\frac{\partial^2 u}{\partial S^2} \\ &= \frac{\partial V}{\partial t} + rV - rS\frac{\partial V}{\partial S} - \frac{\sigma^2}{2}S^2\frac{\partial^2 V}{\partial S^2}. \end{aligned}$$

Finally, recall that time in the Black–Scholes options pricing context is time to completion while time in the heat equation is forward stepping. To accommodate for this, we need to only replace $\partial V/\partial t$ by $-\partial V/\partial t$, bringing us to (5.1.6)

$$\frac{\partial V}{\partial t} = -\frac{1}{2}\sigma^2 S^2\frac{\partial^2 V}{\partial S^2} - rS\frac{\partial V}{\partial S} + rV.$$

To take full advantage of what we have just shown, we should be able to setup a discrete process for an options pricing problem, transform it to a heat equation setting, then execute the numerical solution procedure in the simpler context. There is a problem here as a regular partition of the spot price interval will not yield a homogeneous partition of the spatial variable for the heat equation. Hence, we may not use FDM. Alternatively, if we start with a regular partition of the spatial variable for the heat equation, then the corresponding partition for the spot price interval may not suit our purposes. The solution to the problem is to introduce transient FEM.

In transient FEM, we handle the spatial dimensions as in a finite element process and the temporal as a finite difference. We begin by defining the nodes and elements. First, define a partition of the spot price interval and use (5.2.2) to induce a partition of the corresponding spatial interval $[a, b]$ for the heat equation. Denote the partition $a = x_0 < x_1 < \cdots < x_{n-1} < x_n = b$. We designate the intervals as the elements $[x_{i-1}, x_i]$ and the end points as the nodes, yielding n elements and $n+1$ nodes. Since each element has two nodes, we need two basis functions for each element. These we select as the Lagrange polynomials in one variable,

$$N_1^e(x) = \frac{\beta - x}{\beta - \alpha}, \quad N_2^e(x) = \frac{x - \alpha}{\beta - \alpha},$$

where $E^e = [\alpha, \beta]$. Hence, $N_1^e(\alpha) = 1$, $N_1^e(\beta) = 0$, and $N_2^e(\alpha) = 0$, $N_2^e(\beta) = 1$.

Next, we setup the element matrices via the Galerkin method applied the one-dimensional heat equation. In this case, we use integration by parts:

$$0 = \int \left(\frac{\partial u^e}{\partial t} - \frac{1}{2} \frac{\partial^2 u^e}{\partial x^2} \right) N_j^e(x)$$

$$= \int \frac{\partial u^e}{\partial t} N_j^e(x) - \frac{1}{2} \int \frac{\partial^2 u^e}{\partial x^2} N_j^e(x)$$

$$= \int \frac{\partial u^e}{\partial t} N_j^e(x) - \frac{1}{2} \int \frac{\partial}{\partial x} \left(\frac{\partial u^e}{\partial x} \right) N_j^e(x)$$

$$= \int \frac{\partial u^e}{\partial t} N_j^e(x) - \frac{1}{2} \int \frac{\partial}{\partial x} \left(\frac{\partial u^e}{\partial x} \right) N_j^e(x) + \frac{1}{2} \int \left(\frac{\partial u^e}{\partial x} \right) \frac{\partial}{\partial x} N_j^e(x).$$

Hence, for each j,

$$0 = \sum_{i=1}^{2} \frac{\partial}{\partial t} a_i^e(t) \int_{\alpha}^{\beta} N_i^e(x) N_j^e(x)$$

$$+ \frac{1}{2} a_i^e(t) \int_{\alpha}^{\beta} \frac{\partial}{\partial x} N_i^e(x) \frac{\partial}{\partial x} N_j^e(x) - \frac{1}{2} \left(\frac{\partial u^e}{\partial x} \right) N_j^e(x) \Big|_{\alpha}^{\beta}.$$

Writing this in matrix form, we have

$$0 = M^e \begin{pmatrix} \dfrac{\partial}{\partial t} a_1^e(t) \\[2mm] \dfrac{\partial}{\partial t} a_2^e(t) \end{pmatrix} + \frac{1}{2} K^e \begin{pmatrix} a_1^e(t) \\[2mm] a_2^e(t) \end{pmatrix} - \frac{1}{2} \begin{pmatrix} \dfrac{\partial u^e}{\partial x} N_1^e(x) \\[2mm] \dfrac{\partial u^e}{\partial x} N_2^e(x) \end{pmatrix},$$

where

$$M_{ij}^e = \int_{\alpha}^{\beta} N_i^e(x) N_j^e(x), \quad K_{ij}^e = \int_{\alpha}^{\beta} \frac{\partial}{\partial x} N_i^e(x) \frac{\partial}{\partial x} N_j^e(x).$$

Now, we assemble the local matrices into a single global $n+1 \times n+1$ system,

$$0 = M\frac{\partial}{\partial t}a(t) + \frac{1}{2}Ka(t) - \frac{1}{2}R,$$

where the entries of R, the residual vector, are zero except possibly at the first and last positions, the boundary. Hence, R is where we would set Neumann conditions. But, our intention is to only set Dirichlet conditions. Therefore, without loss of generality, we may suppose that R is zero. We now have

$$0 = M\frac{\partial}{\partial t}a(t) + \frac{1}{2}Ka(t). \tag{5.2.8}$$

Next, we must develop (5.2.8) as a FDM. If we choose forward Euler, then replace the entries of $\partial a(t)/\partial t$ by forward differences, $\left(\frac{1}{\Delta t}\right)(a(t_{m+1}) - a(t_m))$ and expand,

$$Ma(t_{m+1}) = \left(M - \left(\frac{\Delta t}{2}\right)K\right)a(t_m) = Da(t_m). \tag{5.2.9}$$

At the left-hand end point, we set $a_1(t_m) = 0$ for every time step m and set the first row of M and D to the first row of the identity matrix. As with collocation, we set $a_{n+1}(t_m) = b - Ke^{-r(T-t_m)}$. Again, set the last row of M and D to the corresponding row of the identity matrix. Now, M is necessarily non-singular, so (5.2.9) reduces to $a(t_{m+1}) = M^{-1}Da(t_m)$. We now have a standard forward stepping Euler process.

Figure 5.2.1 shows the result of five treatments of the Black–Scholes equation. The parameters are exactly as in Exercise 1 below. The data provided by the closed form solution to the heat equation is generally considered correct. Note that collocation applied to the Black–Scholes is the only numeric treatment that is roughly equivalent to the analytic solution. This would indicate that collocation be used when the analytic solution does not apply. This is important as generally conditions are placed on options making the analytic solution not applicable.

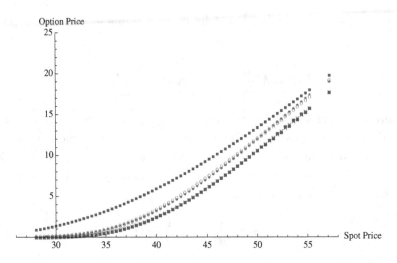

Figure 5.2.1.   Five treatments of the Black–Scholes equation. The closed form solution is plotted with "0", Gaussian collocation "∗", NDSolve Implicit FDM "×".

Exercises

1. Solve the Black–Scholes via forward stepping Euler using the following settings:

$$S \in [1, 100], \quad T = 0.5 \text{ (6 months)},$$
$$K = 40, \quad r = 0.1, \quad \sigma = 0.2.$$

2. Solve the one-dimensional heat equation using transient FEM. As in Section 1.3, set the spatial range to $[-5, 5]$ and the boundary values to zero. Compare the result with the output of *NDSolve*.

3. Some values associated to V are called hedging parameters. One in particular is $\Delta = \partial V / \partial S$.

   a. Plot Δ using the derivative of the collocation solution for V.
   b. Derive the PDE for Δ from the Black–Scoiles equation.
   c. Use Gaussian collocation to estimate the value of Δ.

4. The bond price equation (Ekstrom and Tysk, 2007) is given by

$$\frac{\partial U}{\partial t} = -\frac{1}{2}\sigma^2 \frac{\partial^2 V}{\partial S^2} - \beta \frac{\partial V}{\partial S} + SV, \quad U = 1 \quad \text{at } t = T,$$

where $U = U(t, S)$ denotes the fraction of face value as a function of time, t to maturity T and short term interest rate, S. In this equation, σ and β are functions of t and S. Supposing these parameters are constant, estimate U using Gaussian collocation.

5.3. River Pollution, the Transport–diffusion Equation via Collocation

In this section, we take a second look at collocation as a method for solving differential equations. This time we present the spectral version. This is one of the oldest techniques for generating numerical solutions to differential equations (Hildebrand, 1974). Nevertheless, the term *spectral method* has different meaning for different researchers. In our case, the meaning is to approximate the solution of a PDE with a polynomial of high degree. For other instances of spectral methods, see Xiu (2010); Shen *et al.* (2011). In Bellomo *et al.* (2008), the authors present it as the technique of choice for nonlinear initial value problems. In other circumstances, it refers interpolation with a trigonometric polynomial as might occur when using the discrete Fourier transform. The underlying mathematics that supports this method is polynomial interpolation. We know that polynomial interpolation is not always convergent (see Section 9.6). Hence, the computed results can be questioned. However, in practical terms everything depends on selecting a set of collocation points that works for the problem at hand.

In this section, we also introduce the idea of a dimensionless representation of a problem. We saw something in this direction in Section 4.1. In that context, it served a simplifying function. Briefly, a value in an expression is dimensionless if it is independent of the choice of units chosen for the problem. For instance, if ξ and $\mathcal{L}$ are positions along a curve measured in distance units from the initial point, then $\eta = \xi/\mathcal{L}$ is the relative position for ξ with respect to $\mathcal{L}$. No matter how we choose to measure distance (feet, meters, miles, kilometers, etc.) the value of η is unchanged. Hence, we say that η is a dimensionless quantity. Dimensionless values associated to a particular setting are critical to the development of technology. For instance, if we have

a partially obstructed fluid flow, then the significant values needed
to describe the flow are width of the obstruction, $\mathcal{L}$, the ambient
velocity of the flow, V, the density, ρ, and the viscosity, μ, of the
fluid. We combine these four values to form the Reynolds number,
$\rho V \mathcal{L}/\mu$. Measuring density as kg/m^3, velocity as m/s, width as m,
then the unit of measurement for the numerator of the Reynolds
number is (kg)/m $*$ s. In turn, viscosity is measured as Ns/m^2 =
$((\text{kg}) * \text{m/s}^2)(\text{s}/(\text{m}^2)) = \text{kg}/(\text{m} * \text{s})$. Hence, the Reynolds number is
dimensionless.

The importance of values such as the Reynolds number arises
naturally when we want to design an experiment to predict how a
particular proposed structure will function. For instance, if you want
to design an aircraft then you would like to test the design in an
experimental setting before committing to a full scale prototype. For
instance, you might build a model and measure the characteristics of
the smaller scale object. You would then infer characteristics of the
full scale design from your measurements on the model. But, important characteristics such as lift and drag are not simple functions of
scale. In order to ensure that the experiment is valid, you will need
to design the experiment so that the Reynolds number for the experimental and full scale setting is more or less the same. For air ρ/μ
is about $1.12 * 10^{-6}$. So, if the model scale is 10:1, then you must
increase the velocity by 10-fold in order to maintain the Reynolds
number. If this is not a practical option, you can change the fluid.
For instance, ρ/μ for water is about $1.46 * 10^{-5}$. Setting the velocity
in air as V_a and V_w for the velocity in water, then we must choose
data so that $1.12 \times 10^{-6} V_a = 1.46 * 10^{-5} \times (10^{-1}) V_w$. Hence, we
maintain the Reynolds number by decreasing the velocity by a ratio
of $1.46/1.12 = 0.767$. Aerodynamic designs are frequently tested in
water tunnels, not wind tunnels.

Additionally, since the nature of a flow is relatively constant
within broad Reynolds number intervals, it is not always necessary
to replicate exact values of the Reynolds number in order to have a
valid experimental platform.

We now turn to the problem at hand. We are all familiar with
the havoc caused by a chemical spill in a river or channel. In the

following example, we develop a model that measures the chemical concentrations in a river as the chemical spreads by diffusion and is transported by the current. The independent variables will be time and position along the channel. The primary dependent variable is chemical concentration. Intermediate dependent variables (depending on position) are depth or volume of the channel, velocity of the flow and slope of the river measured against the free surface. In another direction, all chemicals are more or less stable. The rate of decay as a function of time will be important. Finally, a slow but ongoing trickle of chemical pollutant will likely function differently than a fast sudden dump. We will need a function that describes the spill. It will be a function of both time and location.

We model the river or channel as a curve $\theta = \theta(\xi)$ in $\mathbb{R}^2$ which we suppose is parameterized by arc length, ξ. For each position along the channel, ξ, we set a volume function h and velocity function v via

$$h(\xi) = \frac{5}{4}\left(1 + \frac{\xi}{\mathcal{L}}\right)^3,$$

(5.3.1)

$$v(\xi) = \frac{3}{4}\left(1 - \frac{\xi}{\mathcal{L}}\right)^{\frac{1}{4}}.$$

(5.3.2)

Since $\xi/\mathcal{L}$ is also dimensionless, then the right-hand factor of each expression is dimensionless. However, $5/4 = h(0)$ and $3/4 = v(0)$ are dimensioned in length and velocity units, respectively. Note we have modeled volume as a cubic power of location. In turn, the velocity function is more or less level through most of the domain and then rapidly drops to zero at the river mouth. When we actually implement this model, we would not know h or v. Most likely we would take several measurements and then use an interpolation.

Finally, we must set the gradient or the river free surface angle. As the velocity is set to be nearly constant, we will suppose that the surface gradient is also constant. We choose $i_b = 10^{-3}$.

For this problem, the primary dependent variable is the concentration of the chemical. In order to keep the number of variables at a minimum, we consider only the depth average concentration, $C(\tau, \xi)$

at a time τ and a location ξ. The basic equation is

$$\frac{\partial C}{\partial \tau} = \frac{\partial}{\partial \xi}\left(K(\xi)\frac{\partial C}{\partial \xi}\right) - \frac{\partial}{\partial \xi}(v(\xi)C), \qquad (5.3.3)$$

where

$$K(\xi) = k_0 h(\xi)\sqrt{gi_b h(\xi)} = k_0(9.8 \times 10^{-3})^{\frac{1}{2}}h(\xi)^{\frac{3}{2}} \qquad (5.3.4)$$

since the gravitational constant $g = 9.8$ and i_b is taken to be 10^{-3}. Note that this is a diffusion/transport equation similar to the Black–Scholes (Section 5.1) and the herd formation equation (Section 4.3).

If we suppose that the spill is a discrete event and that the chemical is stable within the time period for the observation, then (5.3.3) is sufficient to describe the concentration. Otherwise, we need a decay term and a source term. In particular, decay is written as $-\lambda C(\tau, \xi)^m$ with $\lambda =$ rate of decay and $m \geq 1$. The source term can be written by separating the time variable and the location variable, $cS(\tau)F(\xi)$ where S and F are differentiable and c is the initial value. Adding these terms into (5.3.3), executing the right-hand side differentiation and modifying the expression slightly, we have

$$\frac{\partial C}{\partial \tau} = K(\xi)\frac{\partial^2 C}{\partial \xi^2} + \left[\frac{\partial K}{\partial \xi} - v(\xi)\right]\frac{\partial C}{\partial \xi} - \frac{\partial v}{\partial \xi}C - \lambda C^m + cS(\tau)F(\xi),$$

$$(5.3.5)$$

where K, h, and v are given above, and

$$\frac{\partial h}{\partial \xi} = \frac{3 \times 5}{4\mathcal{L}}(1 + \xi/\mathcal{L})^2,$$

$$\frac{\partial v}{\partial \xi} = -\frac{3}{4 \times 4\mathcal{L}}(1 - \xi/\mathcal{L})^{-\frac{3}{4}}.$$

For convenience, we introduce a more compact notation for K. In particular, we set $a_K = k_0(5/4)^{3/2}(9.8 \times 10^{-3})^{1/2}$ to yield

$$K(\xi) = a_K\left(\frac{1+\xi}{\mathcal{L}}\right)^{\frac{9}{2}}, \qquad \frac{\partial K}{\partial \xi} = \frac{9a_K}{2\mathcal{L}}\left(\frac{1+\xi}{\mathcal{L}}\right)^{\frac{7}{2}}, \qquad (5.3.6)$$

where the independent variables are $\tau \in [0, T_c]$, $\xi \in [0, \mathcal{L}]$ and the concentration is $C \in [0, C_{max}]$.

Next, we transform (5.3.5) into a dimensionless problem. First, set relative position, $x = \xi/\mathcal{L}$, and relative concentration $\varphi = C/C_{max}$ and then set the relative time $t = \tau/T_c$ where $T_c = \epsilon\mathcal{L}^2/a_K$ with $\epsilon \leq 1$. As a result of these variable changes, all independent and dependent variables take values in the unit interval.

To implement the variable change and keep notation under control, we set $f_1(x) = (1+x)^{7/2}$, $f_3(x) = (1+x)^{9/2}$, $\epsilon_1 = 9a_K T_c/2\mathcal{L}^2$ and $\epsilon_3 = a_K T_c/\mathcal{L}^2$. With this notation, we rewrite (5.3.6)

$$K(x) = \frac{1}{T_c}\mathcal{L}^2\epsilon_3 f_3(x), \qquad \frac{\partial K}{\partial \xi} = \frac{\mathcal{L}}{T_c}\epsilon_1 f_1(x). \qquad (5.3.7)$$

In turn, we recast the velocity term using $f_2(x) = (1-x)^{1/4}$, $f_4(x) = -(1-x)^{-3/4}$, $\epsilon_2 = 3T_c/4\mathcal{L}$ and $\epsilon_4 = \epsilon_2/4 = 3T_c/4 \times 4\mathcal{L}$. We recalculate (5.3.5) as

$$C_{max}\frac{\partial \varphi}{\partial t} = C_{max}T_c\frac{\partial \varphi}{\partial \tau}$$
$$= T_c\left[C_{max}K(\xi)\frac{\partial^2 \varphi}{\partial \xi^2} + \left(\frac{\partial K}{\partial \xi} - v(\xi)\right)C_{max}\frac{\partial \varphi}{\partial \xi}\right.$$
$$\left. - \frac{\partial v}{\partial \xi}C_{max}\varphi - \lambda C_{max}^m\varphi^m + cS(\tau)F(\xi)\right]$$
$$= T_c\left[C_{max}K(\xi)\frac{1}{\mathcal{L}^2}\frac{\partial^2 \varphi}{\partial x^2} + C_{max}\left(\frac{\partial K}{\partial \xi} - v(\xi)\right)\frac{1}{\mathcal{L}}\frac{\partial \varphi}{\partial x}\right.$$
$$\left. - C_{max}\frac{\partial v}{\partial \xi}\varphi - \lambda C_{max}^m\varphi^m + c\frac{1}{T_c}S(t)F(\xi)\right].$$

Now, substitute and multiply through by C_{max}^{-1} to get the desired equation,

$$\frac{\partial \varphi}{\partial t} = \epsilon_3 f_3(x)\frac{\partial^2 \varphi}{\partial x^2} + (\epsilon_1 f_1(x) - \epsilon_2 f_2(x))\frac{\partial \varphi}{\partial x}$$
$$- \epsilon_4 f_4(x)\varphi - \mu\varphi^m + \eta s(t)q(x), \qquad (5.3.8)$$

where $\mu = \lambda T_c C_{max}^{m-1}$, $\eta = cT_c C_{max}$ and $s(t)$ is defined so that $so\sigma = S$ with $\sigma(\tau) = \tau/T_c$ represents the conversion of real time to normalized time. In this equation, s represents the temporal extent of the spill, q is the spatial component. Abstractly, this is the Dirac delta. In practical terms, it should be zero except on a small interval located at the spill center. The value of $s(t)q(x)$ near the spill center must be nearly equal to the total spill volume. As noted (5.3.8) has the general form of a diffusion/transport equation. Listing the terms from left to right we have a diffusion term, transport, reaction and finally the source or external force.

With Eq. (5.3.8) done, we turn to the numerical procedure. Our stated aim is to introduce spectral collocation with this example. However, it is important to consider the alternatives. For the case $m = 1$, Eq. (5.3.8) is a linear parabolic PDE. In later chapters, we see that these equations form the core domain of two procedures already developed here, GFEM and the variant, Gaussian collocation. For $m > 1$, the time stepping phase of the solution is not linear in φ, but a polynomial of degree m. This will not cause a problem for forward Euler. On the other hand for implicit time stepping, we must either use Newton's method for functions of several variables (see Gresho and Sani (1998)) or use a time lag procedure (see Section 5.4). From our work with the Black–Scholes equation, we would expect OSC to be the method of choice. It is often the preferred method for nonlinear PDE or settings with singularities on the boundary. Again, the equation is second-order, so we would set a partition of the spatial interval and define basis functions via cubic Hermite polynomials supported by the elements. The collocation points would be placed at the two-point quadrature locations. And there are more, for instance, there is the method of lines and Hermite method.

In any event, we develop spectral collocation. Additionally, we suppose that $m = 1$. We start with an interval $[0, 1]$ for the spatial variable and induce a partition $0 = x_0 < x_1 < \cdots < x_n = 1$. In this case, we refer to the x_i as collocation points. Then, for each i, we define a degree n Lagrange polynomial dual to the corresponding collocation point. As with ordinary one variable polynomial interpolation, we solve the linear system $Aa = e_1$, where A is the usual Vandermonde matrix for degree n interpolation. If we write N_i for the ith polynomial, then $N_i(x_j) = \delta_{ij}$ (see for instance, Section 1.3). Below, we recommend a set of interpolation points that are very close together. The resulting Vandermonde matrix will be very nearly singular. The result will be a large condition number and the Linear-Solve function may fail. In this case, we recommend computing the Lagrange polynomials directly, $N_i(x) = \prod_{j \neq i}(x - x_j) / \prod_{j \neq i}(x_i - x_j)$.

We set $u(t, x) = \sum_{i=0}^{n} a_i(t) N_i(x)$. If we write (5.3.8) in the form $\varphi_t = L(\varphi) + f(t, x)$ where $f = \eta s(t) q(x)$, then both the left-hand side and L are linear. Hence, inserting u into (5.3.8) and expanding, we get $\sum_{i=0}^{n} a_i'(t) N_i(x)$ on the left-hand side and $\sum_{i=0}^{n} a_i(t) L(N_i(x)) + f(t, x)$ on the right. After evaluating this expression at each collocation point x_j, $j = 1, 2, \ldots, n - 1$ and expressing the derivative $a_i'(t)$ as a forward Euler finite difference, $(a_i(t_{m+1}) - a_i(t_m))/\Delta t$, we get the following system of $n - 1$ linear equations in $n + 1$ unknowns:

$$\frac{1}{\Delta t} \sum_{i=0}^{n} N_i(x_j)(a_i(t_{m+1}) - a_i(t_m))$$

$$= \sum_{i=0}^{n} L(N_i(x_j)) a_i(t_m) + f(t_m, x_j),$$

$j = 1, \ldots, n - 1$. By duality this reduces to

$$a_j(t_{m+1}) = a_j(t_m) + \Delta t \sum_{i=0}^{n} L(N_i(x_j)) a_i(t_m) + f(t_m, x_j). \quad (5.3.9)$$

It is reasonable to set boundary values as $u(t, 0) = 0$ and $u(t, 1) = 0$. This amounts to the assumption that the concentration at both spatial end points is zero. In other words, the spill occurs downstream from the left-hand end point and the concentration has dissipated/decayed during the allotted time ahead of the right-hand end point. To be consistent with this decision, we must define f so that both $f(t, x_0) = 0$ and $f(t, x_n) = 0$. In turn, $0 = u(t, 0) = \sum_{i=0}^{n} a_i(t) N_i(0) = a_0(t)$. Similarly $0 = u(t, 1)$ implies that $a_n(t) = 0$. Hence, the effect of the boundary values is to recast

(5.3.9) as a square linear system,

$$a_j(t_{m+1}) = a_j(t_m) + \Delta t \sum_{i=1}^{n-1} L(N_i(x_j))a_i(t_m) + f(t_m n x_j).$$

Next, we rewrite the system in matrix form. We set a^m equal to the column state vector with entries equal to $a_i(t_m)$, f^m to be the column vector with entries $\Delta t f(t_m, x_i)$, and set K as the $n-1 \times n-1$ matrix with i,j entry equal to Δt times $L(N_i(x_j))$. With this notation, we have

$$a^{m+1} = (I + K)a^m + f^m, \tag{5.3.10}$$

where I is the $n-1 \times n-1$ identity matrix. Now, with initial condition a^0, we use forward iteration to calculate values for $u(t_m, x_i)$ at each time state and each collocation point.

Note that the procedure just described is essentially a polynomial interpolation technique. It is well known that there are problems when interpolating functions that are x-axis asymptotic. This is particularly problomatic if you choose a uniformly distributed sequence of points. Since this is case here, we select the collocation points so that they tend to bunch at the end points. The Chebyshev type allocation procedure sets $x_i = [1 - \cos((i-1)\pi/(n-1))]/2$. In Figure 5.3.1, we show 20 collocation points in the unit interval. In this case, 60% of the interpolation points are in 40% of the region. There are similar procedures that place 90% of the points in as little at 10% of the region (see for instance, Chen *et al.* (2011)).

To execute a case, we start by setting the parameters. This particular implementation is modified from the one used in Bellomo *et al.* (2008).

Length: $\mathcal{L} = 33,600$, $C_{\max} = 2 \times 10^{-3}$, $\epsilon = 2.42 \times 10^{-5}$, $\lambda = 10^{-4}$.

To setup the dimensionless form of the transport–diffusion equation, set $k_0 = 6$ and derive $a_K = 0.8301$, $T_C = 32,913$. We next calculate $\epsilon_1 = 1.089 \times 10^{-4}$, $\epsilon_3 = 2.42 \times 10^{-5}$ to get the normalized

Figure 5.3.1.   20 Chebyshev collocation points on the unit interval.

diffusion function and its derivative,

$$K(x) = 0.830098 \times (1 + x)^{4.5},$$

$$\frac{\partial K}{\partial \xi} = 1.11174 \times 10^{-4} \times (1 + x)^{3.5}.$$

In particular,

$$f_1(x) = (1 + x)^{3.5}, \quad f_3(x) = (1 + x)^{4.5},$$

$$f_2(x) = (1 - x)^{0.25}, \quad f_4(x) = -(1 - x)^{-0.75}.$$

Setting $m = 1$ (assuming linearity), we calculate the remaining constants,

$$\epsilon_2 = 2.94 \times 10^{-1}, \quad \epsilon_4 = 7.35 \times 10^{-2},$$

$$\mu = 3.29128, \quad c = 0.0001, \quad \eta = \frac{c}{C_{\max}} T_c = 1645.64.$$

Finally, we set $s(t) = t^2 e^{-20t^2}$, the spill by time and the spill by location, $q(x) = \cos^{50}(2\pi x - 0.5)$ with $0 \leq x \leq 0.33$ and $q(x) = 0$ otherwise. We plot the function $s(t)q(x)$ to get a sense of the spill process without diffusion, transport or decay. Note that the positive time axis is upper right going to the left. Note that the spill time is about 0.5 relative time units. Hence, the duration of the spill is long but the area is small.

Since the choice of s indicates no concentration at $t = 0$, then $u(0, x) = 0$ is a reasonable initial condition. Figure 5.3.3 shows the output. Time stepping was done explicitly. The effect of transport and diffusion/decay is apparent when this output is compared to the spill illustrated in Figure 5.3.2. In natural coordinates (all values are in meters and seconds), the spill has traveled more than 12 km in about 9 hours. Note what appears to be a secondary spill at the same time but downstream from the main spill. This is an artifact of the polynomial interpolation. Since this image is misleading, we correct it (see Figure 5.3.4).

One further word of caution. This version of collocation may not function well for a large number of collocation points. Again, this is an artifact of polynomial interpolation. The problem is that polynomials of large degree have may have several critical points

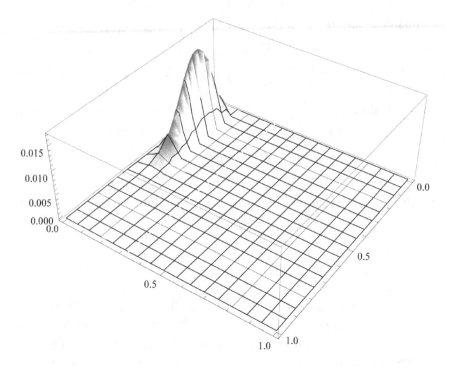

Figure 5.3.2.   Spill by location and time.

or local extrema (see Figure 5.3.3). The resulting bumpiness of the
estimated function is not a true representation of the process. We
used 20 collocation points distributed as in Figure 5.3.1. However,
there are other means of generating collocation points and some of
these will support thousands of collocation points and corresponding
polynomials.

Another approach would be to split the spatial domain into two
segments and then setup the collocation separately in each subdo-
main. For instance, if you split the domain at the spill center, then
you are assured that there will be several collocations in the spill
area. Using the boundary values at the right for the left-hand component and at the left for the right-hand component, we ensure that
the resulting estimated solution is continuous.

The two forms of collocation are related in that they use a polynomial interpolation procedure to estimate the solution to the PDE.

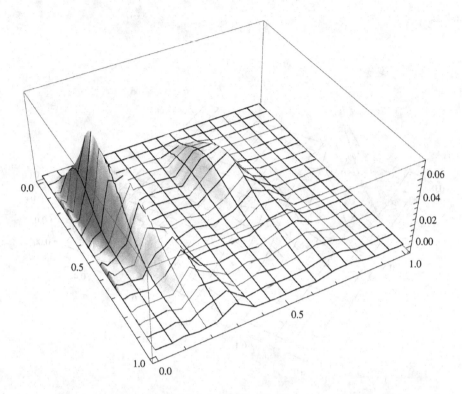

Figure 5.3.3.   Example output.

The first case uses polynomials supported by elements of a partition of a spatial dimension. Hence, the degree of the interpolating polynomials is small and there is local control. Further, the researcher is free to select the partition as he chooses. The second procedure is older. The degree of the interpolating polynomials depends on the number of points of the partition. There is no local control. The number and location of the points in the partition are not controlled by the researcher. Nevertheless, this procedure does yield excellent results and is the procedure of choice of many technicians. In Chapter 10, we consider in detail the mathematical foundation for collocation.

We have presented this solution procedure from the point of view that it provides an interesting vehicle to present this form of collocation. On the other hand, the work by Gresho and Sani (1998) is a

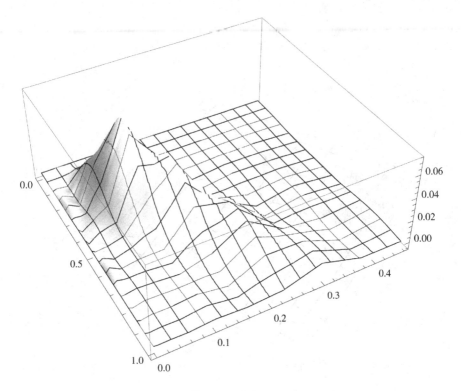

Figure 5.3.4.

two-volume piece where the first volume is dedicated to doing FEM on the transport–diffusion equation. In that connection, they present a variation on implicit FDM time stepping called Trapezoid FDM. We will present this alternative when we look at Stokes flow.

Exercises

1. Solve the River Pollution problem using the collocation technique introduced above. Use Lagrange polynomials with 20 nodes, the given parameters, source functions, boundary and initial conditions.
2. Repeat Exercise 1 with $m = 5$. Compare the output to Figure 5.3.4.
3. Repeat Exercise 1 with $v(\xi) = 0.73 a_v (1 - \xi/\mathcal{L})^{0.2}$.

4. Solve the River Pollution problem using Gaussian collocation introduced in Section 5.1. Use cubic Hermite polynomials and the given parameters, source functions, boundary and initial conditions.
5. Solve the River Pollution problem by spatial FEM and temporal implicit FDM. Setup the model to focus at the spill location. Use linear Lagrange polynomials.

5.4. Traffic Congestion, a Different Flow Problem

We all know the feeling of being stuck in rush hour traffic. Sometimes, there is an accident to reduce the flow (flux) and consequently the mean velocity through the system. But more often the problem is just the number of cars on the road, the system density. In the following problem, we use ideas from fluid flow to model vehicular traffic flow. Since we make the assumption that the road is single lane, then the model may also be applied to a commuter rail system.

In a commuter rail or urban underground system, the spike in ridership at rush hour requires additional rail cars so that the passengers do not over flow the platforms. Since the number of cars per train is fixed by the platform length, then additional trains must be added. But the number of trains is bounded by the required minimal intertrain distance. Because of the increased number of passengers boarding and leaving the trains, the trains spend more time in each station. Also trains that arrive at a station at the onset of rush hour will load more passengers than the train immediately behind. This results in bunching. Altogether, the increased passenger load causes rail line congestion with subsequent decreases in the average velocity along the line.

Modeling traffic flow is an active area of research. There are microscopic models that look at each individual vehicle, there are cumulative microscopic models that describe an entire traffic pattern as the sum of individual vehicles. Alternatively, there are macroscopic models that deal in system flux, mean velocity and density. The macroscopic model treats the traffic problem as a fluid flow. Initially, these models were based solely on conservation laws and gave rise to order two PDE. However, a car with a driver is not a fluid particle. Without any means to incorporate psychological tendencies, the model could not represent the full spectrum of observation (Helburg and Johansson, 2009). As a result, researches began to modify the velocity function to include observed behavior (Aw and Rascle, 2000). This is the type of model we will develop here.

The macroscopic models, though improved, still had short comings (Bellomo and Dogbe, 2011). As a result, current research is mostly microscopic (Mackenzie, 2013). In the microscopic setting, they are able to impose different driving tendencies to each vehicle, and then determine the effect of one style of driving against another. This research tends to conclude that higher levels of automation are necessary to optimize the traffic flow.

Since implementing such a model requires access to a high performance computing facility, we will consider only a macroscopic fluid type model which includes driver tendency parameters.

We set the problem in dimensionless variables. To begin, let $\mathcal{L}$ be the total length, T_C be the characteristic time. We set the latter to $\mathcal{L}/v_M$, the ratio of the system length and the mean maximal allowed vehicle velocity. Hence, T_C as the minimal time to completion. Next, set x_r to be the actual position and t_r the world clock time. Then, the relative position is $x = x_r/\mathcal{L}$, and relative time t is given by $t = t_r/T_C$.

The first dependent variable is the relative density, $u = n/n_M$, where n is the actual number of vehicles and n_M is the maximal number of vehicles. In a rail system n_M, the maximum number of trains in the system is governed by work rules designating the minimum distance between trains and the length of a train. The second dependent variable is the relative velocity, $v = v_r/v_M$, the ratio of the real velocity and the mean maximal velocity. The secondary variable is $q = uv$. This is the mean relative flux as the product of the relative velocity times the relative density. Also, we denote by $q_M = u_M v_M$, the maximum admissible mean flux.

We begin to develop our working PDE. We have already mentioned that we are thinking of the traffic flow as a fluid problem. Using a mass one similar to the conservation argument used to derive the one-dimensional heat equation in Chapter 2, we have the following expression relating the density and flux:

$$\frac{\partial u}{\partial t} + \frac{\partial q}{\partial x} = 0. \tag{5.4.1}$$

Again, q is a function of u and the velocity, v. We would like to write v as a function of u. When $u = 1$ ($n_r = n_M$), then we have gridlock implying that $v = 0$. In turn, $u = 0$ (locally) implies that $v_r = v_M$ or $v = 1$. Now, there are a host of functions that satisfy $v(1) = 0$ and $v(0) = 1$. Any one that we select should be verified by comparing our final result against an actual case. But for the moment, we may choose any one. So, we select a simple one and write $v_e = 1 - u$. We designate the velocity now with v_e as it is an expected relative velocity. Hence, with this selection we may replace q in (5.4.1) by $u - u^2$.

But, before settling on an operative differential equation, there is another factor that we want to include in the model. In Bellomo and Dogbe (2011), it is reported that when we operate a vehicle we tend to over react to local changes in traffic density. Whether the automobile driver sees traffic congestion down road or the train operator is notified by his controller, our tendency is to slow more than necessary. At the other end, the perception of open road ahead influences us to accelerate more than is warranted. Hence, vehicles are operated using a perceived density based on the local situation rather than the actual density. The latter being largely unknown to the vehicle operator. We have noted that there are many means to implement this observation. We will implement the one suggested in Bellomo *et al.* (2008). In this case, we replace actual density by perceived density, u_p, as a function of u,

$$u_p = u \left[1 + \eta(1 - u)\frac{\partial u}{\partial x} \right], \tag{5.4.2}$$

where η is a positive constant. Note that when u is increasing, $\partial u/\partial x > 0$. Given $\partial u/\partial x > 0$, then $\eta v_e(\partial u/\partial x) > 0$ and $u_p > u$. Further, the parameter η measures how much greater. For $\partial u/\partial x < 0$, the opposite is true. Therefore, the function given by (5.4.2) conforms to the observation. Seemingly, each driver has his own value for η.

However, we noted that we cannot manage that level of complexity. Hence, we suppose that η is constant.

Using the perceived relative density in (5.4.2) and doing a few routine manipulations, (5.4.1) becomes

$$\frac{\partial u}{\partial t} = \eta u^2(1-u)\frac{\partial^2 u}{\partial x^2} + \eta u(2-3u)\left(\frac{\partial u}{\partial x}\right)^2 - (1-2u)\frac{\partial u}{\partial x}. \quad (5.4.3)$$

Once we have values for $u = u(t,x)$, then we can derive values for $v = v(t,x)$ via (5.4.2). As in Section 5.3, we will compute values for u using spectral collocation. Since (5.4.3) is nonlinear, then we must execute the time steps with forward Euler. Should we choose to develop implicitly, then the result would be a fourth degree polynomial in 18 variables. We could employ Newton's method for several variables, but the execution time on a personal computer would be prohibitive. Alternatively, we could use a time lag procedure (see Exercise 3).

We set the initial conditions to $u_0 = u(0,x) = 0.2$. Since this setting is constant, then the derivative of u is 0, $u_p = u$ and $v_e = 0.8$ at $t = 0$. We may think of u_0 as the ambient load on the system. For boundary values, take $u(t,0) = 0.5-0.3e^{-t}$ at inflow and $u(t,1) = 0.2$ at outflow. Figure 5.4.1 shows the boundary value at the inflow.

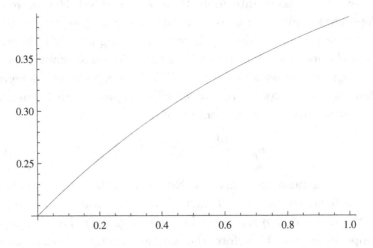

Figure 5.4.1.   Traffic density up tick at the inflow edge.

Our choice indicates most of the increase in system load occurs near $t = 0$ and leveling for large t. This choice highlights the observation that rush hour starts all at once. Everyone is off work at the same time and is on the road at the same time. In other settings, it could be linear, or nonlinear and concave up or S shaped. For instance, using $u(t,0) = 0.2 + t^2 e^{-20t^2}$ would yield a model with inflow density increasing and then decreasing during the time period. The choice should be governed by experience.

The parameter η measures how much the personality or psychology of the drivers affects their behavior. For $\eta = 0$ the vehicles act like fluid particles. The greater we take η, the greater is the psychological factor. We show below the output for $\eta = 0.1$. The reader is encouraged to try a range of values, see Exercise 4.

We implement the spectral collocation method exactly as in Section 5.3. We use the same 20 collocation points, denoted x_j, $j = 1, 2, \ldots, 20$. The discrete form of (5.4.3) setup for explicit FDM time stepping is

$$\frac{u(t_{m+1}, x_j) - u(t_m, x_j)}{\Delta t}$$

$$= \eta u(t_m, x_j)^2 \left(1 - u(t_m, x_j)\right) \frac{\partial^2 u}{\partial x^2}(t_m, x_j)$$

$$+ \eta u(t_m, x_j) \left(2 - 3u(t_m, x_j)\right) \left(\frac{\partial u}{\partial x}(t_m, x_j)\right)^2$$

$$- \left(1 - 2u(t_m, x_j)\right) \frac{\partial u}{\partial x}(t_m, x_j). \tag{5.4.4}$$

As in Section 5.3, we set $u(t, x) = \sum_i a_i(t) N_i(x)$, where $N_i(x)$ are degree 19 polynomials dual to the collocation points, $N_i(x_j) = \delta_{i,j}$. Therefore, $u(t_m, x_j) = a_j(t_m)$. If we set $\xi_{ji} = N_i'(x_j)$ and $\zeta_{ji} = N_i''(x_j)$ and substitute into (5.4.4), we have

$$\frac{u(t_{m+1}, x_j) - u(t_m, x_j)}{\Delta t}$$

$$= \eta a_j(t_m)^2 \left(1 - a_j(t_m)\right) \sum_i \zeta_{ij} a_i(t_m)$$

$$+ \eta a_j(t_m)\,(2 - 3a_j(t_m)) \left(\sum_i \xi_{ij} a_i(t_m) \right)^2$$

$$- (1 - 2a_j(t_m)) \sum_i \xi_{ij} a_i(t_m).$$

Equivalently,

$$u(t_{m+1}, x_j) = u(t_m, x_j) + \Delta t \eta a_j(t_m)^2 \,(1 - a_j(t_m)) \sum_i \zeta_{ij} a_i(t_m)$$

$$+ \Delta t \eta a_j(t_m)\,(2 - 3a_j(t_m)) \left(\sum_i \xi_{ij} a_i(t_m) \right)^2$$

$$- \Delta t \,(1 - 2a_j(t_m)) \sum_i \xi_{ij} a_i(t_m).$$

Setting the vector $a^m = (a_j(t_m))$, we get the linear relation

$$a^{m+1} = a^m + \alpha K a^m + \beta L a^m + \gamma M(a^m), \qquad (5.4.5)$$

where α, β and γ are time-dependent diagonal matrices given by

$$\begin{aligned}
\alpha(i, i) &= \Delta t \eta a_i(t_n)^2 (1 - a_i(t_m)), \\
\beta(i, i) &= -\Delta t (1 - 2a_i(t_m)), \qquad\qquad (5.4.6) \\
\gamma(i, i) &= \Delta t \eta a_i(t_m)(2 - 3a_i(t_m)).
\end{aligned}$$

The matrices K and L have entries

$$K_{i,j} = \zeta_{i,j} = \frac{\partial^2}{\partial x^2} N_j(x_i), \quad L_{i,j} = \xi_{i,j} = \frac{\partial}{\partial x} N_j(x_i).$$

Finally, $M(a^m)$ denotes the vector with entries the squares of the entries of $K a^m$. Hence, we can actualize (5.4.5) by pre-computing the matrices α, β, γ, K, and L, then iteratively composing the result. Note that whereas K and L are time independent, α, β, and γ must be readjusted at each time step.

Figure 5.4.2 shows the development of the system for 2.5 time units.

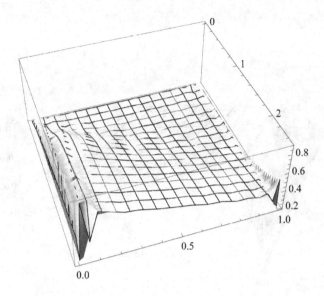

Figure 5.4.2.   Traffic density by time and location.

The choppy quality near the edges is an artifact of the polynomial interpolation. The problem is that a driver is in a traffic jam or he is not. There is no such thing as a semi-traffic jam. In mathematical terms, the traffic density has nearly infinite slope at the two end points. There are insufficient identified locations to approximate the vertical tangent. To correct for this, we multiply by $\sqrt{x}(1 - \sqrt{x})$. Figure 5.4.3 shows the corrected output. Alternatively, we would consider procedures that placed more collocation points near the origin (Chen *et al.*, 2011).

In Figure 5.4.4, we show the modified output for three time values, $t = 1$ is denoted by $\square$, $t = 1.5$ by $\diamond$ and $t = 2$ by *.

Next, we consider a vehicle entering the system at a given time and compute the expected mean velocity. For time t_0 and location x_0, we get the system density $u(t_0, x_0)$ from the model. The expected velocity is $v_e(t_0, x_0) = 1 - u_p(t, 0, x_0)$ and the expected distance is $\delta_0 = v_e(t_0, x_0)\Delta t$. Setting $t_1 = t_0 + \Delta t$ and $x_1 = x_0 + \delta_0$, we repeat the process until the vehicle has left the system, $t_{\max}$ or $t = 2.5$, the end of the execution. When computing u_p from u (Eq. (5.6.2)), the polynomial $\sum_i a_i N_i$ is not a good predictor of $\partial u/\partial x$. You should

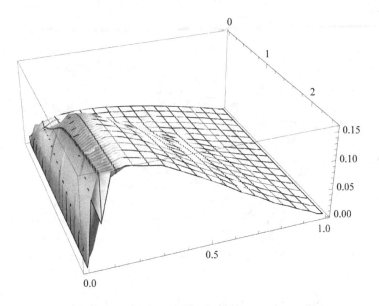

Figure 5.4.3.   Corrected output.

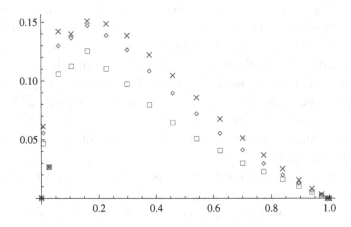

Figure 5.4.4.   3 time slices: $\square\, t = 1$, $\diamond\, t = 1.5$, $\times\, t = 2$.

either use finite differences to estimate the derivative or alternatively use the slope of a spline fit. Then, we compute the mean velocity as $\sum_i \delta_i / t_{\max}$. The results for nine entry times are shown in Figure 5.4.5.

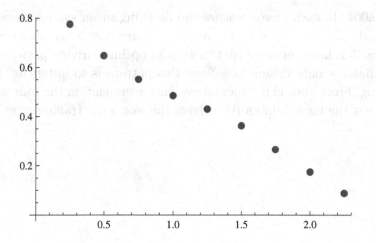

Figure 5.4.5.   Mean expected velocity by system entrance time.

Exercises

1. Execute the traffic flow problem as described in the section.
2. Repeat the traffic flow problem using Gaussian collocation with Hermite cubics (see Section 5.1).
3. Repeat the traffic flow problem using lag time linearization. In particular, write

$$a_j(t_{m+1}) = a_j(t_m) + (\Delta t \eta a_j(t_{m-1})^2)$$
$$\times (1 - a_j(t_{m-1})) \sum_i \xi_{j,i} a_i(t_m)$$
$$+ \Delta t \eta a_j(t_{m-1})(2 - 3a_j(t_{m-1})) \sum_i \xi_{j,i} a_i(t_m)$$
$$- \Delta t (1 - 2a_j(t_{m-1})) \sum_i \xi_{j,i} a_i(t_m).$$

Finally, we take $a_j(t_{-1}) = 0$. The resulting formulation is linear in current time, t_m. With this setup, use Crank–Nicolson time stepping.
4. Use your favorite numerical technique to execute this example for a range of values for the parameter η. For instance, $\eta = 0.5, 0.01$ or

0.001. In each case, measure the resulting mean system through-put. We have noted that recent research involving microscopic models has concluded that there is an optimal driving pattern and that the only means to enforce this pattern is to automate driving. From your experimental evidence comment on the conclusion that the more human the drivers the worse the traffic pattern.

Chapter 6

Transient Problems
in Two Spatial Dimensions

Introduction

In this chapter, we turn to transient PDE with two spatial dimensions. From the point of view of numerical solutions to differential equations, we use these examples to introduce various aspects of transient FEM. From the point of view of applications, we derive the Navier–Stokes equation (NSE) that describes incompressible, viscous fluid flow with momentum. In addition, we look at two biological applications. In one case, we have a mathematical model for chemotaxis. In the other, we continue the population model begun in Section 4.3. From the point of view of differential equations, the equations encountered here are largely nonlinear. Hence, we have the opportunity to introduce more procedures specific to nonlinear PDE.

The NSE is a major event in the history of mathematical modeling. There are books written about it. And there are many open questions associated to it. It is the implied context of Hilbert's sixth problem. It is included in the millennia problems. At this time, we have no proof of the existence of a continuous solution. This result has escaped the best efforts of mathematicians for so long that many now believe that the equation may be missing a term that it may not be the correct model for an incompressible viscous flow.

The NSE looks at fluid flows at the macroscopic level. There are results that connect the NSE to the Lattice-Boltzmann equation which considers fluid flows at the mesoscopic or particle level. However, there is not yet a complete theory that fully describes the connection. At the same time, it has been an active topic in numerical analysis. Much has been written about how to effectively simulate a fluid flow with the NSE. In this chapter, we only touch on this active area. In particular, we show how the equation is derived and some of the basics for flow simulation.

A recent development is the emergence of a special case of the NSE in cell biology. This is the case of incompressible Stokes flow. In its dimension free form (see Section 5.3), the Reynolds number is the coefficient of the momentum term of the NSE. If the Reynolds number is effectively zero, then the flow without momentum is called a Stokes flow. This is the case for fluid motion inside a single cell.

Finally, we conclude Part 1 with a theorem that may be seen as a transition from the applications to the theory. We find the result interesting but most likely not useful. Simply stated, we prove that if a numerical procedure is well behaved through a sequence of domain refinements, then the differential equation has a solution. In the literature, convergence of a numerical process generally requires the existence of a solution. However, there are many cases where numerical processes have been used to produce important information without the benefit of existence theory or convergence for the numerical process. In this context, the successful numerical results may well indicate that a more complete theoretical foundation will one day follow.

The chapter is organized as follows. We begin with the chemotaxis example. In Sections 6.2 and 6.3, we see the derivation of the NSE and the development of the FEM approximation procedure. The Navier–Stokes for viscous, incompressible flows involves two dependent variables, one that describes the flow field and one to describe pressure. This is not the first example involving more than one dependent variable. It is the first time where we treat one variable as primary and use the equilibrium equation to resolve the second. In the next section, we look at the nonlinear second-order PDE that arose from our model for population dynamics for an animal with the tendency to form herds. We will develop the numerical process using spatial FEM with a linearizing time lag. We end the chapter with a theorem that straddles the material in Parts 1 and 2.

6.1. Cell Chemotaxis

Chemotaxis is an important area of current biological research. It is an essential process in the development of multicellular animals and plants. It is the mechanism behind cell-aggregation, and it is cell aggregation that gives rise to the organ structures necessary for complex life forms. In addition, cell aggregation by means of chemotaxis plays a role in tumor formation. For amoeba and many other simple animals chemotaxis gives the animal mobility.

There are many variations of the basic process and several distinct mathematical models. Described at the highest level, there is a chemical attractant which causes changes within the cell. The result is motion. Most often, motion is in the direction of greater attractant concentration. Mathematically, the attractant gradient determines the direction of motion.

The Keller–Segel equation set (Keller, 1980) is the generally accepted mathematical model for chemotaxis at the level of cell densities. Even though this set of simultaneous PDE has been studied for many years it remains largely unresolved. As a result, recent interest has focused on modified versions of the standard model. The version we consider is frequently called cell chemotaxis (Murray, 2003). It is described by the equation pair,

$$\frac{\partial n}{\partial t} = D\nabla^2 n - \alpha\nabla \cdot (n\nabla u) + (sr)n(N - n), \qquad (6.1.1)$$

$$\frac{\partial u}{\partial t} = \nabla^2 u - su + s\frac{n}{1 + n}, \qquad (6.1.2)$$

where $n = n(t, x)$ is the cell density and $u = u(t, x)$ is the concentration of chemical attractant. Both n and u are functions of time and location. The parameters are identified as follows:

N — maximal cell density (usually set to 1).

s — the chemical attractant parameter used to determine the initial distribution of attractant.

For our implementation, we will use a standing wave for the initial chemical concentration. In this case, we set s to the corresponding wave number.

The following two parameters may be considered functionally related, as a high mitosis rate should be accompanied with a large diffusion coefficient:

D — diffusion parameter.

r — mitosis rate.

Note that diffusion is not a directed motion. Here, we treat diffusion as the process for distributing new cells. In some cases, mitosis and diffusion are suspended during chemotaxis. Hence, $D = r = 0$ is an interesting special case. The following parameter is associated to the transport term in (6.1.1). Indeed, it is the sole parameter directly related to chemotaxis:

α — effectiveness parameter.

In Murray (2003), the author uses finite differences to show that the cell chemotaxis pair may account for color patterns on certain reptiles. However, FDM cannot model the complex geometries common in animals and plants. Furthermore, collocation for more than one spatial dimension is not fully resolved over complex geometric domains (Loustau *et al.* (2013); Li *et al.* (2008)). Hence, we are left with FEM.

As the region, we choose a disk which we triangulate with six nodal triangles, see Figure 6.1.1. As in Section 3.2, we use a reference triangle and induce the nodes and polynomial model to the elements by means of the affine transformations. Since the vector space of two-variable polynomials of degree 2 is dimension 6, then this model will allow for FEM degree 2 piecewise polynomial solutions n_h and c_h for the given mesh parameter.

The process proceeds as follows. Since the second equation is the simpler one, we begin there. The weak form of (6.1.2) is

$$\int_U \frac{\partial c}{\partial t}\psi = \int_U \nabla^2 c\psi - s\int_U c\psi + s\int_U \frac{n}{1+n}\psi. \qquad (6.1.3)$$

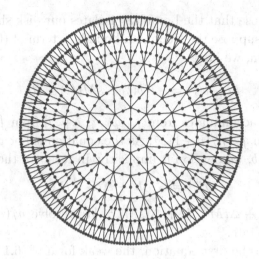

Figure 6.1.1.   Triangulated Disk with 6 nodal triangles.

We apply Green's theorem to the second-order term in (6.1.3) to get

$$\int\limits_U \frac{\partial c}{\partial t}\psi = -\int\limits_U \nabla c \nabla \psi - s\int\limits_U c\psi + s\int\limits_U \frac{n}{1+n}\psi + \int\limits_\Gamma \frac{\partial c}{\partial \mathbf{n}}\psi, \quad (6.1.4)$$

where $\partial/\partial\mathbf{n}$ denotes the normal derivative. We may take (6.1.4) as a statement for piecewise polynomial functions in V^h associated to the partition. Decomposing this equation into its element level constituents, then we may suppose that (6.1.4) is a statement at the element level, E_e. In particular, $\psi \in V_c^h$, $c = c_h = \sum_k b_{e,k}(t)\varphi_k^e$, $\varphi_k^e \in V_e^h$. We substitute this form of c into (6.1.4) to get a linear system, which when assembled and solved will yield the FEM solution c_h:

$$\sum_k \frac{\partial}{\partial t}b_{e,k}(t)\int\limits_{E_e}\varphi_k^e$$

$$= -\sum_k b_{e,k}(t)\int\limits_{E_e}\nabla_j\varphi_k^e\nabla\varphi^e - s\sum_k b_{e,k}(t)\int\limits_{E_e}\varphi_k^e\varphi_j^e$$

$$+ s\sum_k b_{e,k}(t)\int\limits_{E_e}\frac{n}{1+n}\varphi_j^e + \int\limits_{\Gamma_e}\frac{\partial c}{\partial \mathbf{n}}\varphi_j^e. \quad (6.1.5)$$

If we suppose that the boundary isolates our disk shaped region, then we may suppose the Neumann boundary term of (6.1.5) is zero. In matrix form, we have

$$M\frac{\partial}{\partial t}b(t) = -Kb(t) - sMb(t) + R_c,$$

where $M[i,j]$ accumulates from $\int_{E_e} \varphi_i^e \varphi_j^e$, $K[i,j]$ from $\int_{E_e} \nabla\varphi_i^e \cdot \nabla\varphi_j^e$ and $R_c[i]$ from $\int_{E_e} n/(n+1)\varphi_i^e$. As the equation is linear in the coefficient vector b, we develop it implicitly. In terms of the time steps,

$$b^{t+1} = G^{-1}b^t + \Delta t G^{-1} R_c,$$

where $G = I + \Delta t M^{-1}K + \Delta t s M^{-1}$. We resolve $n/(n+1)$ at the current time state.

Turning to the first equation, the weak form of (6.1.1) is

$$\int_U \frac{\partial n}{\partial t}\psi = D\int_U \nabla^2 n\psi - \alpha\int_U (\nabla n \cdot \nabla c)\psi$$

$$- \alpha\int_U n(\nabla^2 c)\psi + sr\int_U n(N-n)\psi.$$

After applying Green's theorem,

$$\int_U \frac{\partial n}{\partial t}\psi = -D\int_U \nabla n\nabla\psi - \alpha\int_U (\nabla n \cdot \nabla c)\psi - \alpha\int_U n(\nabla^2 c)\psi$$

$$+ sr\int_U n(N-n) + D\int_\Gamma \frac{\partial n}{\partial \mathbf{n}}\psi.$$

At the element level with $n = \sum_i a_{e,i}(t)\varphi_i^e$,

$$\sum_i \frac{\partial}{\partial t}a_{e,i}\int_{E_e} \varphi_i^e\varphi_j^e$$

$$= -D\sum_i a_{e,i}\int_{E_e} \nabla\varphi_i^e\nabla\varphi_j^e - \alpha\sum_i a_{e,i}\int_{E_e} (\nabla\varphi_i^e \cdot \nabla c)\varphi_j^e$$

$$- \alpha\sum_i a_{e,i}\int_{E_e} \varphi_i^e(\nabla^2 c)\varphi_j^e + sr\int_{E_e} n(N-n)\varphi_j^e + D\int_{\Gamma_e} \frac{\partial n}{\partial \mathbf{n}}\varphi_j^e,$$

which resolves in matrix/vector notation as

$$M\frac{\partial}{\partial t}a(t) = -DKa(t) - \alpha La(t) - \alpha Na(t) + R_n, \qquad (6.1.6)$$

where M and K are as before, $L[i,j]$ is accumulated from $\alpha \int_{E_e} (\nabla \varphi_i^e \cdot \nabla c)\varphi_j^e$; $N[i,j]$ from $\alpha \int_{E_e} \varphi_i^e (\nabla^2 c)\varphi_j^e$; $R[j]$ from $sr \int_{E_e} n(N-n)\varphi_j^e$. As before, we take the Neumann boundary term to be zero and resolve c at the current time state. As (6.1.6) is nonlinear, then do as in Sections 5.3 and 5.4 and use explicit time stepping. Hence, $n(N-n)$ is also resolved at current time. Now, the time stepping procedure is given by

$$a^{t+1} = (I - \Delta t DM^{-1}K - \Delta t \alpha M^{-1}L - \Delta t \alpha M^{-1}N)a^t$$
$$+ \Delta t M^{-1}R_n.$$

To execute a simulation, we need to only set initial conditions. For n, we choose a uniform distribution and for c, the pattern shown in Figure 6.1.2, the approximate solution for the Helmholtz equation with wave number is 28.2686. As already mentioned, we use backward time stepping for c and forward time stepping for n. To resolve the integrals, we use the quadrature technique described in Section 1.4 for triangles. Figure 6.1.3 shows that cell density after 1,000 time steps ($\Delta t = 0.0001$). The other parameters for this execution are $D = 0.25$, $\alpha = 1.0$, $r = 1.0$, $N = 1$, and s is the wave number.

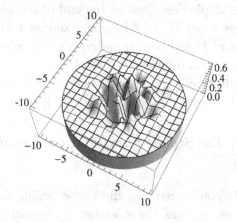

Figure 6.1.2. Initial chemical concentration.

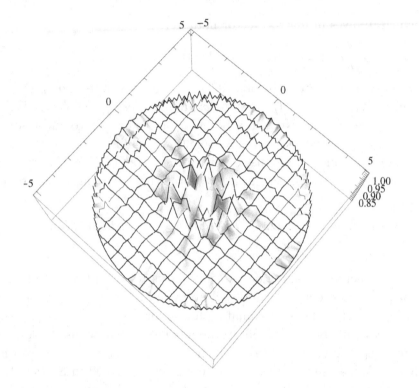

Figure 6.1.3.   Cell density plot at 50 iterations.

Exercises

1. Repeat the example described at the end of section.
2. What happens when $D \to 0$? Repeat Exercise 1 for progressively smaller values of D and correspondingly small values of r. Repeat each case for several different values of α.
3. List some additional experiments with the chemotaxis data parameters that would be interesting to follow-up on.

6.2.   The NSE, the Stokes Flow, and the Bernoulli Equation

We have already considered two fluid flow problems. One was the airfoil and the second was the obstructed channel. In both cases, the fluid was incompressible, steady, and irrotational. In addition, we ignored viscosity or friction. Hence, there was no momentum. With a flow of this sort Bernoulli's equation (6.2.1) holds (Heubner *et al.*, 2001). This equation provides a simple relationship between the velocity field, u, and the pressure P along a streamline. For laminar or two-dimensional flow, this equation states

$$\frac{1}{\rho}P(x,y) = \frac{1}{2}\|u(x,y)\|^2 + g(x,y) + \text{constant} \qquad (6.2.1)$$

or alternatively,

$$\frac{1}{\rho}\big(P(x_1,y_1) - P(x_0,y_0)\big)$$

$$= \frac{1}{2}\big(\|u(x,y)\|^2 - \|u(x,y)\|^2\big) + g(x_1,y_1) - g(x_0,y_0), \quad (6.2.2)$$

where g represents external body forces including gravity. Since all points in the field lie on a streamline starting at the inflow edge, then the pressure is completely resolved from the flow field and the boundary values. Furthermore, if the flow vectors to the left of the inflow edge are horizontal of constant length, then the pressure is also constant to the left of the inflow edge. Therefore, we may take the pressure to be constant on the inflow edge. For a viscous fluid, all of this changes. For such fluids, we must use the NSE to model the flow. In this section, we will derive this important PDE. Our derivation is the standard one, see for instance, Munson *et al.* (2005) or Gresho and Sani (1998).

Consider a small rectangular hexahedron (box shaped) volume of fluid. Suppose that the faces of the volume are parallel to the coordinate planes (xy-plane, xz-plane, or yz-plane). As neighboring regions of fluid move, the volume is deformed. One type of deformation is identified by measuring the change in the normal vector to a face. This effect is called *normal stress*. In addition, vectors on the face parallel to the coordinate axes may also be deformed. These deformations are called *shearing stress*. For the box plane parallel to the yz coordinate plane, the normal stress is denoted σ_{xx}, while the shearing stresses are denoted τ_{xy} and τ_{xz}.

The total force, F, acting on the small fluid volume V is the sum of the surface forces, F_s, and external body force, gm, where mass

$m = \rho V$ as density times volume. Considering a small volume, δV, then for the corresponding force δF and surface force δF_s,

$$\delta F = \delta F_s + g\rho \delta V$$

$$= \left(\frac{\partial \sigma_{xx}}{\partial x} + \frac{\partial \tau_{xy}}{\partial y} + \frac{\partial \tau_{xz}}{\partial z}, \frac{\partial \tau_{yx}}{\partial x} + \frac{\partial \sigma_{yy}}{\partial y} + \frac{\partial \tau_{yz}}{\partial z}, \right.$$

$$\left. \frac{\partial \tau_{zx}}{\partial x} + \frac{\partial \tau_{zy}}{\partial y} + \frac{\partial \sigma_{zz}}{\partial z} \right) \delta V + g\rho \delta V. \qquad (6.2.3)$$

In turn, writing force as mass times acceleration and noting that the flow field u is a function of time and location and location is also a function of time, where the derivative of location with respect to time is velocity,

$$\rho \frac{Du}{Dt}\delta V = \rho \left(\frac{\partial u}{\partial t} + (u \cdot \nabla)u \right) \delta V. \qquad (6.2.4)$$

Equating (6.2.3) and (6.2.4), dividing by δV yields

$$\left(\frac{\partial \sigma_{xx}}{\partial x} + \frac{\partial \tau_{xy}}{\partial y} + \frac{\partial \tau_{xz}}{\partial z}, \frac{\partial \tau_{yx}}{\partial x} + \frac{\partial \sigma_{yy}}{\partial y} + \frac{\partial \tau_{yz}}{\partial z}, \frac{\partial \tau_{zx}}{\partial x} + \frac{\partial \tau_{zy}}{\partial y} + \frac{\partial \sigma_{zz}}{\partial z} \right)$$

$$+ \rho g = \rho \left(\frac{\partial u}{\partial t} + (u \cdot \nabla)u \right). \qquad (6.2.5)$$

A fluid is called *Newtonian* provided the stresses are linearly related to the angular accelerations. If we suppose that the fluid is Newtonian then together with incompressibility, $\nabla \cdot u = 0$, we may write (see Munson *et al.* (2005)),

$$\sigma_{xx} = -P + 2\mu\frac{\partial u_1}{\partial x}, \quad \sigma_{yy} = -P + 2\mu\frac{\partial u_2}{\partial y}, \quad \sigma_{zz} = -P + 2\mu\frac{\partial u_3}{\partial z},$$

$$(6.2.6)$$

where μ designates the viscosity, and

$$\tau_{xy} = \tau_{yx} = \mu \left(\frac{\partial u_1}{\partial y} + \frac{\partial u_2}{\partial x} \right),$$

$$\tau_{yz} = \tau_{zy} = \mu \left(\frac{\partial u_2}{\partial z} + \frac{\partial u_3}{\partial y} \right), \qquad (6.2.7)$$

$$\tau_{xz} = \tau_{zx} = \mu \left(\frac{\partial u_1}{\partial z} + \frac{\partial u_3}{\partial x} \right).$$

Inserting (6.2.6) and (6.2.7) into (6.2.5) yields the NSE

$$\rho \frac{Du}{Dt} + \nabla P = \mu \nabla^2 u + \rho g.$$

Expanding the first term on the left-hand side and dividing by ρ yield the more common form of this equation,

$$\frac{\partial u}{\partial t} + (u \cdot \nabla)u + \nabla \frac{P}{\rho} = \nu \nabla^2 u + g, \tag{6.2.8}$$

where $\nu = \mu/\rho$ is called the *kinematic viscosity*. In the latter formulation, $u \cdot \nabla u$ is called the *momentum* term.

The NSE is a nonlinear second-order PDE that describes an equilibrium between the velocity field and the pressure. When using the NSE to simulate a flow, we need to model both dependent variables. Usually, a higher degree model is used for u, the flow variable. For instance, if we used a domain partition into rectangular elements with degree 2 Lagrange polynomials to model u, then degree 0 (constant) polynomials are used to model P. Alternatively, researchers use degree 6 Hermite polynomials for u and degree 2 Lagrange polynomials for the pressure.

One problem with modeling the NSE with FEM occurs when there is a singularity on the boundary. In this case, the singularity will resonate into the flow field causing chaotic results. Depending on the application, this problem may be resolved using extended finite differences (see Chapter 7). In other cases, researchers use a technique called discontinuous Galerkin FEM (Hesthaven and Warburton, 2008).

The *Stokes* equation is an important spatial case of (6.2.8). If we render (6.2.8) in dimensionless form, the coefficient of the nonlinear or momentum term is the Reynolds number. When the number is very small, momentum is negligible. The resulting flow is called a Stokes flow:

$$\frac{\partial u}{\partial t} + \nabla \frac{P}{\rho} = \nu \nabla^2 u + g. \tag{6.2.9}$$

Stokes flow is usually reserved for highly viscous, slow moving fluids such as glycerin or heavy oil. More generally stated, any case with Reynolds number ≈ 0 (see Guazzelli and Morris (2012)). Very recently cell biologists and biomathematicians have used the Stokes equation to model intracellular flows. See for instance, Shao *et al.* (2012) or Strychalski and Guy (2013). However, it also plays an important role in simulating Navier–Stokes flow. In particular, when simulating a non-steady process you need an initial state. But a flow starting from zero must pass through a Stokes phase as the velocity ramps up. Hence, the initial state for the Navier–Stokes flow is the final state of the Stokes phase. In the next section, we develop an FEM procedure for the Stokes equation with a velocity ramp up function.

Throughout this section, we have restricted our attention to incompressible flows. All along we have ignored the simplest consequence of incompressibility. We have a divergence free flow field. Simply, $\nabla \cdot u = 0$. In the literature, this is called the continuity equation. Using this equation, we can derive an alternate form of the NSE. Indeed, we calculate

$$\nabla \cdot (\nabla u + (\nabla u)^T) = \nabla^2 u + \nabla \cdot (\nabla u)^T = \nabla^2 u + \nabla(\nabla \cdot u) = \nabla^2 u.$$

Substituting into the NSE yields the *stress-divergence* from of the equation,

$$\frac{\partial u}{\partial t} + (u \cdot \nabla)u + \nabla \frac{P}{\rho} = \nu \nabla \cdot (\nabla u + (\nabla u)^T) + g. \qquad (6.2.10)$$

For Stokes flow, we get

$$\frac{\partial u}{\partial t} + \nabla \frac{P}{\rho} = \nu \nabla \cdot (\nabla u + (\nabla u)^T) + g. \qquad (6.2.11)$$

Many authors prefer (6.2.10) over (6.2.8) as the Neumann boundary values are more apparent.

Bernoulli's equation is also a consequence of (6.2.5). Indeed, in the case of an inviscid, Newtonian fluid, (6.2.6) and (6.2.7) reduce to $\sigma_{xx} = \sigma_{yy} = \sigma_{zz} = -P$ and $\tau_{xy} = \tau_{yx} = \tau_{yz} = \tau_{zy} = \tau_{xz} = \tau_{zx} = 0$. Inserting these values into (6.2.5) yields $\nabla P + \rho g = \rho(\partial u/\partial t) + \rho u \cdot \nabla \cdot u$. If we suppose that the flow is steady, we get $\nabla P/\rho + g = (u \cdot \nabla) \cdot u$. Next, we verify $(u \cdot \nabla)u = \nabla \cdot (u^T \cdot u)/2 - u \times (\nabla \times u)$. We insert this relation and write g as $\gamma \nabla z$ to get

$$\frac{1}{\rho} \nabla P + \gamma \nabla z = \frac{1}{2} \nabla \cdot (u^T \cdot u) - u \times (\nabla \times u). \qquad (6.2.12)$$

Next, take the line integral of (6.2.12) along a streamline and (6.2.2) follows.

Exercises

1. Complete the derivation of the NSE. You will need to assume that u has continuous second partial derivatives.
2. Complete the derivation of Bernuolli's equation. Recall that streamlines are parallel to the flow field.
3. Suppose we have a partition of a domain D into elements. Let φ_i^e be a basis for the element velocity field and ψ_i^e be a basis for the pressure field. In addition, we write (6.2.10) as $R_1(u, P)$ and the continuity equation as $R_2(u)$. Writing the weak form as $\int R_1(u, P)\psi_i^e$ and $\int R_2(u)\psi_i^e$ derive the matrix equations for the discrete weak form NSE and continuity equations.

6.3. Applying FEM to the Stokes Equation, TR Time Stepping

There are two common numerical approaches to the Stokes equation. In the first instance, the Stokes equation may be combined with the continuity equation to yield an equivalent fourth-order equation in a single dependent variable, the flow potential. We will demonstrate this for the case of a laminar flow. To begin, recall from (6.2.6) and (6.2.7),

$$\sigma_x = \sigma_{xx} + P = 2\mu\frac{\partial u_1}{\partial x}, \quad \sigma_y = \sigma_{yy} + P = 2\mu\frac{\partial u_2}{\partial y}, \quad (6.3.1)$$

$$\tau_{xy} = \tau_{yx} = \mu\left(\frac{\partial u_1}{\partial y} + \frac{\partial u_2}{\partial x}\right). \quad (6.3.2)$$

If we suppose that the Stokes flow is steady, then the external body forces are in equilibrium and (6.2.5) yields

$$\frac{\partial}{\partial x}(\sigma_x - P) + \frac{\partial \tau_{xy}}{\partial y} = 0, \quad \frac{\partial}{\partial y}(\sigma_y - P) + \frac{\partial \tau_{xy}}{\partial x} = 0. \quad (6.3.3)$$

Additionally, we include the continuity equation,

$$\frac{\partial u_1}{\partial x} + \frac{\partial u_2}{\partial y} = 0. \quad (6.3.4)$$

Using (6.3.4) and results from elementary complex analysis, there is a stream potential ψ so that $u_1 = \partial\psi/\partial y$ and $u_2 = -\partial\psi/\partial x$. If we substitute (6.3.1) and (6.3.2) into (6.3.3), we get

$$0 = \frac{\partial}{\partial x}\left(2\mu\frac{\partial u_1}{\partial x} - P\right) + \frac{\partial}{\partial y}\left(\mu\left(\frac{\partial u_1}{\partial y} + \frac{\partial u_2}{\partial x}\right)\right),$$

$$0 = \frac{\partial}{\partial y}\left(2\mu\frac{\partial u_2}{\partial y} - P\right) + \frac{\partial}{\partial x}\left(\mu\left(\frac{\partial u_1}{\partial y} + \frac{\partial u_2}{\partial x}\right)\right)$$

or

$$0 = 2\mu\frac{\partial^2 u_1}{\partial x^2} - \frac{\partial}{\partial x}P + \mu\left(\frac{\partial^2 u_1}{\partial y^2} + \frac{\partial^2 u_2}{\partial y \partial x}\right),$$

$$0 = 2\mu\frac{\partial^2 u_2}{\partial y^2} - \frac{\partial}{\partial y}P + \mu\left(\frac{\partial^2 u_1}{\partial x \partial y} + \frac{\partial^2 u_2}{\partial x^2}\right).$$

We rewrite in terms of the potential function ψ,

$$0 = 2\mu\frac{\partial^3 \psi}{\partial x^2 \partial y} - \frac{\partial}{\partial x}P + \mu\left(\frac{\partial^3 \psi}{\partial y^3} - \frac{\partial^3 \psi}{\partial y \partial x^2}\right),$$

$$0 = -2\mu\frac{\partial^3 \psi}{\partial y^2 \partial x} - \frac{\partial}{\partial y}P + \mu\left(\frac{\partial^3 \psi}{\partial x \partial y^2} - \frac{\partial^3 \psi}{\partial x^3}\right).$$

Next, we differentiate the first of these equations with respect to y and the second with respect to x,

$$0 = 2\mu\frac{\partial^4 \psi}{\partial y \partial x^2 \partial y} - \frac{\partial^2}{\partial y \partial x}P + \mu\left(\frac{\partial^4 \psi}{\partial y^4} - \frac{\partial^4 \psi}{\partial y^2 \partial x^2}\right),$$

$$0 = -2\mu\frac{\partial^4 \psi}{\partial x \partial y^2 \partial x} - \frac{\partial^2}{\partial x \partial y}P + \mu\left(\frac{\partial^4 \psi}{\partial x^2 \partial y^2} - \frac{\partial^4 \psi}{\partial x^4}\right).$$

Setting these two equations equal, the pressure term cancels, provided it is sufficiently smooth to support interchanging the order of the partials. After dividing by μ, we get

$$2\frac{\partial^4 \psi}{\partial y \partial x^2 \partial y} + \frac{\partial^4 \psi}{\partial y^4} - \frac{\partial^4 \psi}{\partial y^2 \partial x^2} = -2\frac{\partial^4 \psi}{\partial x \partial y^2 \partial x} + \frac{\partial^4 \psi}{\partial x^2 \partial y^2} - \frac{\partial^4 \psi}{\partial x^4}.$$

Finally, supposing that the partials of ψ are interchangeable, we arrive at the desired expression:

$$0 = \frac{\partial^4 \psi}{\partial x^4} + 2\frac{\partial^4 \psi}{\partial x^2 \partial y^2} + \frac{\partial^4 \psi}{\partial y^4}. \tag{6.3.5}$$

Setting up the weak form for Eq. (6.3.5) and applying the divergence theorem twice, yields an expression employing second partial derivatives. Hence, we would use a GFEM model derived from two-variable Hermite polynomials for a rectangular partition or two-variable degree 3 polynomials for a triangular partition (see Section 8.2).

An alternate approach is suggested in [Gresho and Sani (1998)]. This approach deals directly with the Stokes equation as given in (6.2.11),

$$0 = \frac{\partial u}{\partial t} + \nabla\frac{P}{\rho} - \nu\nabla \cdot (\nabla u + (\nabla u)^T) - g. \tag{6.3.6}$$

Suppose we have a domain $D = \bigcup_e E_e$ with a finite element partition. We select a basis $\{\varphi_i^e : i = 1, \ldots, n\}$ of scalar valued functions as a basis for V_e^h. We will use these functions to model the components of the velocity field u. Second, we choose a basis $\{\psi_j^e : j = 1, \ldots, m\}$ of linearly independent scalar valued functions to model P. Because P has only first-order derivatives in (6.3.6), then we may take the ψ_j^e with smaller degree than the φ_i^e.

We begin with the weak form restricted to the element E_e:

$$\int\limits_{E_e} R_1(u_1, u_2, P)\varphi_i^e = 0, \quad \int\limits_{E_e} R_2(u_1, u_2)\psi_j^e = 0, \tag{6.3.7}$$

where R_1 is the right hand expression in (6.3.6) and R_2 is the left hand expression in (6.3.4), the continuity term. Upon application of the divergence theorem only first-order derivatives remain in (6.3.7). Now writing $u_i = \sum_k a_{i,k}^e \varphi_k^e$, $i = 1, 2$, and $P = \sum_j b_j^e \psi_j^e$, then we have a square $2n + m$ linear system for each element.

Suppose we consider the partition of a partially obstructed channel shown in Figure 6.3.1. The elements are rectangles. Hence, we

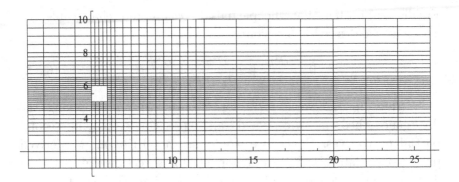

Figure 6.3.1.   Obstructed channel with finite element partition.

may setup four noded rectangles and take the φ_i^e to be first-order
Lagrange polynomials in two variables. Additionally, we may designate a pressure node at the centroid of each rectangle and take
$m = 1$, $\psi_1^e = 1$. With these choices, we have the following linear
system on the element:

$$
\begin{pmatrix} M_1^e & 0 & 0 \\ 0 & M_2^e & 0 \\ 0 & 0 & 0 \end{pmatrix}
\begin{pmatrix} \dot{a}_{11} \\ \dot{a}_{12} \\ \dot{a}_{13} \\ \dot{a}_{14} \\ \dot{a}_{21} \\ \dot{a}_{22} \\ \dot{a}_{23} \\ \dot{a}_{24} \\ \dot{P} \end{pmatrix}
+
\begin{pmatrix} K_1^e & K_{12}^e & L_1^e \\ K_{21}^e & K_2^e & L_2^e \\ (L_1^e)^T & (L_2^e)^T & 0 \end{pmatrix}
\begin{pmatrix} a_{11} \\ a_{12} \\ a_{13} \\ a_{14} \\ a_{21} \\ a_{22} \\ a_{23} \\ a_{24} \\ P \end{pmatrix}
=
\begin{pmatrix} f_{11}^e \\ f_{12}^e \\ f_{13}^e \\ f_{14}^e \\ f_{21}^e \\ f_{22}^e \\ f_{23}^e \\ f_{24}^e \\ \gamma^e \end{pmatrix}.
$$

The dot represents differentiation with respect to t. The matrix
entries are as follows:

$$
M_1^e(i,j) = M_2^e(i,j) = \int_{E_e} \varphi_i^e \varphi_j^e,
$$

$$
K_1^e(i,j) = \frac{\mu}{\rho} \int_{E_e} 2 \frac{\partial \varphi_i^e}{\partial x} \frac{\partial \varphi_j^e}{\partial x} + \frac{\partial \varphi_i^e}{\partial y} \frac{\partial \varphi_j^e}{\partial y},
$$

$$K_2^e(i,j) = \frac{\mu}{\rho} \int\limits_{E_e} \frac{\partial \varphi_i^e}{\partial x} \frac{\partial \varphi_j^e}{\partial x} + 2 \frac{\partial \varphi_i^e}{\partial y} \frac{\partial \varphi_j^e}{\partial y},$$

$$K_{12}^e(i,j) = K_{21}^e(i,j) = \frac{\mu}{\rho} \int\limits_{E_e} \frac{\partial \varphi_i^e}{\partial x} \frac{\partial \varphi_j^e}{\partial y},$$

$$L_1^e(i,1) = - \int\limits_{E_e} \frac{\partial \varphi_i^e}{\partial x},$$

$$L_2^e(i,1) = - \int\limits_{E_e} \frac{\partial \varphi_i^e}{\partial y}.$$

On the right-hand side, we have

$$f_{1,i}^e = \int\limits_{\Gamma_e} \left(2\frac{\partial u_1^e}{\partial x}, \frac{\partial u_1^e}{\partial y} + \frac{\partial u_2^e}{\partial x} \right) \varphi_i^e \cdot \mathbf{n} - \int\limits_{E_e} g_1 \varphi_i^e,$$

$$f_{2,i}^e = \int\limits_{\Gamma_e} \left(\frac{\partial u_1^e}{\partial y} + \frac{\partial u_2^e}{\partial x}, 2\frac{\partial u_2^e}{\partial x} \right) \varphi_i^e \cdot \mathbf{n} - \int\limits_{E_e} g_1 \varphi_i^e,$$

$$\gamma^e = \int\limits_{\Gamma_e} u \cdot \mathbf{n},$$

where Γ_e denotes the boundary of E_e. This then assembles into a global linear system,

$$\begin{pmatrix} M & 0 & 0 \\ 0 & M & 0 \\ 0 & 0 & 0 \end{pmatrix} \begin{pmatrix} \dot{a}_1 \\ \dot{a}_2 \\ \dot{P} \end{pmatrix} + \begin{pmatrix} K & L \\ L^T & 0 \end{pmatrix} = \begin{pmatrix} f \\ \gamma \end{pmatrix}. \qquad (6.3.8)$$

The model shown in Figure 6.3.1 has 976 elements and 1,052 velocity nodes. In this case, (6.3.8) is a 3,080 square system.

We now turn to the boundary value issues. When dealing with viscous fluids, you always take the bounding surfaces as non-penetrating and non-slip. Hence, we set Dirichlet boundary values to zero at the upper and lower edges and around the obstruction. We also suppose that the flow field is horizontal at the inflow and outflow edges. Hence, u_2 is included in the Dirichlet zero boundary at the inflow

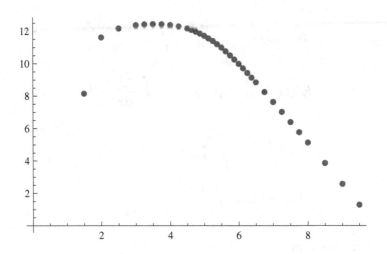

Figure 6.3.2.   Asymmetrical Newman boundary values at the inflow edge.

and outflow. It remains to determine the boundary values for u_1 at the inflow and outflow. We may suppose there is no external force at the outflow. Hence, we set Neumann boundary values to zero here. Finally, there is the Neumann boundary values at the inflow. There are 35 nodes at the inflow edge. We have chosen an asymmetrical pattern (see Figure 6.3.2) to assign Neumann values at these locations.

For the steady-state problem, we solve

$$\begin{pmatrix} K & L \\ L^T & 0 \end{pmatrix} \begin{pmatrix} a_1 \\ a_2 \\ P \end{pmatrix} = \begin{pmatrix} f \\ 0 \end{pmatrix} \tag{6.3.9}$$

for u and P. In Gresho and Sani (1998), the authors prove that if the element partition is regular, then the coefficient matrix is singular. They do this by identifying two linearly independent pressure vectors lying in the kernel. Since neither is possible for an irregular partition, then a partition such as the one used here is essential. Indeed, the coefficient matrix is non-singular for this model.

The resulting steady-state flow is shown in Figure 6.3.3. Besides the effect of asymmetrical forces, note that there is back flow at the outer edges. This is analogous to the splash which occurs when you drop a pebble into a still pool. The effect may be prevented by

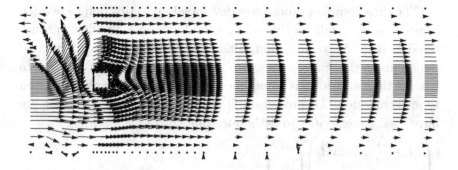

Figure 6.3.3. Steady-state Stokes flow with asymmetrical force.

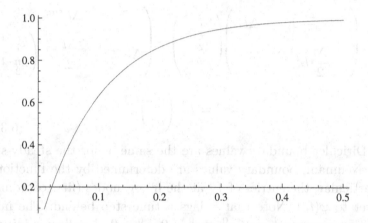

Figure 6.3.4. Burst start up, $\varphi(t) = 1 - Exp[-1/0.1t]$.

setting Dirichlet boundary values at these locations. In Chapter 8, we discuss the relationship between Neumann and Dirichlet boundary values and derive a functional relationship between them. With that result, we understand boundary values that prevent the back splash are unnatural.

As indicated in the previous section, the transient case is the more important. In this case, we start the flow at zero and proceed through a ramp up process that lasts half a second. To simulate this process, any number of ramp up functions may be used to model the external force. We have used a burst type function shown in Figure 6.3.4 which is applied uniformly at each inflow node.

The time stepping procedures developed up to this point are often considered too crude for the complex fluid dynamics problem encountered here (Gresho and Sani, 1998). A technique favored in fluid dynamics is the *trapezoid method* or *TR*. Because it involves data from 3 time states, it is sometimes classified as a Crank–Nicholson type technique. The procedure is shown in (6.3.10). As usual, we have used superscripts to designate time stages.

$$
\begin{pmatrix} M + \dfrac{\Delta t}{2} K & \Delta t L \\[2mm] \dfrac{\Delta t}{2} L^T & 0 \end{pmatrix} \begin{pmatrix} a_1^n \\ a_2^n \\ P^n \end{pmatrix}
$$

$$
= \begin{pmatrix} M - \dfrac{\Delta t}{2} K & -\Delta t L \\[2mm] -\dfrac{\Delta t}{2} L^T & 0 \end{pmatrix} \begin{pmatrix} a_1^{n-1} \\ a_2^{n-1} \\ P^{n-1} \end{pmatrix} + \begin{pmatrix} \Delta t f^n \\[2mm] \Delta t \gamma^{n-1} - \dfrac{\Delta t}{2} L^T \begin{pmatrix} a_1^{n-1} \\ a_2^{n-1} \end{pmatrix} \end{pmatrix},
$$

$$
\gamma^n = L^T \begin{pmatrix} a_1^n \\ a_2^n \end{pmatrix}.
$$
(6.3.10)

The Dirichlet boundary values are the same as for the steady-state case. Neumann boundary values are determined by the function φ. In particular, the entries of f^n at the inflow nodes (first coordinate) are set to $\varphi(t_n)$. Note that γ lags a time step behind. The initial conditions represent rest, $a_1^0 = a_2^0 = 0$, $\gamma^0 = 0$. The flow is initiated by the force f^1. Finally, we set $\Delta t = 0.05$. The resulting flow at $t = 0.5$ is shown in Figure 6.3.5.

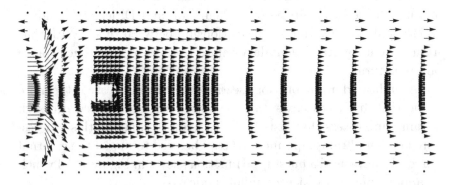

Figure 6.3.5.   Stokes flow with burst start up after 0.5 s.

Note that the flow behind the obstruction is very different from the output in Section 3.1. In this case, the velocity lost while passing the obstruction is not recovered.

Exercises

1. Repeat the flow experiment shown in Figure 6.3.5.
2. Repeat the procedure for transient Stokes flow using a ramp up function that is concave up.
3. Simulate stokes flow from the fourth-order PDE form of the stokes equation (6.3.5).

6.4. Herd Formation Model, FEM

We are continuing the equation developed in Section 3.4 now renamed as

$$\frac{\partial u}{\partial t} = \delta \nabla^2 u - \nabla \cdot ((\beta z + w)u) + uE(u), \qquad (6.4.1)$$

where $E(u) = (1 - u)(u - \zeta)$, $w = \nabla E = (\partial E/\partial u)\nabla u$, and z is a unit vector orthogonal to w. We want to simulate (6.4.1) using finite element method. We will use a rectangle for the domain as the geometry of the region is not an important part of the model. For the elements, we use squares and Lagrange polynomials for the local polynomial space N_i^e in V_e^h. We write

$$u(t, x, y) = \sum_{e,i} \alpha_{e,i}(t) N_i^e(x, y)$$

for the piecewise polynomial FEM solution. Next, we write $\beta z + w = F(u)$ and recast (6.4.1) as

$$\frac{\partial u}{\partial t} = \delta \nabla^2 u - \nabla \cdot (F(u)u) + uE(u).$$

We are going to resolve the nonlinearity via a time lag procedure. For this purpose, we differentiate between the u that is the argument of E and F and the current population density function u

$$\frac{\partial u}{\partial t} = \delta \nabla^2 u - \nabla \cdot (F(\hat{u})u) + uE(\hat{u}). \qquad (6.4.2)$$

If D_e denotes an element, then the weak form of (6.4.2) at the element level is

$$\int_{D_e} \frac{\partial u}{\partial t} N_j^{e'} = \int_{D_e} \delta \nabla^2 u N_j^{e'} - \int_{D_e} \nabla \cdot [F(\hat{u})u] N_j^{e'} + \int_{D_e} u D(\hat{u}) N_j^{e'}.$$

We use D_e to denote an element as E is used here to designate the birth/death function.

With the usual application of Green's theorem, we have

$$\int_{D_e} \frac{\partial u}{\partial t} N_j^{e'} = - \int_{D_e} \delta \nabla u \cdot \nabla N_j^{e'} + \int_{\Gamma_e} \delta \frac{\partial u}{\partial \mathbf{n}} N_j^{e'}$$

$$- \int_{D_e} \nabla \cdot [F(\hat{u})u] N_j^{e'} + \int_{D_e} u D(\hat{u}) N_j^{e'},$$

where Γ_e denotes the boundary of D_e. Substituting the representation of u as a piecewise polynomial function and noting that each N_j^e has support contained in D_e, we have the following linear system:

$$\sum_{e,i} \frac{\partial}{\partial t} \alpha_{e,i}(t) \int_{D_e} N_i^e(x,y) N_j^e(x,y)$$

$$= - \sum_{e,i} \alpha_{e,i} \int_{D^e} \delta \nabla N_i^e(x,y) \cdot \nabla N_j^e(x,y) + \int_{\Gamma_e} \frac{\partial u}{\partial \mathbf{n}} N_j^e(x,y)$$

$$- \sum_{e,i} \int_{D_e} \nabla \cdot [F(\hat{u}) N_i^e(x,y)] N_j^e(x,y)$$

$$+ \sum_{e,i} \int_{D_e} E(\hat{u}) N_i^e(x,y) N_j^e(x,y),$$

$j = 1, 2, \ldots, m$, the number of nodes. This can be rewritten as a matrix equation,

$$L^e \frac{d}{dt} \alpha_{e,i}(t) = -\delta K^e \alpha - M^e(\hat{u}) \alpha + S^e(\hat{u}) \alpha + R^e,$$

where

$$K^e[i,j] = \int_{D_e} \delta \nabla N_i^e(x,y) \cdot \nabla N_j^e(x,y),$$

$$M^e[i,j] = \int_{D_e} N_i^e(x,y) N_j^e(x,y),$$

$$L^e(u)[i,j] = \int_{D_e} \nabla \cdot [F(\hat{u}) N_i^e(x,y)] N_j^e(x,y),$$

$$S^e(u)[i,j] = \int_{D_e} E(\hat{u}) N_i^e(x,y) N_j^e(x,y),$$

$$R^e[j] = \int_{\Gamma_e} \frac{\partial u}{\partial n} N_j^e(x,y)$$

and the calculation of $F(\hat{u})$ and $E(\hat{u})$ is to be determined. As usual, these element level matrices assemble to a global matrix equation,

$$M\frac{d}{dt}\alpha(t) = -\delta K\alpha(t) - L(\hat{u})\alpha(t) + S(\hat{u})\alpha(t) + R. \qquad (6.4.3)$$

To proceed, we look at a particular example. We set the parameter $\zeta = 0.2$ so that E is now fully specified. We will suppose that the spatial dimension is $n = 2$ and take $D = [0, 10] \times [0, 10]$ with uniform partitions $\Delta x = \Delta y = 1.0$. This partition is very sparse. However, for a first example, it is sufficient. For initial condition, we scatter a population about the partitioned domain. For instance, we randomly select 20 of the 121 nodes and then randomly set the initial value for u at these locations. The set of values will be in the range $[0.0, 1.0]$. We suppose that δ is constant and equal to 0.1 as suggested by Grindrod (1991). The deflection parameter β is now the only random coefficient. We suppose that β is independently normally distributed at each location in D. We set the common mean and variance to 0 and 0.1.

Next, we select a time step size, Δt and write $\alpha(t_n) = \alpha^n$ at the nth time state. Resolving the derivative on the right-hand side of (6.4.3) as a forward difference, we get

$$\alpha^{n+1} = \alpha^n - \delta\Delta t M^{-1} K\alpha^n - \Delta t M^{-1} L(\hat{u})\alpha^n$$
$$+ \Delta t M^{-1} S(\hat{u})\alpha^n + \Delta t M^{-1} R.$$

The underlying application is a population problem where u is population density by location and time. Hence, it is reasonable to suppose that there is no population movement across the boundary. Therefore, the normal derivatives $\partial u/\partial \mathbf{n}$ must be zero. In turn, $R = 0$.

To resolve $L(\hat{u})$ and $S(\hat{u})$ we set $\hat{u}(x,y) = \sum_{e,i} \alpha_{e,i}^{n-1} N_i^e(x,y)$, the prior state. Now $E(\hat{u})$ and $\nabla E = (dE/d\hat{u})(\partial\hat{u}/\partial x, \partial\hat{u}/\partial y)$ are easily resolvable and independent of u. This yields a linear system of equations. To further simplify matters, we have also assumed that E is constant on elements. Consequently, the same is true for w and z. We do this by taking the average of the values at the vertices. In particular, at the element level,

$$\int_{D^e} E(\hat{u}) N_i^e(x,y) N_j^e(x,y) = E(\hat{u}) \int_{D^e} N_i^e(x,y) N_j^e(x,y)$$

and

$$\int_{D^e} (\nabla \cdot [F(\hat{u}) N_i^e(x,y)]) N_j^e(x,y) = \int_{D^e} F(\hat{u}) \cdot \nabla N_i^e(x,y) N_j^e(x,y).$$

Figures 6.4.1–6.4.3 illustrate the process implicit in this PDE. The first figure shows the initial state with 20 population centers, followed by output after 100 iterations and then 200 iterations.

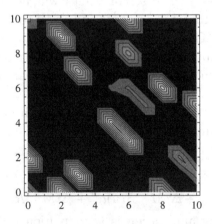

Figure 6.4.1. The initial state.

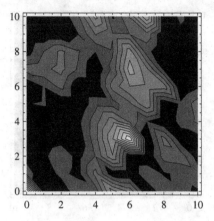

Figure 6.4.2.   After 100 iterations.

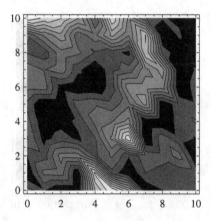

Figure 6.4.3.   After 200 iterations.

The basic pattern is for the population to increase to the 0.8 to 0.9 range. This is followed by a flattening and spreading process to densities in the 0.5 to 0.6 range. The spreading generally occurs in the direction of another population center. This process leads to smaller groups joining together to form larger subpopulations. Figures 6.4.2, 6.4.4 and 6.4.5 show three stages in this process. The maximal density values are for the region in the lower center.

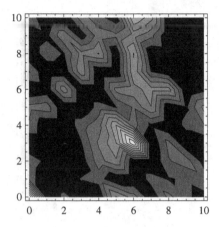

Figure 6.4.4. 80 iterations, $\max u = 0.9$.

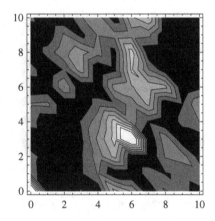

Figure 6.4.5. 90 iterations, $\max u = 0.6$.

Exercises

1. For the model described in Eq. (4.3.4) and parameter values as described in the text.

   a. Setup the nodes and elements for a 1D FEM partition (in one-dimension D is an interval), then elements are subintervals and the nodes are the endpoints.

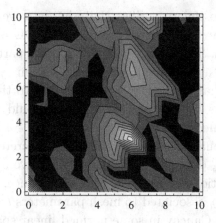

Figure 6.4.6.   100 iterations, max $u = 0.8$.

  b. Take a sample for the diffusion coefficient and then define β on
     each element using two-variable polynomial interpolation.
  c. Setup the spatial FEM for the rectangular partition.

2. For the model described in Eq. (4.3.4) and parameter values as
   described in the text.
   a. Setup the nodes and elements for a rectangular FEM partition.
   b. Take a same for the deflection coefficient and then define β on
      each element using two-variable polynomial interpolation.
   c. Setup the spatial FEM for the rectangular partition.

3. Set $E = (0.8 - u)(u - 0.2)$. Note that this choice for E penalizes
   both under and over populated regions. Derive the corresponding
   equation to (4.3.4) and execute the example.

6.5. A Remark on Numerical Processes, Convergence, and Existence Theory

We present this section as a transition to the second part. In the
previous chapters, we have looked at many examples of numerical
techniques to simulate the process associated to a differential equation. Some of these are supported by a convergence theory, others
are not. Nevertheless, they are all considered successful.

A convergence theory verifies that the data we compute is correct. In particular, suppose we are given a PDE with boundary or initial conditions. Then, we develop a domain partition with mesh parameter h and a discrete approximate solution φ_h. A convergence theorem states that if u is the actual solution to the PDE, then φ_h converges to u as $h \to 0$. Such a theorem would be stated in the context of a normed linear space.

It seems reasonable to ask, if a family of discrete approximations are well behaved, does the PDE necessarily have a solution and do the discrete solutions converge to it. For instance, suppose we had FEM solutions φ_h associated to mesh parameters h. If we know that the solutions are Cauchy in some normed linear space, can we then assert that the PDE has a solution? It is not likely that we would be able to verify that a particular family of solutions was Cauchy, but nevertheless, we can make the supposition. In this section, we consider a particular instance of this and show that the PDE indeed has a solution.

To begin, we will suppose that the discrete process is FEM and that the partition is triangulation. We further suppose that the refinement process is accomplished by joining the midpoints of the sides of each triangular element. The result is to divide each triangle into four congruent triangles. Each of the four is similar to the original by a ratio of 1:2. Hence, if h is the mesh parameter at a given stage, then $h/2$ is the mesh parameter at the next stage. Therefore, the mesh parameter will go to zero as the partition is refined.

Next, we need a normed linear space V that contains every FEM solution, φ_h to the given second-order PDE. We will then suppose that the FEM solutions are Cauchy for this norm. For our puspose, we will need an FEM model based on C^1 piecewise polynomial functions. The Hermite polynomials in two variables (see Section 5.1 for the 1 variable case) on a rectangular partition is a C^1 FEM model. In Section 8.2, we develop a C^1 model on a triangular parttition. This is done with fifth degee polynomials in two variables. In either case, V^h will be a subspace of a linear space denoted $H^2 \subset L^2$ (see Section 9.2). Proceeding, H^2 is a Hilbert space with inner product,

$$\sigma(u,v) = \int_D \left[uv + \sum_i \frac{\partial u}{\partial x_i} \frac{\partial v}{\partial x_i} + \sum_{i,j} \frac{\partial^2 u}{\partial x_i \partial x_j} \frac{\partial^2 v}{\partial x_i \partial x_j} \right].$$

In Chapter 9, we develop a special notation in order to separate weak differentiability from ordinary differentiability. In any event, we are assured that the sequence of FEM solutions converges to a function $u \in H^2$.

We want to prove that u is a solution to the PDE. We write the second-order PDE $Lv - f = 0$ with L linear. Since we are concerned with FEM, we pass to the weak form of the PDE

$$0 = \int_D [Lu - f]v = \tau[u,v] - \int_D fv, \qquad (6.5.1)$$

where τ is a bilinear function. A solution is a function u in H^2 so that (6.5.1) is satisfied for every v in H^2. In particular, for any piecewsie C^1 polynomial function v,

$$\tau[u - \varphi_h, v] = \tau[u,v] - \int_D fv - \tau[\varphi_h, v] + \int_D fv = \tau[u,v] - \int_D fv,$$

by the definition of the FEM solution.

Next, we need to add a boundedness assumption. If we suppose that L is bounded with norm C, then by Cauchy Schwarz, $\tau[u,v] = \int_D (Lu)v \leq C\|u\|\|v\|$ for any u and v in H^2. We now have

$$\tau[u,v] - \int_D fv = \tau[u - \varphi_h, v] \leq C\|u - \varphi_h\|\|v\| \to 0,$$

as $h \to 0$.

Next, take w in H^2. The Bramble–Hilbert theorem (see Theorem 9.6.7) applies to the given partition and refinement process. Therefore, w is the L^2-limit of a sequence of piecewise polynomial functions w_h again associated to the partitions. Now, since τ is continuous in each argument, we have for the limit as $h \to 0$,

$$\tau[u, w] - \int_D fw = \lim \tau[u, w_h] - \int_D fw_h = 0.$$

Therefore, u is the weak solution. We have proved the following:

Theorem 6.5.1. *Consider a PDE $Lv - f = 0$ with L linear and bounded with respect to the Sobolev space H^2. Suppose we have an FEM triangular partition with the regular refinement process described above. If the FEM solutions are Cauchy, then the PDE has weak solution u and the FEM solutions converge to the actual solution in L^2.*

Similar results hold for other discrete processes. The reader is asked to prove the corresponding result for Gaussian collocation in Exercise 1. The technique was introduced in Section 5.1. The convergence theory is developed in Chapter 10.

Exercise

1. Consider Gaussian collocation in one spatial dimension. For the normed linear space, take the Banach space of continuous functions with sup norm. Suppose that the PDE is given by $Lv - f = 0$ where L is linear and bounded, f is continuous. Suppose further that the partition refinement process divides each interval element in half. Let u_C^h be a Cauchy sequence of collocation solutions, then the PDE has solution u and $u_C^h \to u$ as $h \to 0$ (*Hint*: The union of the collocation point set is a dense subset of the domain).

PART 2
Methods and Theory

Chapter 7

Finite Difference Method

Introduction

In Chapter 2, we used finite difference method (FDM) to simulate
the solutions to the diffusion (heat) equation in one and two spatial
variables. At that time, we introduced stability and Neumann
stability. We followed by developing the discrete Fourier transform
as the preferred means to resolve Neumann stability. We saw that
without stability, the accumulated FDM error was so great as to render
the computed data unreliable. Stated otherwise, the computed
data could not possibly approximate or converge to the actual solution
of the PDE. We begin the chapter with the Lax–Richtmyer or
Lax equivalence theorem. This result clearly states the relationship
between stability and convergence.

For the Lax theorem, we begin with a formal development of
an FDM scheme followed by definitions for convergence and consistency.
Intuitively, convergence means that the FDM computed values
converge to the actual values as the mesh parameter goes to
zero. Consistency is more subtle. The idea is that an FDM scheme
is consistent, if the finite difference rendering of the PDE is a fair
approximation of the actual equation. The usual development of the
elementary finite differences, first-order forward and backward differences
and the first and second-order central differences, are based on
the Taylor expansion. Hence, for a sufficiently smooth function, we

189

may suppose that the finite difference operators converge uniformly to actual differential operators. This, we see, goes a long way toward resolving consistency.

The importance of the Lax theorem is that consistency and stability do not require knowledge of the solution of the PDE. Convergence is different. To verify convergence we need specific knowledge of the limit function, the actual solution to the PDE. However, if we knew the solution, we would not need a discrete numerical technique to approximate it.

Our theorem does not just apply to FDM. The Lax result applies to cases where we develop the spatial variables via FEM or collocation and the time variable is developed with FDM. Recall that time stepping is nearly always based on an FDM rendering of the d/dt term. Therefore, stability is an issue even when the spatial rendering is not FDM.

Following the Lax theorem, we develop the extended differences. These difference formula are extensions of the ones we have been using up to now. The extended differences converge at higher-order than the elementary differences, and they are more sensitive to stability issues. Even though they are not often used, researchers do prefer them when there is a boundary or initial value singularity, for instance, in the Black–Scholes equation. (See Section 5.1.).

In Chapter 2, we looked at FDM as applied to parabolic equations. In this chapter we look at the FDM as applied to hyperbolic initial value problems in one or two spatial variables. The development is analogous but different. For instance, unconditional instability will arise for the order 1 wave equation. Also in this context, we will have occasion to introduce two new FDM techniques. Both are associated to Peter Lax, the Lax–Wendroff method, and the Lax–Friedrichs method.

We continue with a short discussion of stability and convergence. Recall that in Chapter 2 we stated that Neumann stability is equivalent to stability in the case of a symmetric FDM transformation. In this chapter, we will prove that stability implies Neumann stability. Whereas, the converse is false. In another direction, there is a general theory for stability associated to the discrete Laplace transform.

We will discuss some aspects of this development. We also use this section to introduce the Courant–Friedrichs–Lewy convergence theorem.

We end the chapter with the theory of FDM applied to elliptical PDE. We prove the convergence theorem for elliptical PDE with Dirichlet boundary values.

The chapter is organized as follows. In Section 7.1 we prove our major result on convergence for FDM schemes. In Section 7.2 we develop the extended differences. We continue with a section devoted to introducing FDM techniques for the hyperbolic initial value problem. Next, we discuss stability and convergence for initial value problems. We end the chapter with a section presenting some of the theory of FDM applied to elliptical PDE.

7.1. Convergence, Consistency, and Stability

In this section, we prove the Lax convergence theorem, sometimes referred to in the literature as the Lax–Richtmyer theorem or the Lax equivalence theorem. Whichever the case, it largely justifies the use of FDM as a solution technique for PDE.

We begin with several simplifying assumptions. First, we restrict attention to differential equations involving only first-order time derivatives and using only two time states in the FDM scheme. These are called single step processes. Secondly, we will suppose the coefficients of the PDE are constant and the equation is linear and homogeneous. Further, we restrict attention to functions of a single spatial variable. This is the context of the original theorem, see Lax and Richtmyer (1956). There are extension of this basic result, see Atkinson and Han (2009); Strikwerda (2004). As in Part 1, we suppose that the spatial domain of the PDE is compact. For notational simplicity, we consider only scalar-valued functions.

We begin with a Banach space V and a dense subspace V_0. This context will present no real restriction as any normed linear space has a completion, the completion is a Banach space and the original normed space is isometric to a dense subspace of the completion. Moreover, any bounded linear operator on V_0 is uniquely extendable to a bounded linear operator on V. Throughout this section, all norms refer to the norm of V or the norm of a bounded operator on V. Furthermore, if $f(t, x)$ is a real valued function of time and space which is differentiable in each variable, then $u(t_0)(x) = f(t_0, x)$ may be considered an element of the normed linear space of bounded functions of the spatial variable with sup norm.

We define our differential equation as follows. Given L linear on V_0 and u_0 in V_0, we seek a $u : [0, T] \to V_0$ which is differentiable in t and satisfies

$$\frac{\partial u}{\partial t} = L(u), \quad u(0) = u_0. \tag{7.1.1}$$

It is usual to refer to u_0 as the *initial value*.

Our goal here is to prove a convergence theorem for finite difference method. But in order to prove that a discrete process converges as the mesh parameter goes to zero, we must have a target, a solution to the PDE. Existence of a solution is often referred to as well-posed.

Definition 7.1.1. The initial value problem (7.1.1) is *well-posed* provided it has unique solution $u : [0, T] \to V_0$. Furthermore the solution must vary continuously from the initial value. In particular, there is a constant c such that if u and v are solutions associated to initial values u_0 and v_0, then

$$\sup_{0 \le t \le T} \|u(t) - v(t)\| \le c\|u_0 - v_0\|.$$

We may suppose that the solution evolves from the initial state, or stated formally, there is a function S defined on $[0, T]$ so that for each t, $S(t)$ is an operator on V_0, given by $u(t) = S(t)(u_0)$. The following result resolves the iterated application of S.

Lemma 7.1.1. *Given a well-posed PDE and* $t_1, t_2 \in [0, T]$, $t_1 + t_2 \le T$, *then* $S(t_1 + t_2) = S(t_1)S(t_2)$.

Proof. We set $u(t, \cdot) = S(t)u_0$ and $\bar{u}(t, \cdot) = u(t_1 + t, \cdot) = S(t_1 + t)u_0$. But $\bar{u}(t, \cdot) = S(t)\bar{u}_0$, where $\bar{u}_0 = S(t_1)u_0$. Putting this together, the result follows. $\square$

There are two ways to go about proving an FDM convergence theorem. We can use interpolation to resolve the FDM data as a function and prove the theorem as a linear space convergence result, or we can reduce the solution to a sequence of values and prove the theorem as a linear algebra result. In this case, we choose the former approach. At the end of the chapter, we prove convergence for elliptical PDE. In that case, we use the latter approach.

Our next task is to relate an FDM operator R to the continuous solution process that we have already introduced. The idea here is to define the finite difference operators as functions on V associated to a uniform partition of the time interval with mesh parameter k. If we were to take $R_k = S(k)$, then by the lemma, $S(t_m) = S(mk) = R_k^m$, or $u(t_m, \cdot) = S(t_m)u_0 = R_k^m u_0$. Hence, iterating R_k would be the same as evaluating S at the points in the partition. However, this would accomplish nothing. We need to define R_k in terms of the FDM scheme for the PDE and then resolve the relationship to S. Hence, we define R independent of S and seek conditions under which R_k converges to S as the partition is refined.

In particular, we take a positive constant c and a maximal mesh parameter k_0 and define for each non-negative $k \leq k_0$ a bounded linear operator of V so that the norm $\|R_k\| < c$. We will refer to the operators R_k as a *finite difference scheme*. We justify this terminology in two stages.

Suppose we have a linear PDE representing a single step process. Then, we may define uniform partitions for the spatial x_i and temporal t_n variables and then render the PDE in finite differences using forward Euler time stepping. The result is a matrix C_k that relates two consecutive state vectors, $a^{n+1} = C_k a^n$, or $a^{n+1} = C_k^{n+1} a^0$.

Alternatively, consider u in V_0 and the vector of function values $\{u(x_i)\}$ associated to the spatial partition X. Define $\mathcal{V}_X(u) = \{u(x_i)\}$ and set $\mathcal{F}_X(u)$ to the discrete Fourier transform of u at the spatial partition, or equivalently, the interpolation of $\{u(x_i)\}$ with trigonometric polynomials (see Section 2.3). Now, we define $R_k = \mathcal{F}_X C_k \mathcal{V}_X$. Note that R_k operates on u_0 the same as C_k acts on a^0. Furthermore, R_k is linear and bounded on the image of $\mathcal{F}_X$. In addition, using results of Section 2.3, we can prove that R_k has the same norm as C_k.

In another direction, we can use the current finite difference scheme to represent the time stepping process that arises after the spatial variable has been rendered with FEM or CM. In this case we have $\partial a/\partial t = Ka$. If we develop this with forward Euler FDM, we get a statement for successive time states $a^{n+1} = a^n + kKa^n$ or $a^{n+1} = (I + kK)a^n$. Hence, we define $C_k = (I + kK)$ and set R_k accordingly. In this context there is a separate convergence theory for the spatial treatment. This convergence theory plays the role of consistency defined below.

With this discussion in mind, there is no need to redefine stability for this setting. We see R_k as the continuous domain extension of C_k. Definition 2.3.1 will suffice. Once we have R_k we have the corresponding C_k.

We now turn to the formal development of consistency. Notice that the R_k are defined with little or no connection to the PDE. Consistency imposes that connection. Specifically, it states that the truncation error (see Section 2.3) is zero at the limit. We develop this statement after the definition.

Definition 7.1.2. The finite difference scheme is *consistent* if for every solution u

$$\lim_{k \to 0} \frac{1}{k} \|R_k u(t, \cdot) - u(t + k, \cdot)\| = 0. \qquad (7.1.2)$$

At the end of this section, we will verify consistency in a particular case. At this time, it is interesting to look closer at Eq. (7.1.2). Recall that the PDE is written as $\partial u/\partial t = Lu$. Starting with truncation error $R_k u(t, \cdot) - u(t+k, \cdot)$, we divide by k, and add and subtract first with $Lu(t, \cdot)$ and then with $u(t, \cdot)/k$ to get

$$\left[\frac{R_k u(t, \cdot)}{k} - L(u(t, \cdot)) \right] - \left[\frac{u(t + k, \cdot)}{k} - L(u(t, \cdot)) \right]$$

$$= \left[\frac{R_k}{k} - L \right] u(t, \cdot) - \left[\frac{u(t + k, \cdot)}{k} - L(u(t, \cdot)) \right]$$

$$= \left[\frac{R_k}{k} - L \right] u(t, \cdot) - \frac{I}{k} u(t, \cdot) + \frac{u(t, \cdot)}{k}$$

$$- \left[\frac{u(t+k, \cdot)}{k} - L(u(t, \cdot)) \right]$$

$$= \left[\frac{R_k - I}{k} - L \right] u(t, \cdot) - \left[\frac{u(t+k, \cdot) - u(t, \cdot)}{k} - L(u(t, \cdot)) \right].$$

Given that u is the solution to the PDE, then $[u(t+k, \cdot) - u(t, \cdot)]/k - Lu(t, \cdot) \to 0$ as $k \to 0$. Therefore consistency implies that

$$\left[\frac{R_k - I}{k} - L \right] u(t, \cdot) \to 0 \quad \text{as } k \to 0.$$

Hence, consistency implies that we can approximate the PDE using R_k.

It now remains to define convergence and recall stability from Chapter 2. Referring back to our earlier discussion, for convergence, we want the powers of R_k to act like $S(t)$ at the limit. However, the precise limit statement is complex. To start, given a time t, we will need a sequence of time steps sizes k_i with $m_i k_i \to t$ where m_i is an integer sequence and $k_i \to 0$.

Definition 7.1.3. An FDM scheme is *convergent* as the mesh parameter $k \to 0$ provided for any t,

$$\lim_{k_i \to 0} \|[R_{k_i}^{m_i} - S(t)]u_0\| = 0, \tag{7.1.3}$$

where k_i is a sequence of mesh parameters (Δt), m_i is a sequence of integers with $m_i k_i \to t$.

We defined stability in Definition 2.3.1. The supposition is that $\|C_k^m\|$ is bounded independent of m for all $k < k_0$ and $mk \le T$. We note now that this is equivalent to the corresponding statement for the operators R_k (see Exercise 5). The following result is the main event of this section. The notation used in the proof is the one developed in this section.

Theorem 7.1.1. (*Lax and Richtmyer*) *Any well-posed, consistent initial value problem is stable if and only if it is convergent.*

Proof. We first show that stability and consistency imply convergence. In particular, we must verify the limit statement in Definition 7.1.3. For this purpose we telescope the difference in (7.1.3),

$$R_k^m u_0(x) - u(t, x)$$

$$= \sum_{j=1}^{m-1} R_k^j \left[R_k u((m - (j+1))k, x) - u((m-j)k, x) \right]$$

$$+ [u(mk, x) - u(t, x)].$$

Next, we take the norm of both sides, apply the triangle inequality and stability with uniform bound M.

$$\left\| R_k^m u_0(x) - u(t, x) \right\|$$

$$\leq M \sum_{j=1}^{m-1} \| R_k u((m - (j+1))k, x) - u((m-j)k, x) \|$$

$$+ \| u(mk, x) - u(t, x) \|.$$

Given $\epsilon > 0$, then setting $t = m - (j+1)k$, we have

$$\| R_k u((m - (j+1))k, x) - u((m-j)k, x) \|$$

$$= \| R_k u(t, x) - u(t+k, x) \|$$

$$= k \left[\frac{1}{k} \| R_k u(t, x) - u(t+k, x) \| \right] < k\epsilon,$$

as R_k is consistent. Since there are $m - 1$ summands and since $mk \leq T$,

$$\| R_k^m u_0(x) - u(t, x) \| < Mk(m-1)\epsilon + \| u(mk, x) - u(t, x) \|$$

$$\leq MT\epsilon + \| [S(mk) - S(t)] u_0 \|.$$

As S is continuous, the right-hand summand goes to zero as $mk \to t$, and convergence is verified.

Conversely, suppose that the FDM scheme is not stable. We proceed to prove that the FDM scheme cannot be convergent. By instability, there are sequences k_j, m_j with $k_j m_j < T$, $k_j \leq k_0$ so that $\| R_{k_j}^{m_j} \|$ is unbounded. Since k_j is bounded, we may suppose that it is convergent. By definition of the FDM operator, there is a positive constant c with $\|R_{k_j}\| \leq c$ for all k_j. If m_j is bounded, then $\|R_{k_j}^{m_j}\| \leq \|R_{k_j}\|^{m_j}$ is also bounded. But this contradicts our supposition. Therefore, we conclude that m_j is unbounded, $k_j \to 0$ and convergence applies.

By the definition of convergence and the triangle inequality, given $\epsilon > 0$, $\|R_{k_j}^{m_j} u_0\| \leq \epsilon + \|S(t)u_0\|$, for j sufficiently large and for each u_0. By the uniform boundedness principle (UBP — see next paragraph), $\|R_{k_j}^{m_j}\|$ must also be bounded and we have reached the required contradiction.

For completeness, we include the proof of the UBP now. Note that the proof requires the Baire category theorem (Rudin, 1986). Suppose that L_k is a sequence of bounded operators of a Banach space V so that $\|L_k(v)\|$ is bounded for every k in $\mathbb{Z}$ and every v. Note that it does not matter that we have these statements on a dense subspace as they extend to the entire space. We set $A_n = \{v \epsilon V : \|L_k(v)\| \leq n\}$. This is a nested increasing sequence of closed subsets of V. Furthermore, by hypothesis, every element of V must belong to at least one A_n. But then by the Baire category theorem, at least one A_{n_0} must contain an open set B.

Select v in V with $\|v\| = 1$ and select v_0 in B, so that a sphere about v_0 is contained in B. Set the radius of the sphere to n_0. Take ρ sufficiently small so that $\rho v + v_0$ lies in the sphere contained in B. We calculate

$$\|L_k(v)\| = \frac{1}{\rho} \|L_k(\rho v + v_0 - v_0)\|$$

$$\leq \frac{1}{\rho} [\|L_k(\rho v + v_0)\| + \|L_k(v_0)\|] \leq \frac{2n_0}{\rho}.$$

Since we have a Banach space, there is a minimal ρ over all v. It now follows that the L_k are bounded independent of k. This completes the proof. □

We have seen that stability may be resolved using the discrete Fourier transform. No information regarding the solution of the PDE is required. Likewise, verifying consistency requires no specific knowledge of the solution. We emphasize this point by ending the section with an example, explicit FDM for the 1D heat equation. The idea is to express the truncation error, the error between the exact process and the discrete process, in terms of the Taylor expansion and then reduce that expression to one involving only the remainder terms. In which case, the existent but unknown solution to the PDE cancels from the calculation.

We verify consistency for the 1D heat equation with forward time stepping and second central difference applied to the spatial variable. In particular, we begin with $\partial u/\partial t = \alpha \partial^2 u/\partial x^2$ and write $\lambda = \alpha k/h^2$, where k is the time partition mesh parameter and h is the spatial mesh parameter.

$$R_k u - u_l^{j+1} = -\lambda u_{l+1}^j - (1 - 2\lambda)u_l^j - \lambda u_{l-1}^j \qquad (7.1.4)$$

for $m > l > 0$ with the first and last equations to be determined via the boundary values. Next, we rewrite this expression using the Taylor expansion for each of the state values (Note that when we write the Taylor expansion, we make a smoothness assumption for u.). In particular,

$$u_l^{j+1} = u_l^j + \frac{\partial u}{\partial t}(t_j, x_l)k + C_1 \frac{k^2}{2},$$

$$u_{l+1}^j = u_l^j + \frac{\partial u}{\partial x}(t_j, x_l)h + \frac{\partial^2 u}{\partial x^2}(t_j, x_l)\frac{h^2}{2} + \frac{\partial^3 u}{\partial x^3}(t_j, x_l)\frac{h^3}{6} + C_2 \frac{h^4}{24},$$

$$u_{l-1}^j = u_l^j - \frac{\partial u}{\partial x}(t_j, x_l)h + \frac{\partial^2 u}{\partial x^2}(t_j, x_l)\frac{h^2}{2} - \frac{\partial^3 u}{\partial x^3}(t_j, x_l)\frac{h^3}{6} + C_3 \frac{h^4}{24},$$

where the C_i are constants arising from the remainder terms. Upon substitution into (7.1.4), the coefficient of u_l^j evaluates to zero, leaving

$$R_k u = \frac{\partial u}{\partial t}(t_j, x_l)k + C_1 \frac{k^2}{2}$$

$$- \lambda \left(\frac{\partial u}{\partial x}(t_j, x_l)h + \frac{\partial^2 u}{\partial x^2}(t_j, x_l)\frac{h^2}{2} + \frac{\partial^3 u}{\partial x^3}(t_j, x_l)\frac{h^3}{6} + C_2 \frac{h^4}{24} \right)$$

$$- \lambda \left(-\frac{\partial u}{\partial x}(t_j, x_l)h + \frac{\partial^2 u}{\partial x^2}(t_j, x_l)\frac{h^2}{2} - \frac{\partial^3 u}{\partial x^3}(t_j, x_l)\frac{h^3}{6} + C_3 \frac{h^4}{24} \right).$$

Next, collect the remainder terms on the right and factor out the h from the denominator to get

$$R_k u = \frac{\partial u}{\partial t}(t_j, x_l)k - \lambda\left(\frac{\partial u}{\partial x}(t_j, x_l)h + \frac{\partial^2 u}{\partial x^2}(t_j, x_l)\frac{h^2}{2} + \frac{\partial^3 u}{\partial x^3}(t_j, x_l)\frac{h^3}{6}\right)$$

$$- \lambda\left(-\frac{\partial u}{\partial x}(t_j, x_l)h + \frac{\partial^2 u}{\partial x^2}(t_j, x_l)\frac{h^2}{2} - \frac{\partial^3 u}{\partial x^3}(t_j, x_l)\frac{h^3}{6}\right)$$

$$+ C_1\frac{k^2}{2} - \alpha k(C_2 + C_3)\frac{h^2}{24}.$$

The first-order and third-order spatial derivatives cancel, leaving

$$R_k u = \frac{\partial u}{\partial t}(t_j, x_l)k - \lambda\frac{\partial^2 u}{\partial x^2}(t_j, x_l)h^2 + C_1\frac{k^2}{2} - \alpha k(C_2 + C_3)\frac{h^2}{24}.$$

Since u is a solution to the differential equation, then

$$\lambda\frac{\partial^2 u}{\partial x^2}(t_j, x_l)h^2 = \alpha k\frac{\partial^2 u}{\partial x^2}(t_j, x_l) = \frac{\partial u}{\partial t}(t_j, x_l)k.$$

Therefore, we have

$$R_k u = C_1\frac{k^2}{2} - \alpha k(C_2 + C_3)\frac{h^2}{24},$$

which goes to zero as $k \to 0$.

For implicit schemes, the process iterates only with the inverse of the tridiagonal matrix. Hence, similar to stability, we must consider the norm of this matrix.

Exercises

1. Consider the differential equation $\partial u/\partial t = \alpha(\partial^2 u/\partial x^2) + \beta u$.

   a. Verify that the stiffness matrix for the FTCS FDM scheme yields a symmetric tridiagonal matrix.

   b. Determine the norm of the matrix.

2. Consider the differential equation $\partial u/\partial t = \alpha(\partial^2 u/\partial x^2 + \partial u/\partial x)$.

   a. Determine sufficient conditions for consistency for the FTCS FDM scheme.

   b. Repeat (a) for BTCS.

3. The functions $f_n - x^n$ are square integrable and continuous on $[0, 1]$. Use this sequence of functions to show that $L = d/dx$ is bounded on $L^2[0, 1]$ and unbounded on $C[0, 1]$ with sup norm (*Hint*: Compute $\lim_m \|Lf_n\|$ for the two cases).

4. Prove that the operators R_k for the 1D heat equation with forward Euler time stepping is a uniformly bounded family of bounded linear operators of the Banach space V (see Section 2.2).

5. Prove that $\|C_k^m\|$ is bounded independent of m for all $k < k_0$ and $mk \leq T$ if and only if the same holds for R_k.

7.2. Extended Differences

The difference formulations derived below are more complicated than the elementary ones we have seen up to now. They are important because they converge faster. Recall that in Section 2.2 we looked at the FDM output data for the heat equation against the values of the analytic solution. In that case, the FDM output was not particularly good. In addition, the FDM results may not vary significantly with Δx or Δt. In such a situation, the analyst might consider extended differences, finite difference formulations which converge faster. In addition, the *Mathematica* function NDSolve supports extended differences via options which allow the programmer to specify extended differences (see *Mathematica* documentation).

In the exercises to the section, we see that extended differences also tend to be less stable. NDSolve is able to get around this problem by employing adaptive time stepping. Even so, extended differences are not often used. Nevertheless, there are specific settings where they are used. For instance, situations where the second derivative of the solution is known to be discontinuous, for non-uniform partitions they are often the technique of choice or when the boundary or initial conditions are not smooth, see Chung (2002); Hildebrand (1974). The non-uniform partition case may also be considered as discontinuous, since if $x_i - x_{i-1}$ differs largely from $x_{i+1} - x_i$, then the forward and backward differences will be significantly different at x_i.

Consider a real-valued function, $u(x)$, which is defined on an interval $[a, b]$. In order to have free access to the truncated Taylor expansions, we suppose that u is n times continuously differentiable on $[a, b]$ and $n+1$ times differentiable on the open interval (a, b). Suppose that we have a uniform partition of the interval and set $x_{i+1} - x_i = \Delta x$ for each i. We set $D = d/dx$, and $(Du)_i = (du/dx)(x_i)$. Now, we use this notation to write the n term Taylor expansion (with remainder term) for $u(x_{i+1})$ around x_i.

$$u(x_{i+1})$$

$$= u(x_i) + \Delta x (Du)_i + \frac{\Delta x^2}{2}(D^2 u)_i + \cdots + \frac{\Delta x^n}{n!}(D^n u)_i + R_n(u, x_i)$$

$$= \left[\left(I + \Delta x D + \frac{1}{2!}(\Delta x D)^2 + \cdots + \frac{1}{n!}(\Delta x D)^n + R_n \right)(u) \right](x_i),$$

$$(7.2.1)$$

where the second version of the right-hand side of (7.2.1) suggests an operator that maps the pair (u, x_i) to $u(x_{i+1})$. We formalize this operator in the following definition. For this purpose, we need to develop the notation. We let V denote the vector space of all n-times continuously differentiable functions on $[a, b]$ and $n + 1$ times differentiable functions on (a, b). Further, we let P denote a uniform partition of $[a, b]$.

Definition 7.2.1. The operator $E_n^{\rightarrow} : V \times P \to \mathbb{R}$ given by

$$E_n^{\rightarrow}(u, x_i) = \left[\left(I + \Delta x D + \frac{1}{2!}(\Delta x D)^2 \right. \right.$$

$$\left. \left. + \cdots + \frac{1}{n!}(\Delta x D)^n + R_n \right)(u) \right](x_i)$$

is called the *forward displacement operator*. In addition, the operator $\delta_n^{\rightarrow} = E_n^{\rightarrow} - I$ is called the *forward difference operator*.

By (7.2.1), $u(x_{i+1}) = E_n^{\rightarrow}(n, x_i)$ and $\delta_n^{\rightarrow}(u, x_i) = u(x_{i+1}) - u(x_i)$. These two equations motivate the terminology. Since the expression for $E_n^{\rightarrow}$ is reminiscent of the truncated exponential series, it is commonly written $E_n^{\rightarrow}(u) = \exp_n(u)$. It is easy to verify that repeated application of the displacement operator moves along the partition. In particular, $(E_n^{\rightarrow})^m(u_i) = u_{i+m}$ provided $i + m$ is not larger than the length of the partition. Here, we have written $u(x_{i+m})$ as u_{i+m}. We will continue to use this standard FDM notation.

We now get an expression for the iterated forward difference operator.

Lemma 7.2.1. *For any $j \leq n$, the iterated forward difference operator satisfies*

$$(\delta_m^{\rightarrow})^j(u_i) = \sum_{k=0}^{j}(-1)^k C_k^j u_{i+j-k}, \qquad (7.2.2)$$

where $C_k^j = j!/k!(j-k)!$ is the usual binomial coefficient.

Proof. Indeed, since $(\delta_m^{\rightarrow})^j(u_i) = (E_n^{\rightarrow} - I)^j(u_i)$, then by the binomial theorem,

$$(\delta_m^{\rightarrow})^j(u_i) = \left[\sum_{k=0}^{j}(-1)^k C_k^j (E_n^{\rightarrow})^{j-k}\right](u_i) = \sum_{k=0}^{j}(-1)^k C_k^j u_{i+j-k}. \qquad \square$$

The following result yields expressions for the derivative of u for each $m \leq n$. These formulations converge to the derivative of u order Δx^m. As they are based on the forward difference operator, they are referred to as extended forward differences.

Theorem 7.2.1. *If u has n continuous derivatives on the closed interval $[a, b]$ and $n + 1$ derivatives on the open interval (a, b), then for any $m \leq n$,*

$$\frac{1}{\Delta x}\sum_{j=1}^{m}\frac{(-1)^{j-1}}{j}\sum_{k=0}^{j}(-1)^k C_k^j u_{i+j-k}, \qquad (7.2.3)$$

converges to $(-1)^{(m-1)}Du_i$ order Δx^m.

Proof. We rewrite the expression in (7.2.3) using the forward difference operator,

$$\frac{1}{\Delta x} \sum_{j=1}^{m} \frac{(-1)^k}{j} \sum_{k=0}^{j} (-1)^k C_k^j u_{i+j-k}$$

$$= \frac{1}{\Delta x} \left[\overrightarrow{\delta_n} - \frac{1}{2}(\overrightarrow{\delta_n})^2 + \cdots + \frac{(-1)^{m-1}}{n}(\overrightarrow{\delta_n})^m \right] (u_i)$$

$$= \frac{1}{\Delta x} \left[(\overrightarrow{E_n} - I) - \frac{1}{2}(\overrightarrow{E_n} - I)^2 + \cdots + \frac{(-1)^{m-1}}{n}(\overrightarrow{E_n} - I)^m \right]$$

$$\times (u_i).$$

From (7.2.1),

$$\overrightarrow{E_n} - I = \Delta x D + \frac{1}{2!}(\Delta x D)^2 + \cdots + \frac{1}{n!}(\Delta x D)^n + R_n.$$

Whereas the expression,

$$(\overrightarrow{E_n} - I) - \frac{1}{2}(\overrightarrow{E_n} - I)^2 + \cdots + \frac{(-1)^{m-1}}{n}(\overrightarrow{E_n} - I)^m$$

is formally just the truncated Taylor expansion of $\log(1 + x)$ evaluated at $\overrightarrow{E_n} - I$. In turn, the expression (7.2.1) for $\overrightarrow{E_n}$ is formally the truncated Taylor expansion for the exponential. Whether these are formal expansions or actual Taylor polynomials, the algebraic manipulation is the same. Therefore, evaluating the formal *log* at the formal exponential yields

$$(\overrightarrow{E_n} - I) - \frac{1}{2}(\overrightarrow{E_n} - I)^2 + \cdots + \frac{(-1)^{m-1}}{n}(\overrightarrow{E_n} - I)^m$$

$$= (-1)^{m-1}\Delta x D + \mathcal{O}(\Delta x^{m+1}).$$

The term $\mathcal{O}(\Delta x^{m+1})$ arises from R_n and the terms of the exponential series after the mth in (7.2.1). Finally, we divide through by $(-1)^{m-1}\Delta x$ to conclude

$$\frac{1}{\Delta x} \sum_{j=1}^{m} \frac{(-1)^k}{j} \sum_{k=0}^{j} (-1)^k C_k^j u_{i+j-k} = (Du)_i + \mathcal{O}(\Delta x^m).$$

The proof is now complete. $\qquad\square$

For $m = 1$, $j = 1$ and we have

$$\frac{1}{\Delta x} \sum_{j=1}^{1} \frac{(-1)^0}{1} \sum_{k=0}^{1} (-1)^k C_k^1 u_{i+1-k}$$

$$= \frac{1}{\Delta x} [(-1)^0 C_0^1 u_{i+1-0} + (-1)^1 C_1^1 u_{i+1-1}] = \frac{1}{\Delta x} [u_{i+1} - u_i],$$

which is the usual expression for the forward difference. Setting $m = 2$, $j = 1$ and 2 and calculating yields

$$\frac{1}{\Delta x} \sum_{j=1}^{2} \frac{(-1)^{j-1}}{j} \sum_{k=0}^{j} (-1)^k C_k^j u_{i+j-k}$$

$$= \frac{1}{\Delta x} \frac{(-1)^{1-1}}{1} \sum_{k=0}^{1} (-1)^k C_k^1 u_{i+1-k}$$

$$+ \frac{1}{\Delta x} \frac{(-1)^{2-1}}{2} \sum_{k=0}^{2} (-1)^k C_k^2 u_{i+2-k}$$

$$= \frac{1}{\Delta x} \sum_{k=0}^{1} (-1)^k C_k^1 u_{i+1-k} - \frac{1}{2\Delta x} \sum_{k=0}^{2} (-1)^k C_k^2 u_{i+2-k}$$

$$= \frac{1}{\Delta x} \left[(-1)^0 C_0^1 u_{i+1-0} + (-1)^1 C_1^1 u_{i+1-1} \right]$$

$$- \frac{1}{2\Delta x} \left[(-1)^0 C_0^2 u_{i+2-0} + (-1)^1 C_1^2 u_{i+2-1} + (-1)^2 C_2^2 u_{i+2-2} \right]$$

$$= \frac{1}{\Delta x} u_{i+1} - \frac{1}{\Delta x} u_i - \frac{1}{2\Delta x} u_{i+2} - \frac{1}{2\Delta x} u_{i+1} + \frac{1}{2\Delta x} u_i.$$

Collecting terms and multiplying by -1, we have

$$\frac{1}{2\Delta x} (u_{i+2} - 4u_{i+1} + 3u_i). \tag{7.2.4}$$

And for $m = 3$,

$$\frac{1}{6\Delta x} (2u_{i+3} - 9u_{i+2} + 18u_{i+1} - 11u_i). \tag{7.2.5}$$

Both (7.2.4) and (7.2.5) are expressions for the forward difference. They are convergent order Δx^2 and Δx^3, respectively. Note that

(7.2.4) involves three consecutive values of u, (7.2.5) uses four. This is referred to as the *span* or *reach* of the finite difference operator.

We next turn to the extended backward differences. Again, the Taylor expansion is the starting point.

$$u(x_{i-1}) = u(x_i) - \Delta x (Du)_i$$
$$+ \frac{\Delta x^2}{2}(D^2 u)_i + \cdots + \frac{(-1)^n}{n!}\Delta x^n (D^n u)_i + R_n(u, x_i).$$

$$(7.2.6)$$

We use (7.2.6) to define the operators which give rise to the extended backward differences.

Definition 7.2.2. The operator $E_m^{\leftarrow} : V \times P \to \mathbb{R}$ given by

$$E_n^{\leftarrow}(u, x_i) = \left[\left((I - \Delta x D + \frac{1}{2!}(\Delta x D)^2\right.\right.$$
$$\left.\left. + \cdots + \frac{(-1)^n}{n!}(\Delta x D)^n + R_n)(u)\right)(x_i)\right]$$

is called the *backward displacement* operator. In addition, $\delta_n^{\leftarrow} = I - E_n^{\leftarrow}$ is called the *backward difference operator*.

In this case, the backward displacement operator is formally the exponential of $-\Delta x D$. By repeated application of (7.2.6), we have $(E_n^{\leftarrow})^m(u_i) = u_{i-m}$ provided there are sufficient elements of the partition to the left of x_i and $m \leq n$. As with the case of the forward difference, $(\delta_n^{\leftarrow})^j(u_i) = \sum_{k=0}^j (-1)^k C_k^j u_{i-j+k}$. We now state the backward difference version of Theorem 7.2.1.

Theorem 7.2.2. *If u has n continuous derivatives on the closed interval $[a, b]$ and $n + 1$ derivatives on the open interval (a, b), then for any $m \leq n$,*

$$\frac{1}{\Delta x}\sum_{j=1}^m \frac{(-1)^j}{j}\sum_{k=0}^j (-1)^k C_k^j u_{i-j+k}, \qquad (7.2.7)$$

converges to $(Du)_i$ order Δx^m.

Proof. Since $E_n^{\leftarrow}$ is defined analogously to the exponential of $-\Delta x D$, then we look at the polynomial in $\delta_n^{\leftarrow}$ given by

$$\delta_n^{\rightarrow} - \frac{1}{2}(\delta_n^{\rightarrow})^2 + \cdots + \frac{(-1)^m}{n}(\delta_n^{\rightarrow})^m. \tag{7.2.8}$$

On the one hand, expanding (7.2.8) by means of $(\delta_n^{\leftarrow})^j(u_i) = \sum_{k=0}^{j}(-1)^k C_k^j u_{i-j+k}$ yields (7.2.7). On the other hand, arguing analogously to the prior case, the given polynomial is $(Du)_i$ plus a term that converges to 0 order Δx^m.   □

We now list the second and the third backward differences. The first difference is the usual backward difference.

$$\frac{1}{2\Delta x}(3u_i - 4u_{i-1} + u_{i-2}), \tag{7.2.9}$$

$$\frac{1}{6\Delta x}(11u_i - 18_{i-1} + 9u_{i-2} - 2u_{i-3}). \tag{7.2.10}$$

As before, these are expressions for the backward difference which converge faster than the simple backward difference. Note that if we order the terms in ascending order, then the corresponding coefficients for the backward and forward differences are the same. Finally, we mention that $E_n^{\leftarrow}$ is often denoted as the inverse of $E_n^{\rightarrow}$. This is because in some sense it acts like a reciprocal. But, it is not the case nor is it meaningful. Hence, we have chosen notation that emphasizes the forward/backward viewpoint.

For the central difference, we proceed in a manner identical to the usual means for deriving the expression from the two Taylor expansions. We start with $\delta_n^{\leftrightarrow} = (1/2)(E_n^{\rightarrow} - E_n^{\leftarrow})(u_i) = (1/2)(u_{i+1} - u_{i-1})$.

Definition 7.2.3. The central difference operator is defined as

$$\delta_n^{\leftrightarrow}(u, x_i) = \frac{1}{2}(E_n^{\rightarrow} - E_n^{\leftarrow})u_i.$$

We can evaluate the right-hand side of the expression for $\delta_n^{\leftrightarrow}$ or recall that $e^x - e^{-x} = 2\sinh(x)$. In any case, we get the following expression:

$$\delta_n^{\leftrightarrow}(u, x_i) = \left[\Delta x D + \frac{1}{3!}(\Delta x D)^3 \right.$$

$$\left. + \frac{1}{5!}(\Delta x D)^5 + \cdots + \frac{1}{(2m-1)!}(\Delta x D)^{2m-1} \right](u_i),$$

where m is the greatest integer of $n/2$. But the second alternative has the added value that we can derive the central difference using the expansion for the inverse function, $\sinh^{-1}$.

Theorem 7.2.3. *If u has n continuous derivatives on the closed interval $[a, b]$ and $n + 1$ derivatives on the open interval (a, b), then for any $m \leq n$, the expression*

$$\frac{1}{\Delta x} \left(\delta_n^{\leftrightarrow} + \sum_{j=1}^{m} (-1)^j \frac{(2j-)^2}{(2j+1)!} (\delta_n^{\leftrightarrow})^{2j+1} \right) u_i, \qquad (7.2.11)$$

converges to $(Du)_i$ order Δx^{2m}.

We leave the proof of this result to the reader. As should be expected, the order of accuracy for the extended central difference is more than either the forward or backward case:

$$\frac{1}{2\Delta x}(u_{i+2} - 8u_{i+1} + 8u_{i-1} - 2u_{i-2}), \qquad (7.2.12)$$

$$\frac{1}{60\Delta x}(u_{i+3} - 9u_{i+2} + 45u_{i+1} - 45u_{i-1} + 9u_{i-2} - u_{i-3}). \qquad (7.2.13)$$

The rate of convergence is order Δx^4 and Δx^6.

For the second and higher-order extended differences, we set the following notation:

$$Ln_m \delta_n^{\rightarrow} = \delta_n^{\rightarrow} - \frac{1}{2}(\delta_n^{\rightarrow})^2 + \cdots + \frac{(-1)^{m-1}}{m}(\delta_n^{\rightarrow})^m, \qquad (7.2.14)$$

$$Ln_m \delta_n^{\leftarrow} = \delta_n^{\leftarrow} - \frac{1}{2}(\delta_n^{\leftarrow})^2 + \cdots + \frac{1}{m}(\delta_n^{\leftarrow})^m. \qquad (7.2.15)$$

We saw that applying these operators to u_i yields a finite difference expression for the derivative of u at x_i. Hence, applying any of them multiple times will generate an expression for a higher-order derivative. The first two of these operators will give rise to higher-order forward and backward extended differences. In particular, applying the operator $Ln_3\delta_n^{\rightarrow}$ twice to u_i will yield the expression for Δx^2 times the third forward second difference. The reader should note that we had no prior means of generating an expression for the forward or backward second difference.

Since the second central difference is most often used, we focus on this case. We begin with a variation on the central difference operator, $\gamma^{\leftrightarrow}(u_i) = u_{i+(1/2)} - u_{i-(1/2)}$. This is called the half step central difference operator. Applying this operator twice to u_i and dividing by $2/\Delta x^2$ yields the usual central second difference:

$$\frac{1}{\Delta x^2}(\gamma^{\leftrightarrow})^2(u_i) = \frac{1}{\Delta x^2}(u_{i+1} - 2u_i + u_{i-1}) = (D^2 u)_i + \mathcal{O}(\Delta x^2).$$

$$(7.2.16)$$

As in the case of the central difference, we use the formal series for $\sinh^{-1}$ to generate the extended second central differences. In particular,

$$\left(\frac{2}{\Delta x}\sinh^{-1}\frac{1}{2}\gamma_n^{\leftrightarrow}\right)^2$$

$$= \frac{1}{2\Delta x^2}\left[(\gamma^{\leftrightarrow})^2 - \frac{1}{12}(\gamma^{\leftrightarrow})^4 + \frac{1}{90}(\gamma^{\leftrightarrow})^6 - \frac{1}{560}(\gamma^{\leftrightarrow})^8 + \cdots\right].$$

$$(7.2.17)$$

Using the truncated series on the right with only one term, we have the second central difference in (7.2.16). If we use two terms from the right-hand side of (7.2.17), we get

$$\frac{1}{\Delta x^2}(\gamma^{\leftrightarrow})^2\left[I - \frac{1}{12}(\gamma^{\leftrightarrow})^2\right](u_i)$$

$$= \frac{1}{\Delta x^2}\left[I - \frac{1}{12}(\gamma^{\leftrightarrow})^2\right](u_{i+1} - 2u_i + u_{i-1})$$

$$= \frac{1}{12\Delta x^2}[-u_{i+2} + 16u_{i+1} - 30u_i + 16u_{i-1} - u_{i-2}],$$

or

$$\frac{1}{2\Delta x^2} \left[-u_{i+2} + 16u_{i+1} - 30u_i + 16u_{i-1} - u_{i-2} \right]$$

$$= (D^2 u)_i + \mathcal{O}(\Delta x^4). \tag{7.2.18}$$

In the following theorem, δ represents either of the three difference operators, $\delta_n^{\rightarrow}$, $\delta_n^{\leftarrow}$ or $\gamma^{\leftrightarrow}$ and D_m denotes the *log* or *arcsinh* operator truncated to m terms. Using analogous arguments to the one just given, we have the following theorem:

Theorem 7.2.4. *For any positive integer p no larger than m, $(1/\Delta x^p)(D\delta)^p u_i$ yields an expression which converges to the pth derivative of u at x_i. The order of convergence for the forward and backward differences is $m+1$. The order of convergence for the central difference is $2m$.*

Keep in mind that when doing Neumann stability analysis for the extended differences, it is best to use a maximum value process. Compute the ratio for two successive discrete Fourier coefficients on the left. In the case of the 1D heat equation, the right-hand side will be an expression in trigonometric functions in $\lambda = \alpha \Delta t / \Delta x^2$. After this function is differentiated, set to zero and solved, we have an expression for maximal value in terms of λ. Set the maximum to 1 and solve. Even if this expression appears intractable, *Mathematica* can resolve it.

Exercises

1. Use the Expand and Simplify functions in *Mathematica* to compute the expression for the forward and backward differences for $m = 3$ and 4.
2. Fix an integer m and prove that if $\alpha_j u_{i+j}$ occurs in the expression for the first forward difference and $\beta_j u_{i-j}$ occurs in the expression for the first backward difference, then $|\alpha_j| = |\beta_j|$.
3. Use *Mathematica* to derive the expressions for the extended first central difference for $m = 1$ and 2.
4. Prove Theorem 7.2.4.

5. Use the Expand and Simplify in *Mathematica* to calculate the expression for the second central difference with $m = 3$.
6. Do the following FDM example with extended differences.

   a. Setup explicit FDM with extended differences in time and space to solve the one-dimensional heat equation on the spatial interval $[-5, 5]$ and time interval $[0, 20]$ with the following assumptions.
   (IC) $u(0, x) = 0$ for every x,
   (BC) $u(t, -5) = 0$, $u(t, 5) = 40$ for all $t > 0$.
   Set $\Delta x = 0.2$ and $\alpha = 2$.
   b. Do a stability analysis to determine Δt.
   c. Execute the explicit FDM.

7. In Section 2.3, we saw that the FDM using standard difference formula do not produce results that are particularly close to the solution of the PDE that is derived from the Fourier transform. Is the situation improved when extended differences are used?
8. Calculate the forward second difference for $m = 1$ and 2. How do these compare to the corresponding central second differences?

7.3. Difference Schemes for Hyperbolic PDE

We begin by looking at the first-order wave equation,

$$\frac{\partial u}{\partial t} + \alpha \frac{\partial u}{\partial x} = 0. \tag{7.3.1}$$

If we are given an initial value function $g(x) = u(0, x)$, then it follows that $g(x - \alpha t) = u(t, x)$. Hence, u is constant on lines $x = \alpha t + x_0$ of slope α. In particular, $x_0 = x_0 - \alpha \times 0$ and $x_0 - \alpha = x_0 - \alpha \times 1$ lie on a line. Therefore $u(0, x_0) = u(1, x_0)$ or $u(0, x_0) = u(0, x_0 - \alpha)$. Hence, u is periodic in x for fixed t or vice versa. In particular, the values of u at a fixed x repeat indefinitely as $t \to \infty$. This is very different from the diffusion equation where the values of u go to zero as $t \to \infty$.

As is usual, we take spatial and temporal partitions with mesh parameters Δx and Δt. We set $C = \alpha \Delta t / \Delta x$. This quantity is called the *Courant number*. It plays a similar role to λ in the parabolic cases. If we setup the FDM scheme FTFS, then we have

$$u_i^{n+1} - u_i^n = -C\left(u_{i+1}^n - u_i^n\right).$$

Continuing with the Neumann stability analysis, we get the following expression:

$$\sum_{k=-N}^{N} \left(c_k^{n+1} e^{iki\Delta x} - c_k^n e^{iki\Delta x}\right) = -C \sum_{k=-N}^{N} \left(c_k^n e^{ik(i+1)\Delta x} - c_k^n e^{iki\Delta x}\right)$$

for the discrete Fourier representation. Next, we compute $\sigma(\cdot, e^{ij\Delta x})$ for both the left-hand and right-hand sides to isolate the Fourier coefficients.

$$c_j^{n+1} - c_j^n = -C\left(c_j^n e^{i\Delta x} - c_j^n\right).$$

Dividing through by c_j^n and expanding $e^{i\varphi} = \cos\varphi + \mathbf{i}\sin\varphi$, $\varphi = \Delta x$, we get the following representation for the ratio of two successive amplitudes,

$$\frac{c_j^{n+1}}{c_j^n} = 1 - C(\cos\varphi - 1) - \mathbf{i}C\sin\varphi = 1 + C - C\cos\varphi - \mathbf{i}C\sin\varphi.$$

The modulus squared of the right-hand side is equal to

$$(1+C)^2 - 2C\cos\varphi + C^2 = 1 + 2C(1 - \cos\varphi) + 2C^2.$$

Since $1 - \cos\varphi \geq 0$ then the FDM scheme is unconditionally unstable provided $\alpha > 0$. If $\alpha < 0$, we differentiate this last expression as a function in φ to get $-2C(\sin\varphi)$. Setting this to zero yields critical points at $\varphi = 0$ or the interval end points, $\varphi = \pm\pi$. For the case $\varphi = 0$, stability will require $1 + 2C^2 \leq 1$. As $C < 0$, we conclude that this case is impossible. At $\varphi = \pm\pi$, $1 \geq 1 + 4C + 2C^2 = (1 + 2C)^2$, or $1 \geq |1 + 2C|$. Hence, $0 > C \geq -1$.

It is important to know that unconditional instability is possible. This helps us to orient our thinking about time stepping. For instance, we know that Δt occurs in the numerator of a time stepping procedure. When faced with obvious instability, our first reaction is to decrease this parameter. This often works. However, we now know that this may not help. On the theoretical side, this case will provide examples of non-convergent FDM.

We proceed with forward time and central space and backward space. We will collect all three cases into Theorem 7.3.1. If we consider the FTCS scheme, forward time and central space, then we get

$$u_i^{n+1} - u_i^n = -\frac{1}{2}C(u_{i+1}^n - u_{i-1}^n).$$

In turn, for this case, the ratio of two successive Fourier amplitudes is

$$\frac{c_j^{n+1}}{c_j^n} = 1 - \mathbf{i}\frac{C}{2}(2\sin\varphi) = 1 - \mathbf{i}C\sin\varphi.$$

Again taking the absolute value of both sides, we conclude that $1 \geq 1 + C^2\sin^2\varphi$. This is unconditionally unstable since $\varphi = \Delta x > 0$ independent of sign of C.

Finally, we look at FTBS. In this case the FDM formulation is

$$u_i^{n+1} - u_i^n = -C(u_i^n - u_{i-1}^n)$$

and the amplitude ratio is

$$\frac{c_j^{n+1}}{c_j^n} = 1 - C(1 - \cos\varphi + \mathbf{i}\sin\varphi).$$

Now, the real part of the right-hand side is $(1 - C) + C\cos\varphi$ and the imaginary part is $-C\sin\varphi$. This describes the parametric form of a circle with center at $(1 - C, 0)$ and radius $|C|$. The circle lies within the unit circle provided $|C| \leq 1$.

Theorem 7.3.1. *For the order 1 wave equation given by (7.3.1), the FDM scheme FTBS is conditionally stable depending on the Courant number. In particular, stability is assured provided $|C| \leq 1$. Additionally FTFS is unconditionally unstable for $C > 0$ and stable for $0 > C \geq -1$. Finally, FTCS is unconditionally unstable.*

With these cases set aside, we introduce a new technique, Lax–Wendroff. We begin with (7.3.1) and differentiate with respect to t

to get for sufficiently smooth u

$$\frac{\partial^2 u}{\partial t^2} = \alpha \frac{\partial}{\partial t} \frac{\partial u}{\partial x} = \alpha \frac{\partial}{\partial x} \frac{\partial u}{\partial t} = \alpha^2 \frac{\partial}{\partial x} \frac{\partial u}{\partial x} = \alpha^2 \frac{\partial^2 u}{\partial x^2} \qquad (7.3.2)$$

the order 2 wave equation. We expand u with respect to the time variable and insert (7.3.2).

$$u_k^{n+1} = u_k^n + \frac{\partial u}{\partial t}(t_n, x_i)\Delta t + \frac{\partial^2 u}{\partial t^2}(t_n, x_i)\frac{\Delta t^2}{2} + \mathcal{O}(\Delta t^3)$$

$$= u_k^n - \alpha \frac{\partial u}{\partial x}(t_n, x_i)\Delta t + \alpha^2 \frac{\partial^2 u}{\partial x^2}(t_n, x_i)\frac{\Delta t^2}{2} + \mathcal{O}(\Delta t^3).$$

We resolve the two spatial derivatives with central differences,

$$u_k^{n+1} = u_k^n - \frac{C}{2}\left(u_{k+1}^n - u_{k-1}^n\right) + \frac{C^2}{2}\left(u_{k+1}^n - 2u_k^n + u_{k-1}^n\right) + \mathcal{O}(\Delta t^3).$$
$$(7.3.3)$$

We now prove the following theorem.

Theorem 7.3.2. *The Lax–Wendroff FDM scheme given by* (7.3.3) *applied to* (7.3.1) *is conditionally stable for* $|C| \le 1$.

Proof. Neumann stability analysis yields

$$\frac{c_j^{n+1}}{c_j^n} = 1 - 2C^2 \sin^2 \frac{\varphi}{2} + iC \sin \varphi,$$

$$\left|\frac{c_j^{n+1}}{c_j^n}\right|^2 = 1 - 4C^2 \sin^4 \frac{\varphi}{2} + 4C^4 \sin^4 \frac{\varphi}{2} + C^2 \sin^2 \varphi.$$

The extreme points for this function are at $\pm\pi$. We substitute into the last equation and factor to get $(1 - 2C^2)^2 \le 1$. This inequality implies $-1 \le 1 - 2C^2 \le 1$, or $1 \ge C^2$. The result now follows. $\square$

Continuing, there are backward time and central time cases that are similar to those just presented. And the results are again distinct from the results we saw with parabolic equations. Several of these cases are covered in the exercises to this section. One particularly interesting situation arises with the Crank–Nicholson method. In this case the Courant number must be exactly equal to 1. This means that the method is not practical as a small round off error yields a Courant number close to one but not exactly equal to one. In particular, we start with the FTCS and BTCS rendering of (7.3.1).

$$u_k^{n+1} - u_k^n = \frac{C}{2}\left(u_{k+1}^n - u_{k-1}^n\right), \quad u_k^{n+1} - u_k^n = \frac{C}{2}\left(u_{k+1}^{n+1} - u_{k-1}^{n+1}\right).$$

We add these and divide by 2 to get

$$-\frac{C}{4}u_{k+1}^{n+1} + u_k^{n+1} + \frac{C}{4}u_{k-1}^{n+1} = \frac{C}{4}u_{k+1}^n + u_k^n - \frac{C}{4}u_{k-1}^n. \qquad (7.3.4)$$

Equation (7.3.4) may be separated with the aid of a half-time step. In this case, there is an explicit step followed by an implicit step.

$$u_k^{n+1} - u_k^{n+1/2} = \frac{C}{4}\left(u_{k+1}^{n+1} - u_{k-1}^{n+1}\right),$$

$$u_k^{n+1/2} - u_k^n = \frac{C}{4}\left(u_{k+1}^n - u_{k-1}^n\right),$$

We know the discrete Fourier series process applied to the implicit step. (See Exercise 2.) We have previously calculated the explicit step.

$$\frac{c_j^{n+1}}{c_j^{n+1/2}} = 1 - \mathbf{i}\frac{C}{2}\sin\varphi, \quad \frac{c_j^{n+1/2}}{c_j^n} = \left(1 - \mathbf{i}\frac{C}{2}\sin\varphi\right)^{-1}.$$

Iterating the two steps is equivalent to multiplying the two equations. Hence, for the absolute value, we have

$$\frac{|c_j^{n+1}|}{|c_j^n|} = \frac{\left|1 - \mathbf{i}\dfrac{C}{2}\sin\varphi\right|}{\left|1 - \mathbf{i}\dfrac{C}{2}\sin\varphi\right|} = 1.$$

We have proved the following result.

Theorem 7.3.3. *Crank–Nicolson FDM for the 1D wave equation (7.3.1) is unconditionally stable with $|c_j^{n+1}| = |c_j^n|$ for each j.*

We now turn to the two-dimensional case, $u(t, x, y)$ with

$$\frac{\partial u}{\partial t} + \alpha \left(\frac{\partial u}{\partial x} + \frac{\partial u}{\partial y} \right) = 0, \qquad (7.3.5)$$

with initial condition $u(0, x, y) = g(x, y)$. We now have two independent uniform spatial partitions and hence, two Courant numbers $C_x = \alpha \Delta t / \Delta x$ and $C_y = \alpha \Delta t / \Delta y$. Setting up FTCS, we have

$$u_{j,k}^{n+1} - u_{j,k}^n = -C_x \left(u_{j+1,k}^n - u_{j-1,k}^n \right) - C_y \left(u_{j,k+1}^n - u_{j,k-1}^n \right).$$

The result of the two variable Neumann stability analysis is

$$\frac{c_{j,k}^{n+1}}{c_{j,k}^n} = 1 - C_x \left(1 - e^{-i\varphi} \right) - C_y (1 - e^{-i\psi}).$$

Taking the absolute value squared and then differentiating, we must solve

$$0 = -4 \cos \frac{\xi}{2} \sin \frac{\xi}{2} \left(1 - 2 \sin^2 \frac{\xi}{2} - 2 \sin^2 \frac{\eta}{2} \right),$$

$$0 = -4 \cos \frac{\eta}{2} \sin \frac{\eta}{2} \left(1 - 2 \sin^2 \frac{\xi}{2} - 2 \sin^2 \frac{\eta}{2} \right).$$

For positive Courant numbers these equations are nearly intractable. The reader is invited to execute Solve in *Mathematica* to see the derived solution. In any event, it has been proved that the criterion for stability is $0 < C_x + C_y \leq 1$. Recall that FTCS is unconditionally unstable for the one-dimensional case.

We end this section with another contribution by Peter Lax, the Lax–Friedrichs method. We begin with FTCS

$$u_{j,k}^{n+1} = u_{j,k}^n - C_x \left(u_{j+1,k}^n - u_{j-1,k}^n \right) - C_y (u_{j,k+1}^n - u_{j,k-1}^n)$$

and replace $u_{j,k}^n$ with the mean of the surrounding values,

$$u_{j,k}^{n+1} = \frac{1}{4} (u_{j+1,k}^n + u_{j-1,k}^n + u_{j,k+1}^n + u_{j,k-1}^n)$$
$$- C_x \left(u_{j+1,k}^n - u_{j-1,k}^n \right) - C_y (u_{j,k+1}^n - u_{j,k-1}^n).$$

In this case

$$\frac{c_{j,k}^{n+1}}{c_{j,k}^n} = \frac{1}{2} (\cos \xi + \cos \eta) - i(C_x \sin \xi + C_y \sin \eta).$$

Taking the absolute value squared, the right-hand side becomes

$$1 - (\sin^2 \xi + \sin^2 \eta) \left[\frac{1}{2} - C_x^2 - C_y^2 \right]$$

$$- \frac{1}{4}(\cos \xi + \cos \eta)^2 - (C_x \sin \xi + C_y \sin \eta)^2$$

$$\leq 1 - (\sin^2 \xi + \sin^2 \eta) \left[\frac{1}{2} - (C_x^2 + C_y^2) \right]$$

since the omitted terms are negative. In turn, this expression is less than 1 provided

$$C_x^2 + C_y^2 \leq \frac{1}{2}.$$

Exercises

1. Complete the stability analysis for Eq. (7.3.1) using BTFS when
   a. $C < 0$,
   b. $C > 1$.
2. Complete the stability analysis for Eq. (7.3.1) using BTCS.
3. Prove that the order 1 wave equation rendered with forward time and central space is consistent provided u is sufficiently smooth.
4. Use Lax–Wendroff to solve (7.3.1) when $\alpha = 300$ for the following setting:

   Region: $[0, 300]$; Time interval $[0, 5]$,

   Initial Conditions:

   $$u(0, x) = 0, \quad 0 \leq x \leq 50;$$

   $$u(0, x) = 100 \sin \left[\pi \frac{(x - 50)}{60} \right], \quad 50 \leq x \leq 110;$$

   $$u(0, x) = 0, \quad 110 \leq x \leq 300.$$

   Boundary Values: $u(t, 0) = 0$ and $u(t, 300) = 0$,
   $\Delta x = 5$; $\Delta t = 0.0015$.

   Use B-Splines to draw the curve at 5 given time steps, 0, 1, 2, 3, 4.

5. Use FTBS FDM to solve the first-order wave equation with the same partitions, initial and boundary conditions as in Exercise 4.

6. Use Lax–Friedrichs method to solve the two-dimensional analogue of Exercise 4. The equation is now (7.3.5), we take $\alpha = 300$ for the following setting

Spatial Region: $[0, 300] \times [0, 300]$; Time interval $[0, 5]$,

Initial Conditions:

$$u(0, x, y) = 0, \quad 0 \le x \le 50 \text{ or } 0 \le y \le 50,$$

$$u(0, x, y) = \left[100 \sin \frac{\pi(x - 50)}{60} \right] \left[100 \sin \frac{\pi(y - 50)}{60} \right],$$

$$(x, y) \in [50, 110] \times [50, 110],$$

$$u(0, x, y) = 0, \quad 110 \le x \le 300 \text{ or } 110 \le y \le 300.$$

Boundary Values:

$$u(t, 0, y) = 0, \quad u(t, 300, y) = 0,$$

$$u(t, x, 0) = 0, \quad u(t, x, 300) = 0,$$

$$\Delta x = 5; \quad \Delta y = 5; \quad \Delta t = 0.0015.$$

Use *Mathematica* Plot3D to draw the surface at 5 given time steps, 0, 1, 2, 3, 4.

7. Use FTCS to resolve the order 2 wave equation in two spatial dimensions,

$$\frac{\partial^2 u}{\partial t^2} - \alpha^2 \left(\frac{\partial^2 u}{\partial x^2} + \frac{\partial^2 u}{\partial y^2} \right) = 0. \tag{7.3.6}$$

Repeat the setting given for the prior exercise.

In order to resolve this problem, you will need an initial estimate of $\partial u / \partial t$ at t_1. Consider the Taylor expansion for u with respect to the time variable,

$$u(t_1, x, y) = u(t_0, x, y) + \frac{\partial u}{\partial t}(t_0, x, y) + \frac{1}{2} \frac{\partial^2 u}{\partial t^2}(t_0, x, y) + \mathcal{O}(\Delta t^3),$$

and use the initial value for u at $t = t_0$ to resolve the time derivative at $t = t_0$.

7.4. Remarks on Stability and Convergence

Before moving on to elliptical PDE, we include a final discussion of stability and convergence for initial value problems. Our current status is as follows. From the Lax equivalence theorem, we know that convergence of an FDM scheme applied to a differential equation is resolved using consistency and stability. Moreover, in most cases, consistency may be routinely resolved using the Taylor expansion provided the solution is sufficiently smooth. So, the question of convergence rests primarily on stability.

Stability was introduced in Definition 2.3.1, in terms of the norm of the FDM time state transformation. At the same time, we defined Neumann stability, Definition 2.3.2. In this context we introduced the discrete Fourier transform and developed a technique for verifying Neumann stability. In addition, we proved that Neumann stability was equivalent to stability provided the matrix C_k is symmetric, where C_k is the linear transformation that maps time state vectors, the discrete form of the FDM scheme transformation R_k. (See Section 7.1.) In this section, our first task will be to prove stability implies Neumann stability. This proof relies heavily on the Jordan form of a matrix. For a fast development of the Jordan form see Loustau and Dillon (1993).

We begin with the following result of Gelfand.

Theorem 7.4.1. *Suppose that A is an $n \times n$ complex matrix with spectral radius $\sigma(A)$. Then*

$$\lim_{k \to \infty} \|A^k\|^{1/k} = \sigma(A). \qquad (7.4.1)$$

Proof. Without loss of generality, we may suppose that A is in Jordan form. In fact, it suffices to prove the result for a Jordan block matrix. Let λ be the single eigenvalue of A. To prove the theorem, we take $\epsilon > 0$ and verify that for k sufficiently large,

$$\sigma(A) - \epsilon < \|A\|^{1/k} < \sigma(A) + \epsilon.$$

Set $\hat{A} = (\sigma(A) + \epsilon)^{-1}A$, so that $\sigma(\hat{A}) < 1$. Hence, $\hat{A}^k \to 0$ as $k \to \infty$. (See Exercise 1.) Therefore,

$$(\sigma(A) + \epsilon)^{-k}\|A^k\| = \|\hat{A}^k\| < 1$$

for every k sufficient large. Hence, $\|A^k\|^{1/k} < \sigma(A) + \epsilon$.

Conversely, set $\bar{A} = (\sigma(A) - \epsilon)^{-1}A$, so that $\sigma(\bar{A}) > 1$. Again using Exercise 1, $\|\bar{A}^k\|$ is unbounded. Hence, for large k,

$$(\sigma(A) - \epsilon)^{-k}\|A^k\| = \|\bar{A}^k\| > 1.$$

In particular, $\|A^k\|^{1/k} > \sigma(A) - \epsilon$. The result follows. $\square$

We now prove the basic result that relates stability to Neumann stability.

Theorem 7.4.2. *Suppose C_k maps FDM time states, $C_k u^n = u^{n+1}$. Then $\|C_k^n\|$ is bounded for all m implies that $\|u^{n+1}\| \leq \|u^n\|$ for n sufficiently large.*

Proof. We use the contrapositive. If the FDM is not Neumann stable, then there is the sequence of integers n_j with $\|C_k^{n_j} u^0\| > 1$. Without loss of generality, we may suppose that this sequence is bounded. Otherwise, it is not stable and the proof is complete.

By Theorem 7.4.1, $\sigma(C_k) \geq 1$. If $\sigma(C_k) > 1$, then by a Jordan form argument, $\|C_k^{n_j}\|$ is unbounded. Hence, the result follows. Finally, if $\sigma(C_k) = 1$, then there are two cases. In one case, all eigenvalues with absolute value $= 1$ occur only in 1×1 Jordan blocks and $\|C_k^n\| = 1$ for every n sufficiently large. The alternative is that there is an eigenvalue of absolute value 1 which belongs to a Jordan block bigger than 1×1. In this case $\|C_k^n\| > 1$ also for n sufficiently large. In any case, the result of proved (see Exercise 2). $\square$

Next, we consider the Courant–Friedrichs–Lewy theorem. We start with an initial value problem associated to a PDE with initial condition and boundary values. Suppose that we have a single spatial variable. The extension to more variables is analogous (see Exercise 3). If u is the solution, then the value $u(t, x)$ will depend on certain of the boundary values and initial values. For instance, in the case of Eq. (7.3.1), the initial value of g at $x_0 = x - at$ completely determines $u(t, x)$. More generally, given u, then the set of all points in the domain of the PDE that determine a given value $u(t, x)$ is called the *analytic domain of definition* of (t, x).

We consider a familiar example. For the 1D heat equation, the solution is given by (2.2.10). Suppose further that u_0, the initial value function is zero for $|x| > 5$. In addition, suppose that there is a partition of the interval $[-5, 5]$ and that u_0 is constant on subintervals associated to the partition. It is immediate that any two choices of u_0 will yield distinct values of $u(t, x)$, $x \in [-5, 5]$. Therefore, in this case the analytic domain of definition of any point (t, x) is $[-5, 5]$.

Moving on, we consider the discrete FDM process. We start with a rectangular domain in time and space and a family of uniform partitions with mesh parameters h_i, k_i, $i \in \Lambda$. We represent the points of any partition as (nk_i, mh_i) in $\mathcal{P}_i$. Further, we suppose that $\bigcup_{i \in \Lambda} \mathcal{P}_i$ is dense. Note that any element of a uniform partition does belong to a family of partitions whose union is dense in the domain of the PDE.

We continue with the discrete case. On the discrete side, suppose that (t, x) is a point in the domain. If this point lies in a partition $\mathcal{P}_i$, then we may take $(t, x) = (nk_i, mh_i)$ in the partition. In turn we can write $u_{nh_i}^{mk_i}$, the FDM solution at the point, as a linear combination of FDM generated values at the $(n - 1)k_i$ time states. Each of these in turn may be written in terms of the $(n - 2)k_i$ time stage. Continuing in this manner, we have a set of FDM generated values that result in the FDM solution at $(nk_i, mh_i) = (t, x)$. We call this set the *numerical domain of definition* of (nk_i, mh_i) associated to the partition and denote it by K_i. We then extend this idea to the closure of the union of all K_i. The latter set is the *numerical domain of definition* for $o(t, x)$.

It is important to note that the size of the numerical domain of definition will vary depending on the spread (see Section 7.2) of the finite difference operators that we are using. Certainly, the extended differences must be taken into account whenever we develop theoretical results based on FDM.

Now, suppose the (t_0, x_0) lies in the analytic domain of definition of (t, x). If (t_0, x_0) is isolated from the numerical domain of definition, then we can change the value of $u(t_0, x_0)$ while preserving the smoothness of u and without changing the FDM computed value $u_{mh_i}^{nk_i}$. This is in effect the Courant–Friedrichs–Lewy theorem.

Theorem 7.4.3. *We are given an initial value problem with FDM rendering. Suppose that (t, x) is a point in a uniform partition. If the analytic domain of definition for (t, x) does not contain the numerical domain, then the FDM rendering is not convergent.*

Proof. The notation is as above. Since the analytic domain of definition is closed, then any point of the numerical domain not lying in the analytic domain is isolated from the analytic domain. Therefore, as noted above, it is possible to have two distinct solutions to the PDE satisfying distinct boundary or initial values that give rise to the same FDM approximation. Hence, it is impossible for the FDM scheme to be convergent. □

The condition in Theorem 7.4.3 is necessary but not sufficient for convergence. In Exercise 4, we ask the reader to consider FTFS for the wave equation with parameter $\alpha > 0$ and verify that the condition in Theorem 7.4.3 is not sufficient.

There are later results that do provide necessary and sufficient conditions for stability. This development is included in Gustofsson *et al.* (2013). The approach uses a definition for stability that is particularly suited to their purpose. We will briefly discuss this approach and leave the serious study to the interested reader.

The basic time stepping process (simple forward or backward) can be written as $u_i^{n+1} = \beta u_i^n + \Delta t G$. However, in Section 7.2, we saw that the time step may be governed by an extended first difference. Then, in Section 7.3, we saw the use of Lax–Friedrichs where the term u_i^n is further developed using finite differences. Hence, the time step may look like $u_i^{n+1} = Q + \Delta t G$, where Q is a complex expression of difference terms at the mth time state, $m \leq n$. It is interesting that stability depends only on Q and not on $\Delta t G$. Next, write Q as a sum $\sum_{j=0}^{s} Q_j u_i^{n-1}$, where each Q_i is a finite difference operator. Hence, each Q_i can be written as a linear combination of powers of forward and backward displacement operators (see Section 7.2).

The stability statement starts with Q written as a discrete Laplace series,

$$u_i^{n+1} - \sum_{i=0}^{s} z_{-j-1} Q_j u_i^{n-j} = 0, \qquad (7.4.2)$$

where $z = \alpha e^{k\varphi}$. The statement is that if z exists satisfying (7.4.2), then the time-stepping procedure is unstable. The necessary and sufficient condition is that the left-hand side of (7.4.2) is continuous in z and the solution to (7.4.2) has absolute value 1 and is the limit of values at w, $|w| > 1$.

Exercises

1. Let A be an $n \times n$ Jordan block matrix, $A_{i,i} = \lambda$, $A_{i,i+1} = 1$ and $A_{i,j} = 0$ otherwise.

   a. Prove that for any k, the non-zero entries of A^k are integer multiples of powers of the eigenvalue, $C_{k-m}^k \lambda^m$, $m = k - n, \dots, k$ and C_{k-m}^k is the usual binomial coefficient.

   b. Prove that $C_{k-m}^k \leq k^n$, conclude that the non-zero entries of A^k converge to zero as $k \to \infty$, if $|\lambda| < 1$.

   c. If $|\lambda| > 1$, then the entries of A^k, $k = 1, 2, \dots$ are unbounded.

2. Given an $n \times n$ matrix A with $\sigma(A) = 1$, prove that there is an m_0 so that $\|A^m\| > 1$ for every $m \geq m_0$ precisely when there is an eigenvector v of eigenvalue λ, $|\lambda| = 1$ which associates to a Jordan block of size > 1.

3. Extend the Courant–Friedrichs–Lewy result to two spatial dimensions. Develop the concepts of analytic and numerical domain of definition for this case.

4. Consider the wave equation with $\alpha > 0$. Select a point (t, x) and verify that $(0, x_0)$ with $x_0 = x - \alpha t$, lies in the numerical domain of definition. Conclude that the condition in Theorem 7.4.3 is not sufficient to imply convergence.

7.5. Difference Schemes for Elliptical PDE

In this section, we consider the elliptical Poisson equation, $\nabla^2 u = f$ over a domain in the real plane. Recall that elliptical PDE are steady-state second-order equations. The steady-state part of a parabolic equation is elliptical. In particular, elliptical PDE often arise from a conservation law. In Part 1, we saw several examples of this including the Laplace equation (the steady-state part of the diffusion equation). Other examples include the Navier–Stokes and Stokes equations of Sections 6.2 and 6.3. In Section 3.2, we saw how the elliptical Helmholtz equation is derived from the second-order wave equation after separating the spatial from the temporal variable.

We begin this section by setting notation. Because of the restrictions inherent in FDM, the domain D will be rectangular with uniform partitions in each direction. We denote the mesh parameters as h and l. For notational simplicity, we may suppose that $D = [0, 1] \times [0, 1]$. We derive the discrete form of the Laplace operator using second central differences. For a function u defined on the domain, we set as usual $u(x_i, y_k) = u_{j,k}$.

For the FDM realization of the PDE, we write $C_{j,k} v_{j,k} = f_{j,k}$, $0 \le j \le M_x$. $0 \le k \le M_y$, where

$$
\begin{aligned}
C_{j,k} v_{j,k} &= \frac{1}{h^2}(v_{j+1,k} - 2v_{j,k} + v_{j-1,k}) \\
&\quad + \frac{1}{l^2}(v_{j,k+1} - 2v_{j,k} + v_{j,k-1})
\end{aligned}
\tag{7.5.1}
$$

and $f_{i,j} = f(x_j, y_k)$.

In turn, if we write $\bar{v} = (\bar{v}_{j,k})$ as an $(M_x + 1)(M_y + 1)$-tuple with entries ordered in lexicographic order, then (7.5.1) determines a linear transformation C given by $C\bar{u} = (C_{j,k}\bar{u}_{j,k})$. If we set this equal to $f_{j,k}$, then we have the FDM form of the Poisson equation. In particular, it is a linear system that may be solved for each partition. We will denote the vector solution as $\bar{u} = (\bar{u}_{j,k})$.

Our goal is to prove convergence for FDM solutions for a family of partitions with $h, l \to 0$. For this purpose, we suppose that u is a solution to the PDE and prove that $\bar{u} \to (u(x_j, y_k))$ in L^∞ or *sup* norm. We proceed as follows. We prove two technical results. With this done, the proof of the convergence theorem is routine.

Note that on this occasion, convergence is proved without interpolating the FDM solution but rather by restricting the actual solution of the points in the partition. This is the approach used in Thomas (1995) and Thomas (1999).

Let P denote the set of points of the partition, P^o the points of P that are interior to D and ∂P the points of P that are on the boundary. The first result is the discrete version of a standard extrema result.

Theorem 7.5.1. *If $C_{j,k}v_{j,k} \leq 0$ ($C_{j,k}v_{j,k} \geq 0$) for every j, k, then the minimum (maximum) entry of $v_{j,k}$ occurs at the boundary.*

Proof. The inequality $C_{j,k}v_{j,k} \leq 0$, implies that

$$\left(\frac{1}{h^2} + \frac{1}{l^2}\right) v_{j,k} \geq \frac{1}{2}\left[\frac{1}{h^2}\left(v_{j+1,k} + v_{j-1,k}\right) + \frac{1}{l^2}\left(v_{j,k+1} + v_{j,k-1}\right)\right].$$

$$(7.5.2)$$

If we have an interior minimum at $v_{j,k}$, then

$$\frac{1}{2}\left[\frac{1}{h^2}\left(v_{j+1,k} + v_{j-1,k}\right) + \frac{1}{l^2}\left(v_{j,k+1} + v_{j,k-1}\right)\right]$$

$$\geq \frac{1}{2}\left[\frac{1}{h^2}\left(v_{j+1,k} + v_{j,k}\right) + \frac{1}{l^2}\left(v_{j,k} + v_{j,k}\right)\right]$$

$$\geq \left(\frac{1}{h^2} + \frac{1}{l^2}\right) v_{j,k}.$$

$$(7.5.3)$$

Therefore, equality holds in (7.5.2) and (7.5.3). In particular, $v_{j,k} = v_{j+1,k}$. Using similar argument $v_{j,k} = v_{j-1,k} = v_{j,k+1} = v_{j,k-1}$. Hence, either $(v_{j,k})$ is constant or there is a true minimum on the boundary. The proof of the maximum result is analogous. $\square$

The next result is an example of an *inverse inequality*. The terminology refers to the fact that we expect the application of a linear transformation to decrease norms. There are other examples of this sort of inequality in Chapters 9 and 10.

Lemma 7.5.1. *If v is zero on the boundary ∂P, then*

$$\|v\|_\infty \le \frac{1}{8}\|Cv\|_{\infty,o}, \qquad (7.5.4)$$

where $\|v\|_{\infty,o}$ denotes the sup norm taken over the interior points.

Proof. Let $w_{j,k} = (1/4)[(x_j - 1/2)^2 + (y_k - 1/2)^2]$ and $w = (w_{j,k})$. It is routine to verify that $C_{j,k}w_{j,k} = 1$ for each (x_j, y_k) in P.

We take $v_{j,k}$ at an interior point and notice that

$$C_{j,k}v_{j,k} - \|Cv\|_{\infty,o} \le 0 \le C_{j,k}v_{j,k} + \|Cv\|_{\infty,o}.$$

Therefore, by the definition of w,

$$\begin{aligned}
C_{j,k}\,&(v_{j,k} - \|Cv\|_{\infty,o}w_{j,k}) \\
&= C_{j,k}v_{j,k} - \|Cv\|_{\infty,o}C_{j,k}w_{j,k} \\
&= C_{j,k}v_{j,k} - \|Cv\|_{\infty,o} \le 0 \le C_{j,k}v_{j,k} + \|Cv\|_{\infty,o} \\
&= C_{j,k}v_{j,k} + \|Cv\|_{\infty,o}C_{j,k}w_{j,k} = C_{j,k}\,(v_{j,k} + \|Cv\|_{\infty,o}w_{j,k}).
\end{aligned}$$

Theorem 7.5.1 implies that the minimum of $v - \|Cv\|_{\infty,o}w$ and the maximum of $v + \|Cv\|_{\infty,o}w$ are at the boundary. Since v is zero there, then the minimum is $-\|Cv\|_{\infty,o}\|w\|_{\infty,\partial}$ and the maximum is $\|Cv\|_{\infty,o}\|w\|_{\infty,\partial}$. Since $w_{j,k}$ is always positive, then

$$v_{j,k} - \|Cv\|_{\infty,o}w_{j,k} \le v_{j,k} \le v_{j,k} + \|Cv\|_{\infty,o}w_{j,k}.$$

Hence,

$$-\|Cv\|_{\infty,o}\|w\|_{\infty,\partial} \le v_{j,k} \le \|Cv\|_{\infty,o}\|w\|_{\infty,\partial}.$$

Since $\|w\|_{\infty,\partial} = 1/8$, Eq. (7.5.4) follows. $\qquad\square$

We now state and prove the convergence theorem. But first we emphasize the notation we are using. As noted above, $\bar{u}$ denotes the solution to the linear system $Cv = (f_{j,k})$ while u denotes the designated solution to the PDE. In turn u and the current partition determine a vector of values $(u(x_j, y_k))$.

Theorem 7.5.2. *Let u be a solution to the Poisson equation $\nabla^2 u = f$ defined on a real rectangle D with Dirichlet boundary values. If $\bar{u}$ is the solution to the linear system $Cv = (f_{j,k})$ as developed* *from* (7.5.1), *then there is a constant K depending on u so that,*

$$\|u - \bar{u}\|_\infty \leq K \left(\Delta x^2 + \Delta y^2\right) \|\partial^4 u\|_{\infty,o}, \qquad (7.5.5)$$

provided $u \in C^4[D]$.

Proof. Note that by $\|\partial^4 u\|_{\infty,o}$ we mean the sup of all terms of the form $|\partial^4 u/\partial x^p \partial y^q(x,y)|$ taken over all (x,y) in the interior of D and for every non-negative p, q satisfying $p + q = 4$.

By hypothesis, we may assume that the finite difference operators are derived from the Taylor expansion. Therefore, $C_{j,k} u(x_j, y_k) - f(x_j, y_k)$ goes to zero $\mathcal{O}(\Delta x^2) + \mathcal{O}(\Delta y^2)$. Since $C_{j,k} \bar{u}_{j,k} = f(x_j, y_k)$, we may write

$$C_{j,k} \left(u(x_j, y_k) - \bar{u}_{j,k}\right) \to 0, \quad \text{as } \mathcal{O}\left(\Delta x^2\right) + \mathcal{O}\left(\Delta y^2\right).$$

Taking the sup norms over the interior and applying Taylor's theorem,

$$\|C(u - \bar{u})\|_{\infty,o} \leq K \left(\Delta x^2 + \Delta y^2\right) \|\partial^4 u\|_{\infty,o}.$$

Since u and $\bar{u}$ are equal on the boundary, then Lemma 7.5.1 applies to $u - \bar{u}$. In particular,

$$\|u - \bar{u}\|_{\infty,o} \leq \frac{K}{8} \left(\Delta x^2 + \Delta y^2\right) \|\partial^4 u\|_{\infty,o}.$$

This completes the proof of Theorem 7.5.2. $\qquad\qquad\square$

Exercises

1. Consider the domain $D = [0, 2] \times [0, 1]$ partitioned by a mesh with parameters $h = 0.05$ and $k = 0.05$. Use the FDM scheme (7.5.1) to approximate the solution to the Laplace equation, $\nabla^2 u = 0$ for the boundary values $u(x, 0) = 200$, $u(x, 1) = 0$, $u(0, y) = 0$, $u(2, 0) = 0$.
2. Con tinue problem 1 by using B-splines to draw the level curves for u at 50, 100 and 150.
3. Repeat Exercises 1 and 2 for the Poisson PDE, $\nabla^2 u = x^2 + y^2$.
4. Repeat Exercises 1 and 2 for the Poisson PDE, $\nabla^2 u = \sin x$.
5. Repeat Exercises 1 and 2 for the Helmholtz PDE, $\nabla^2 u + u = 0$.

Chapter 8

Finite Element Method, the Techniques

Introduction

In Part 1, Chapters 3 and 6, we introduced the Finite Element Method by means of examples. At that time, we noted that a primary distinction between FDM and FEM is that the latter produces a function and that this function provides an approximation to the solution of the target differential equation. We noted that there were advantages to having a function rather than values. In the first example, we developed a flow field by modeling the flow potential with FEM. Subsequently, we derived the field from the gradient of the potential. On the other hand, we also saw that FEM is significantly more difficult to implement.

In this chapter, we develop the method in some detail and leave the supporting theory to Chapter 9. We present FEM away from any particular example. Without a particular instance, we are free to investigate the breadth of possible choices with emphasis on common practice. At the same time, we want to lay the foundation for the theory that comes later. The theory is the foundation for the method. Without it, there is no assurance that the computed values will correspond to anything. Moreover, as we describe the implementation, we encounter restrictions to what may or may not be done.

These requirements, imposed on the practitioner of FEM, are indeed assumptions needed to prove the theorems.

In order to have the list at hand, we restate the steps of FEM:

Step 1. Set the geometric model (define the elements and nodes).
Step 2. Determine the polynomial model.
Step 3. Determine the element linear systems.
Step 4. Determine the global linear system.
Step 5. Set boundary values.
Step 6. Solve the linear system.
Step 7. Do post-processing.

Keep in mind that FEM is only applied to the spatial variables. For transient problems, the FEM is used solely to setup the spatial side while time stepping uses finite difference method. In practice, we ignore the time variable when developing the FEM rendering. Or more precisely, we suppose steady state, $\partial u/dt = 0$. At the end of the process, the researcher will then employ any of the FDM techniques developed in previous chapters.

It is perhaps helpful to recall what you do when you setup a discrete process such as FEM. The process of modeling some naturally occurring phenomenon often results in a differential equation. The solution of the differential equation subject to boundary values and initial conditions (for transient problems) would produce a function u that describes the event. However, most likely the differential equation has no known solution. We have no choice but to use a numerical method such as FEM. Now, FEM produces a linear system of equations. The solution to the system determines a function φ_h, the approximate solution. Assuming that the original problem is deterministic, then u and φ_h must be uniquely determined. However, the differential equation alone is insufficient to impose uniqueness. Indeed, the same differential equation will model many very different phenomena. The defining characteristics of any particular setting are imposed by the boundary conditions. In mathematical terms, the derivative with respect to an independent variable x annihilates all terms which are independent of x. Hence, we cannot know u merely from its derivatives. We need additional information.

This information usually comes in the form of data measured at the boundary. Indeed, this particular issue arose with Definition 7.1.1. If the problem is not well-posed, then the numerical technique cannot produce any meaningful information.

The first two steps in the FEM are to define the model, setup the elements, nodes and basis functions. If D is the spatial domain for the problem, then the basis functions define a vector space of piecewise polynomial functions on D. We denote this space by V^h, where h is the mesh parameter, the size of the largest element relative to the size of D. The FEM solution is defined in Section 3.3, Definition 3.2.1. It is a continuous element φ_h of V^h that satisfies

$$0 = \int\limits_D (L\varphi_h - f)\psi$$

for every ψ in V^h. As usual, we set the PDE as $Lv - f$. We also do a formal presentation of degrees of freedom in terms of the dual space to V^h.

Section 8.3 is of particular interest. Here, we present an interesting technique that allows elements with curved edges.

In subsequent sections, we develop steps 3–6. In Section 8.4, we present the Galerkin weak form analysis, which gives rise to the local and global linear systems. At this time, we relate these linear systems to the local and global basis functions developed in Sections 8.2 and 8.3. We show that the assembly process introduced in Chapter 3 produces a linear combination of the global basis functions from the linear span of local basis functions. We see at this point that φ_h is a piecewise polynomial function. It is important to realize that the FEM solution is not the polynomial interpolation of u. Rather, it is determined by the condition designated in Definition 3.2.1.

Finally, we consider boundary value issues. In particular, we look at boundary values as they are applied in the discrete or FEM setting. As with FDM, in FEM, we can apply Dirichlet conditions as specified values of u. In addition, we can apply Neumann conditions which represent external forces on the system. We discuss both in Section 8.5. Moreover, we will demonstrate the relation between them. Time stepping has been discussed in the prior chapters. We do not consider it here.

For future reference, we include a chart listing several FEM models and their properties. This material is developed in this chapter.

Dim	Geometry	Polynomials	DoF	Cts
1	2 Node Intervals	deg 1, Lagrange	2	0
1	2 Node Intervals	deg 3, Hermite	4	1
2	4 Node Rectangles	deg 2, Lagrange	4	0
2	10 Node Rectangles	deg 3	16	0
2	4 Node Rectangles	deg 6, Hermite	16	1
2	3 Node Triangles	deg 1	3	0
2	6 Node Triangles	deg 2	6	0
2	11 Node Triangles	deg 3	10	0
2	6 Node Triangles	deg 5	21	1
2	8 Node Curved Quads	deg 4 Serendipity	8	0
2	12 Node Curved Quads	deg 6 Serendipity	12	0

8.1. The Model 1: Elements and Nodes

We refer to setting the elements, nodes and basis functions as setting the model. It is at this step that you define the space (V^h). In this section, we develop the elements and nodes and hold consideration of basis functions for the following two sections.

Let $U \subset \mathbb{R}^n$ denote a bounded, connected, open set with closure $\bar{U} = D$. Denote the boundary by Γ. From the point of view of FEM, it is necessary that U be bounded. If the problem setting is naturally unbounded, then you must set a practical boundary, as was done for the options pricing model and the river pollution problem in Chapter 5. Since D is closed and bounded, it is compact. For problems arising from mechanical engineering, n is usually 1, 2, or 3. Problems with $n > 3$ do occur in finance, biology and industrial engineering. However, our knowledge and intuition of geometric shapes for $n > 3$ is severely limited. Therefore, this form of structured FEM is usually restricted to cases with $n \leq 3$.

For $n = 1$, elements are intervals with boundary equal to two distinct points. For $n = 2$, elements are bounded sets whose boundary is a finite union of smooth arcs. These arcs are individually referred to as sides or edges. Their union forms a simple closed curve. Most often two-dimensional elements are rectangles or triangles. For the $n = 3$ case, elements are most often rectangular polyhedra (hexahedra) or tetrahedra. The latter case arises as it is possible to pack a cube with 6 tetrahedra. Hence, it is possible to create a variety of shapes using tetrahedra. When $n = 3$, the sides or edges are two parameter smooth surfaces. Their intersection is empty, a smooth arc or a point. It is possible to present FEM with a very general definition of element. However, most commonly, it is one of the cases just mentioned.

Definition 8.1.1. A *finite element decomposition* or *partition* of U is a finite family of open, connected subsets U_e, $e = 1, 2, \ldots, m$, called *elements* and a finite set of points in D, $N = \{P_i : i = 1, 2, \ldots, p\}$, called *nodes*. We denote the closure of U_e by E_e and the boundary by Γ_e. The elements and nodes must satisfy

(i) $D = \cup_e E_e$,
(ii) $E_e \cap E_{e'} = \Gamma_e \cap \Gamma_{e'}$,
(iii) The cardinality of $E_e \cap N$ is constant.
   The partition is a *finite element partition* with polyhedral elements provided each E_e is a smooth polyhedron and
(iv) Each E_e has the same number of faces ($n = 3$), edges ($n = 2, 3$) and vertices,
(v) For each e, the vertices of E_e are contained in N,
(vi) $\Gamma_e \cap \Gamma_{e'}$ is empty, a vertex, an edge or a face.

We also refer to the E_e as *elements* and $E_e \cap N$ as the *element nodes*. Our primary concern is with the finite element method using polyhedral elements. Hence, the reference to FEM will refer to this case (the material in Section 8.3 is an exception to this statement.).

The first two properties state that the elements form a partition of D. The partition is similar to the partition used to define the integral of a function over D. The remaining properties ensure geometric uniformity. Since nodes are defined for D, then we have (without specific mention) that if P is a node lying in $E_e \cap E_{e'}$, then P is a node with respect to both elements.

We have not specifically identified nodes with elements but rather to the partition. However, nodes are commonly element vertices or edge midpoints. Sometimes, it is useful to have nodes that are interior to the elements (see Figure 8.2.3). In Section 8.4, we will consider a technique that allows us to remove interior nodes.

In the next chapter, we will need a different characterization for the idea of a piecewise continuous boundary. Recall that a real valued function f defined on $\mathbb{R}^n$ is *Lipschitz continuous* provided there exists a constant c depending on f with $|f(x) - f(y)| \leq c\|x - y\|$ for any x and y in the domain of f.

Definition 8.1.2. We say that a domain U in $\mathbb{R}^n$ with boundary Γ (parameterized as a subset of $\mathbb{R}^{n-1}$) is a *Lipschitz domain* provided for every x in Γ there is a neighborhood $B \subset \mathbb{R}^n$ of x and a Lipschitz continuous function f on $\mathbb{R}^{n-1}$ with (after a suitable variable change) $U \cap B = \{(x_1, \ldots, x_{n-1}, x_n) \in B : x_n > f(x_1, \ldots, x_{n-1})\}$.

This technical definition generalizes the following familiar idea. A line in the plane, a plane in space can be identified as the set of zeros of a function in one (respectively two) variables. In turn, the function divides the space into two disjoint parts with respect to the line or plane. If the line or plane is properly oriented (the variable change referred to in the definition) with respect to the axes, then the two sides may be characterized by $x_n > 0$ or $x_n < 0$. Specifically, after the change of variables, the line is the x-axis or the xy-plane. For the case at hand, given a point on the boundary, then the boundary divides the space so that nearby interior points are all on one side or the other.

It is important to note that polyhedral elements are Lipschitz domains. In addition, in any case of interest, the domain U of the FEM is also Lipschitz. More generally, domains with piecewise smooth boundaries are Lipschitz.

There are other possible choices for elements. In the plane, any polygonal shape that tessellates the plane will do. However, the technical cost associated to implementing the FEM for other shapes is prohibitive. For instance, regular hexagons would be an appealing choice. However, they are never used. Since the boundary of U may not be a polygon, then it is necessary to allow the sides of the elements to be smooth curves. In practice, curved boundary segments are often rendered as polygons of short line segments.

Regular arrangements of equal sized elements often impose a bias on the computed solution. For instance, the arrangement of triangles given in Figure 8.1.1 should not be used when modeling a fluid flow. The arrangement of triangles in Figure 8.1.2, called the *Union Jack Pattern*, is preferred.

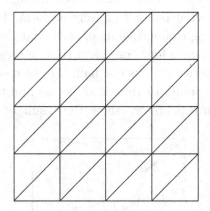

Figure 8.1.1.   Uniformly oriented triangles.

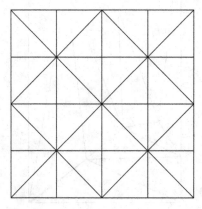

Figure 8.1.2.   Union Jack pattern.

Delaunay triangulation is the standard technique to triangulate a convex planar region. Mathematica has a built in function that executes Delaunay triangulation. The procedure takes a set of boundary and interior points and returns the triangulation. There is one problem. This process commonly produces a *sliver triangle*. A sliver triangle is one with area much smaller than the maximal side length. Generally, there is computational error at sliver triangles. Figure 8.1.3 shows an example of Delaunay output with a sliver triangle at the top edge.

Alternatively, if we start with a regular boundary point distribution then the Delaunay procedure yields a triangulation with triangles that are somewhat but not entirely uniform. This partition can be refined by joining the midpoints of each triangle. The result is to divide each triangle into four congruent triangles, each similar to the original. The similarity ratio is one half. We get a similar refinement by adding the midpoints to the list of points and then re-executing Delaunay. In Figure 8.1.5, we show the result of two refinements

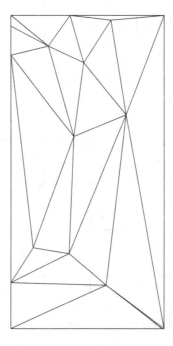

Figure 8.1.3.   Delaunay triangulation with random nodes.

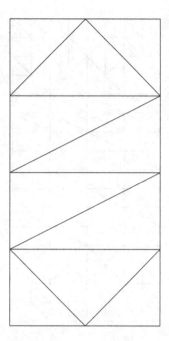

Figure 8.1.4.   Delaunay triangulation with uniform boundary partition.

of the partition in Figure 8.1.4. There are now 160 elements. In Figure 8.1.6, we make small perturbations in the uniform edge distribution and then do the triangulation and refinement. The result is a non-uniform triangulation without sliver triangles.

In the remainder of the section, we consider some of the technical properties of an FEM partition. These will come up in Chapter 9. For the first definition, recall that the elements E_e are compact.

Definition 8.1.3. Let E be a compact subset of $\mathbb{R}^n$, the *diameter* of E is given by $diam(E) = \max_{u,v \in E} \|u - v\|$.

Note that for a segment, triangle or tetrahedron, the diameter is always the length of the longest edge. Therefore, the iterative refinement procedure described above and illustrated in Figures 8.1.4–8.1.6 will diminish the diameter of each element by a factor of $1/2$. If the maximal diameter of any element is h, then after n iterations of the refinement procedure the maximal diameter of any element is $(1/2^n) \times h$.

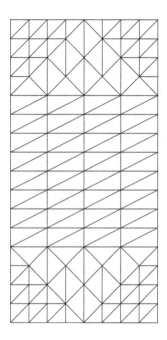

Figure 8.1.5.   Refinement with Delaunay Triangulation.

We next formally define a concept that has come up in several examples in Part 1.

Definition 8.1.4. Let U be a connected, open and bounded set in $\mathbb{R}^n$ with partition $\mathcal{P}$ of U with elements E_e. The *mesh parameter* $h = h_{\mathcal{P}}$ is $h = \max_e[diam(E_e)]$.

When we consider a family of partitions of a domain U, our intention is that the family support a limiting process for the related FEM. For this purpose, we expect that the elements of a sequence of successive refinements to be smaller. In terms of Definition 8.1.4, we expect $h_{\mathcal{P}} \to 0$. However, there will be problems, if the elements become too thin or if the size of the elements varies too much. We have already mentioned that sliver triangles may be computationally problematic. To prevent the later case, we want a lower bound for the ratio $\min_e diam(E_e) / \max_e diam(E_e)$ as $h_{\mathcal{P}} \to 0$. For the former, we consider $r_{\max}(E_e)$ the largest radius of an open ball or circle that fits within the triangle or rectangle or polyhedron. This is called

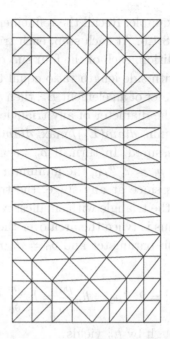

Figure 8.1.6.   Delaunay triangulation with non-uniform edge partition.

the *in-circle* or *in-sphere* for the element. Since U is bounded, radius $r_{\max}(U)$ for the in-circle, in-sphere is finite. In general, we expect each $r_{\max}(E_e)$ to go to zero as $diam(E_e)$ does. However, we can require that the ratio $r_{\max}(E_e)/diam(E_e)$ be bounded below as $h_\mathcal{P} \to 0$. We provide formal definitions now. For this purpose, we use the notation just developed.

Definition 8.1.5. Let U be a connected, bounded open set in $\mathbb{R}^n$, and suppose that we have a family of partitions of U into finite elements E_e. The family is called *quasi-uniform* provided $\inf_\mathcal{P}[\min_e diam(E_e)/ \max_e diam(E_e)] > 0$. Further, the family of partitions is *regular* if it is quasi-uniform and there exists a positive real ρ such that for every partition, $\rho \le r_{\max}(E_e)/diam(E_e)$.

Note that for a triangle, the radius of the in-circle is $2 \times Area/ Perimeter$. We are focusing on elements that are convex. In more general settings, authors often add a property related to convexity. For instance, they may assume star shaped.

We consider an example. Recall the refinement process illustrated in Figures 8.1.5 and 8.1.6. At each step, each triangle is divided into four triangles. In addition, each of the four successor triangles is similar to the original with side length ratio equal to one half. Hence, if we set h_e to the diameter of an element E_e, then at the nth step in the refinement process, the diameter of any triangle contained inside E_e is $2^{-n}h_e$. Therefore, $\min_e diam(E_e)/\max_e diam(E_e)$ is constant and quasi-uniformity is assured. In turn, if $\rho_n = \min_e[r_{max}(E_e)]$ at the nth refinement, then $\rho_n = 2^{-n}\rho_0$, the minimum radius for the initial partition. This holds since at any stage, the triangle with smallest r_{max} is derived from the triangle with smallest r_{max} at the prior stage and the in-circle radius is twice the area over the perimeter. If we set h_0 to the initial mesh parameter and $\delta_0 = diam(U)h_0$, then at the nth step,

$$\min_e[r_{max}] = 2^{-n}\rho_0 = h_n\frac{\rho_0}{h_0} = h_n\frac{\rho_0}{\delta_0}diam(U).$$

Hence, dividing through by h_n yields

$$\frac{\min_e[r_{max}]}{\max_e[diam(E_e)]} = \frac{\min_e[r_{max}]}{h_n} = \frac{\rho_0}{\delta_0}diam(U).$$

Therefore, the family of refinements is regular.

Finally, we mention an important means of associating elements of an FEM partition. The idea came up in Sections 3.2 and 6.1 where it is used to define the polynomial model associated to the partition. In $\mathbb{R}^n$, an *affine map* is a function A of n-space to itself that is a composition of a non-singular linear transformation and a translation. That is, there is a bijective linear map T and a vector α so that for x in $\mathbb{R}^n$, $Ax = Tx + \alpha$. It is well known that affine transformations act transitively on intervals (in 1 space), triangles or rectangles (in 2-space) or tetrahedra, rectangular polyhedra (in 3-space). For instance, given any two triangles, there is an affine transformation that maps one to the other.

Given a compact connected set D with a triangular FEM partition $\mathcal{P}$ in the real plane, select a fixed triangle independent of D as a model triangle. For instance, the triangle with vertices $(0,0)$, $(1,0)$ and $(0,1)$. We denote it by R and refer to it as the *reference element*. Now, for element E_e there is an affine mapping A_e that maps R to E_e. If the nodes are geometrically defined, then A_e will also associate points on the reference element to nodes on the FEM partition element. Conversely, we can define a set of points on the reference element and then use the affine maps to define the nodes in the FEM model. We will see later in this chapter and then in Chapter 9 that the idea of a reference element is very useful.

Exercises

1. Execute Delaunay Triangulation on the following point set. $S = \{(0,0), (1,1), (2,2), (3,3), (4,4), (5,5), (6,6), (2,4), (1,5), (0,6), (4,2), (5,1), (6,0)\}$. Plot the set and the resulting triangulation. Note that the Mathematica function returns an adjacency list, a list that identifies pairs of points that are joined by a triangle side.

2. In the section, we described a refinement process for a triangular finite element partition based on dividing each triangle into four congruent triangles similar to the original. Apply this refinement process to the triangulation that results from Exercise 1.

3. Let D be a subset of $\mathbb{R}^2$ and suppose that Γ is a union of line segments parallel to the coordinate axes. Write a Mathematica program that starts with a partition of Γ into line segments and results in a partition of D into rectangular elements. See for instance, Figure 3.1.1.

4. It is possible to divide a triangle into three parts by joining the triangle centroid to each of the vertices. If you have a planar region with a triangular finite element partition and a sequence of refinements formed by dividing the triangular elements into three parts as just described, is the sequence of refinements quasi-uniform? Is it regular? (Recall that the centroid of a triangle lies at the intersection of the three medians.)

5. Suppose D is a compact domain with a finite element partition. Set R to be the triangle with vertices $(0,0)$, $(1,0)$ and $(0,1)$ and take A_e to be the affine map sending R to the element E_e. Further suppose that Z is the set consisting of the vertices and edge midpoints. Prove that D with the triangular partition and $N = \cup_e A_e(Z)$ is a finite element partition.

6. With the notation of Exercise 5, give an example of a set Z, so that $N = \cup_e A_e(Z)$ is not the set of nodes for any FEM partition.

8.2. The Model 2: Polynomial Basis Functions

The next stage in the FEM model is to determine the local basis functions and the function space associated to the element partition. We consider only polynomial functions. In the literature, there are cases that use trigonometric polynomials or even B-spline parametric curves or surfaces. We begin the section with general comments and a preliminary lemma. Next, we consider the two simplest FEM models, segments and rectangles. This juxtaposition is natural as polynomial spaces on rectangles can be developed as the tensor product of the one-dimensional case. Subsequently, we look at polynomial spaces over triangular partitions.

The setting is the same as in Section 8.1. We have a domain D with FEM partition $\mathcal{P}$ and elements denoted E_e. We call the functions *basis functions* since these functions form a basis for V_e^h and subsequently are used to construct the basis for V^h. However, they are also referred to in the literature as interpolating functions or model functions. This terminology arises from the method for numerically solving differential equations via polynomial interpolation. In FEM, the idea is to do polynomial interpolation locally, then stitch together the local results to approximate the actual solution to the PDE. Joining polynomial functions yields a *piecewise polynomial* function.

In this section, we ignore vector valued models. Indeed, a vector valued model is just the direct product of scalar valued polynomials.

We first recall the standard terminology for degree and order of a polynomial. Let p be a polynomial in variables $x_1, \ldots, x_r$. If $p = \prod_{i=1}^r x_i^{s_i}$, for non-negative integers s_i, then p is a monomial of degree $\sum_{i=1}^r s_i$. More generally, a polynomial is a linear combination of monomials and the *degree of the polynomial* is the maximum of the degrees of the constituent monomials. Alternatively, if the maximal value of s_i for any variable x_i is ρ_i, then we say that p is degree ρ_i

in x_i. If p has the same degree, ρ, for each variable, then p is *order* ρ. For instance, $a + bx_i + cx_2 + dx_1x_2$ is degree 2 and order 1. On the other hand, $a + bx_1 + cx_2 + dx_1^2 + ex_2^2$ is degree 2 and order 2. We caution the reader that the terms degree and order are often interchanged in the literature.

There are three concerns associated to the choice of basis functions. First, the functions should be chosen to achieve the desired level of continuity. Second, the highest derivative that appears in the Galerkin weak formulation must not annihilate the basis functions. We discuss this in a later section and concentrate here on levels of continuity.

Next, we define degrees of freedom. We have already encountered degrees of freedom in Section 5.1. At that time, we were developing the Hermite polynomial model for collocation method. We might have mentioned the idea in Section 3.1 or 3.2. However, at the time there was only one degree of freedom per node. The idea would have been superfluous. With Hermite polynomials, we had two per node. It was reasonable to mention the concept and provide a definition. We now repeat the definition. A general discussion follows.

Definition 8.2.1. A local *degree of freedom* (denoted *DoF*) is an element of the dual space $(V_e^h)^* = Hom[V_e^h, \mathbb{R}]$. The number of (local) degrees of freedom is the dimension of the space. In turn, a *global degree of freedom* is an element of the dual space of V^h. The number of global degrees of freedom is the dimension of $(V^h)^*$.

In Section 1.3, we already noted that each point of E_e defines a linear functional in $(V_e^h)^*$. Also, the same is true for any derivative evaluated at a given point in E_e. Suppose that there are m nodes denoted p_j, then from Section 1.3, we know that we can identify polynomials φ_k^e so that $F_j(\varphi_k^e) = \varphi_k^e(p_j) = \delta_{j,k}$. Hence, the polynomials φ_k^e are linearly independent and the F_j form a basis of functionals (degrees of freedom) dual to the span of the φ_k^e. We now extend this setting to include the derivatives taken at nodes.

A finite sequence $\alpha = (\alpha_1, \ldots, \alpha_n)$ of non-negative integers is called a *multi-index*. We use multi-indices to provide a compact notation for derivatives. For instance, for functions of two independent variables,

$$\partial^\alpha \varphi = \frac{\partial^{\alpha_1}}{\partial x_1^{\alpha_1}} \frac{\partial^{\alpha_2}}{\partial x_2^{\alpha_2}} \varphi.$$

We extend the notation F_j to include $F_{j,\alpha}(\varphi) = \partial^\alpha \varphi(p_j)$. Note $F_{j,\alpha} = F_j$ when $\alpha = (0,0)$. There is a natural extension of the Vandermonde process which produces bases dual to $F_{j,\alpha}$, $\alpha = (0,0)$, $(1,0)$, $(0,1)$ and any number of nodes. And indeed, we are not restricted by the order of the derivatives or the dimension of the underlying vector space. Note that the more degrees of freedom, the greater the dimension of V_e^h and the greater the degrees of the polynomials in the model. For now, we will restrict attention to DoF associated to linear functionals of the form $F_{j,\alpha}$.

Since each basis function for V_e^h is dual to the degrees of freedom and since the degrees of freedom are identified with nodes, then each basis polynomial evaluates zero at all but one node. And at that node, one of its derivatives is not zero. In turn, consider an adjacent element $E_{e'}$. If $\varphi_i^e \in V_e^h$ and $\varphi_j^{e'} \in V_{e'}^h$ and if they are dual to a degree of freedom at a common node, then we want to stitch the two functions together to form a single piecewise polynomial function. Further, we want to consider this process as the natural way to build a basis for V^h. This is the purpose of the following lemma. The lemma is stated for the two-dimensional case. There are analogous results for the line and for three space.

Lemma 8.2.1. *Let $D = \cup_e E_e$ be a partition of D into finite elements. Fix an element E_e and a degree of freedom $F_{i,\alpha}$, and designate the elements sharing $F_{i,\alpha}$ by E_{e_j}, $j = 1, \ldots, s$. For each j, $1 \le j \le s$, let φ_{e_j} be a polynomial function in $V_{e_j}^h$ dual to $F_{i,\alpha}$ and suppose that*

(i) *φ_{e_j} restricted to an edge of Γ_{e_j} is completely determined by the degrees of freedom associated to nodes lying on that edge of Γ_{e_j},*

(ii) *for each j and k, $\varphi_{e_j} = \varphi_{e_k}$ on $E_{e_j} \cap E_{e_k}$,*

(iii) *there is a neighborhood of p_i contained in $\cup_j E_{e_j}$.*

If ψ is defined on D to be equal to φ_{e_j} on E_j and equal to 0 outside of $\cup_{k=1}^s E_{e_k}$, then ψ is a continuous piecewise polynomial function on D. Furthermore, ψ has compact support provided p_i is an interior node of D.

Proof. The result follows from three remarks. For any j between 1 and s, ψ is a polynomial on E_{e_j}. Therefore, it is continuous. For any $1 \leq j < k \leq s$, since $\psi = \varphi_{e_j} = \varphi_{e_k}$ on $E_{e_j} \cap E_{e_k}$ then ψ is continuous on $E_{e_j} \cup E_{e_k}$. Therefore, ψ is continuous on $\cup_{j=1}^s E_{e_j}$. By (iii) ψ is identically zero on the element edges included in the boundary of $\cup_{j=1}^s E_{e_j}$. Hence, ψ is continuous on D. The final assertion is immediate. $\square$

We now turn to the basis functions for interval and rectangular elements. If the spatial dimension is one, then D is an interval and the elements are sub-intervals. The end points of the elements are determined by a partition, $x_0 < x_1 < \cdots < x_m$ of D. We take the interval end points for the nodes. If we set the unit interval as the reference element, then $x \rightarrow (x_e - x_{e-1})x + x_{e-1}$ is the affine map $A_e : R \rightarrow E_e = (x_{e-1}, x_e)$. Further, the affine map sends end points to end points. Therefore, it maps reference nodes to element nodes. Taking the two Lagrange polynomials for the unit interval, N_0 and N_1, we set $\varphi_i^e = N_i A_e^{-1}$. Hence, we have a basis for V_e^h that is dual to the two degrees of freedom associated to the interval endpoints. The basis for the piecewise polynomial functions is ψ_j where ψ_j arises from the join of φ_1^e and φ_0^{e+1}. We collect this information in the following theorem. The proof uses Lemma 8.2.1 and basic properties of Lagrange polynomials.

Theorem 8.2.1. *Suppose that the spatial dimension of D is one with nodes $x_0, \ldots, x_m$ and elements $E_1, \ldots, E_m$ where $E_e = [x_{e-1}, x_e]$. Then $\{\varphi_0^e, \varphi_1^e\}$ forms a basis for the element space V_e^h. Further, for each node p_i, we define a piecewise polynomial function on D by joining the (up to two) local basis Lagrange polynomials that are not zero at p_i. The functions ψ_i are linearly independent. Setting V^h to the function space spanned by the ψ_i, then V^h is $m+1$ dimensional and the elements of V^h are C^0 continuous on D.*

In turn, we can establish the Hermite model from Section 5.2 on the same geometric domain. Alternatively, we can define the Hermite polynomials on the unit interval as the reference element. The element polynomials are again defined via the element's affine transformation. In this case, each V_e^h is four-dimensional with the Hermite polynomials dual to the degrees of freedom, $F_{i,\alpha}$ for $\alpha = (0)$ or (1) and $i = 0$ or 1, identifying the left hand or right hand endpoint. To form the piecewise polynomial basis functions, we join Hermite polynomial functions associated to the same multi-index and opposite end points. Hence, there are two piecewise polynomial functions at each node or a total of $2(m + 1)$.

Theorem 8.2.2. *If the spatial dimension of D is one and the element decomposition is as given in Theorem 8.2.1, then the four cubic Hermite polynomials form a basis for the element space V_e^h. The set $\{\psi_{i,\alpha}(x) : \alpha = (0), (1); i = 0, 1, \ldots, m\}$ is a basis for $V^h \subset C^1$, the continuously differentiable functions on D.*

Proof. It remains to prove that the model is C^1. This question reduces to whether $\psi_{i,\alpha}$ is C^1 at x_i, x_{i-1} and x_{i+1}. For $\alpha = (0)$, the derivative of $\psi_{i,\alpha}$ from the right or left is zero. For $\alpha = (1)$, the derivative is 1 from either direction. There are still two cases to consider, the level of continuity at x_{i-1} and x_{i+1}. We leave this to the reader (Exercise 1). □

Next, we turn to the case of two independent spatial variables and rectangular elements, $E_e = [x_{i-1}, x_i] \times [y_{j-1}, y_j]$ for a given x-axis and y-axis partition. We take the unit square as the reference element and get the affine maps by setting $\alpha = x_i - x_{i-1}$ and $\beta = y_i - y_{i-1}$ so that

$$A_e \begin{pmatrix} x \\ y \end{pmatrix} = \begin{pmatrix} \alpha & 0 \\ 0 & \beta \end{pmatrix} \begin{pmatrix} x \\ y \end{pmatrix} + \begin{pmatrix} x_{i-1} \\ y_{i-1} \end{pmatrix}.$$

The two polynomial models that extend the one-dimensional case are the two variable Lagrange polynomials and the two variable Hermite polynomials. We get the Lagrange polynomials from the Vandermonde matrix using the rectangle vertices. This can be done for the reference element or for each element individually. In any case, the resulting polynomials are dual to the functionals associated to the vertices. Alternatively, using the notation developed for the one dimensional case, the four degree 2 Lagrange polynomials in two variables on the unit square are $N_0(x)N_0(y)$, $N_0(x)N_1(y)$, $N_1(x)N_0(y)$, $N_1(x)N_1(y)$. In any case V_e^h is 4D. We describe this model formally in the following theorem.

Theorem 8.2.3. *Suppose that D has two spatial dimensions and is partitioned into rectangular elements with four nodes per element and elements denoted E_e. Then the first-order, two variable Lagrange polynomials are linearly independent. We denote the span by V_e^h, the local polynomial model. For each node p, we associate the (up to 4) local basis polynomials that evaluate to 1 at p. The resulting piecewise polynomial function ψ_p is continuous. Setting V^h to the linear span of the ψ_p, we have a C^0 model of dimension m, the total number of nodes.*

Proof. It remains to prove the assertions for ψ_p and V^h. There are up to four elements that have vertex node p and each element has a single basis polynomial that has the value one at p. Suppose that φ_p^e is the Lagrange polynomial for one of the elements E_e. As φ_p^e has value zero at each of the other three vertices, then φ_p^e is zero on two edges of E_e. Further, it is easy to verify that φ_p^e has degree 1 on the two remaining edges. Hence the graph is a line connecting a point on the yx-plane to a point of height 1 at the node p. Since the same is true for each element sharing the node, then the join of the polynomials ψ_p is continuous (see Figure 8.2.1).

Finally, if the nodes are identified as p_i, $i = 1, \ldots, m$, then $\sum_i \alpha_i \psi_{p_i} = 0$ implies that $0 = \sum_i \alpha_i \psi_{p_i}(p_j) = \alpha_j$ for each node j. The result now follows. $\qquad\square$

In Section 5.2, we defined the 4 one variable Hermite polynomials. For the unit interval, we have

$$H_{00}(x) = 3(1-x)^2 - 2(1-x)^3, \quad H_{01}(x) = 3x^2 - 2x^3,$$
$$H_{10}(x) = (1-x)^2 - (1-x)^3, \quad H_{11}(x) = -x^2 + x^3.$$

We next define $H_{a,b}(y)$ and then define 16 degree 6 polynomials in two variables. It is easy to verify that these polynomials are dual to the 16 degrees of freedom, $F_{p,\alpha}$ where α is one of the 4 multi-indices

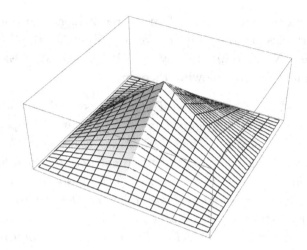

Figure 8.2.1.   Four Lagrange polynomials from adjacent elements.

$(0,0)$, $(0,1)$, $(1,0)$, and $(1,1)$ and p is a vertex of the unit square. Note that when $\alpha = (1,1)$, the derivative ∂^α is the mixed second partial.

For an element E_e, the 16 two variable polynomials are given by φA_e^{-1} where φ is a two variable Hermite polynomial on R. If q is a vertex of E_e, then $q = A_e p$ where p is a vertex of the unit square. It is immediate that φ is dual to $F_{p,\alpha}$ if and only if φA_e^{-1} is dual to $F_{q,\alpha}$. Most of the following theorems have been proved. The remaining parts are left as an exercise.

Theorem 8.2.4. *Suppose that D has two spatial dimensions and is partitioned into rectangular elements. If D has s nodes, then using third-order, two variable Hermite polynomials, V_e^h is $16D$ and V^h is s-dimensional. Further, the elements of V^h are C^1 continuous.*

Proof. See Exercise 3.   □

We now turn our attention to triangular elements. Rectangular partitions are preferred for rectangular domains or infinite domains that are bounded in order to apply FEM. The most important reason is that the integration over a rectangle is less machine intensive than integration over of a triangle (see Section 1.4). However, for irregular domains or domains that are naturally triangular or trapezoidal, we have no choice but to triangulate the domain.

Further, the idea of a reference element is particularly useful for triangular partitions. We have already developed the idea of an affine transformation, a bijective mapping of the plane that maps a reference triangle to a given element triangle. In particular, we set R to be the triangle with vertices $(0,0)$, $(1,0)$, and $(0,1)$. For an element E_e with vertices (α, β), (γ, δ), and (μ, ν), then $(\gamma - \alpha, \delta - \beta)$ and $(\mu - \alpha, \nu - \beta)$ is a basis for $\mathbb{R}^2$ and therefore

$$T_e = \begin{pmatrix} \gamma - \alpha & \mu - \alpha \\ \delta - \beta & \nu - \beta \end{pmatrix}$$

is non-singular and maps $(1,0)$ to $(\gamma - \alpha, \delta - \beta)$ and $(0,1)$ to $(\mu - \alpha, \nu - \beta)$. Hence, the corresponding affine map is $A_e : \mathbb{R}^2 \to \mathbb{R}^2$, $A_e(x, y) = T_e(x, y) + (\alpha, \beta)$. Using the mappings A_e, we can associate polynomials and DoF on R to elements E_e.

We look at three triangular models. If we take the vertices as nodes we have a 3 nodal model. If we take the vertices and side midpoints then we have a 6 nodal triangle. Finally, we may take the vertices, two edge points at distance $1/3$ and $2/3$ from an end point and the triangle centroid. The latter case yields a 10 nodal model. It is easy to verify that these nodal configurations yield triangular sets of points for the reference element. We apply Theorem 1.3.8 to get a set of linear independent polynomials on R. In particular, each polynomial is dual to the degree of freedom associated to a given node. We designate these polynomials as φ_i associated to reference node p_i. Using the affine maps, we induce a polynomial model on the triangular FEM partition. For three nodes, V_e^h is the degree 1 polynomials, for six nodes V_e^h is all degree 2 polynomials and for 10 nodes it is the space of all degree 3 polynomials.

Next, we consider a set of elements $E_{e_1}, \ldots, E_{e_s}$ each sharing a node p_i. Setting ψ to the join of the $\varphi_i^{e_k} = \varphi_i A_e^{-1}$, we claim the ψ_i is continuous. Hence, the model is C^0. Referring to Lemma 8.2.1, it suffices to prove that ψ_i is continuous across common element edges. Suppose that E_j and E_k share a common side, $\Gamma_{j,k}$. Now, $\varphi_i^{e_j}$ and $\varphi_i^{e_k}$ take the same value at every node on the common edge.

Furthermore, when there are 2 nodes on the edge, the two functions are degree one and so may be parameterized as degree 1 polynomials on $\Gamma_{j,k}$. When there are three nodes, they are degree 2 and may both be parameterized as quadratics. In turn, they are parameterizable as cubic polynomials when there are four nodes. Since $r + 1$ points determine a unique degree r polynomial, then $\varphi_i^{e_j}$ and $\varphi_i^{e_k}$ are equal on the common edge. Therefore, ψ_i is continuous by Lemma 3.2.1. We have proved the following theorem.

Theorem 8.2.5. *Suppose that U has two spatial dimensions and is partitioned into triangular elements with any of three models, 3 nodal, 6 nodal, or 10 nodal triangles. Each V_e^h is 3, 6, or $10D$ with basis polynomials $\varphi_i A_e^{-1}$ of degree 1, 2, or 3 respectively. Each basis polynomial is dual to the degree of freedom associated to the corresponding node. Let p_j be a node and designate the elements containing p_j as $E_{e_1}, \ldots, E_{e_s}$. In turn, denote the local basis function in $V_{e_k}^h$, which is dual to p_j by $\varphi_j^{e_j}$. Then the global function space, V^h is m-dimensional (where m is the number of nodes) with basis functions ψ_j defined by joining the local basis functions $\varphi_j^{e_j}$ that do not vanish at p_j. Further, the elements of V^h are C^0 continuous.*

The following four illustrations refer to the 6 nodal, degree 2 model. Figures 8.2.2 and 8.2.3 show the 6 modal triangle and 10 nodal

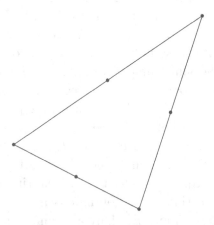

Figure 8.2.2.   Triangle with 6 nodes.

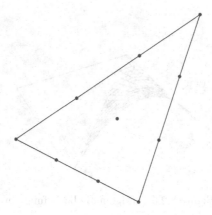

Figure 8.2.3.   Triangle with 10 nodes.

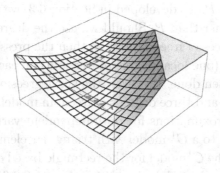

Figure 8.2.4.   Vertex node basis function.

triangles. Figures 8.2.4 and 8.2.5 show plots of elements of V_e^h for the degree 2 model. In the first case, the polynomial is dual to a vertex node. In the second, it is dual to a midpoint node.

It is natural to ask why you might want to use this six degrees of freedom model over the previous one with three degrees of freedom. The primary answer lies in the order of convergence. We will see that the order of convergence depends, in part, on the degree of the polynomials used to model the process. This is largely due to the standard error estimation procedure for polynomial interpolation. The topic was considered in Section 1.3. An alternative answer lies in the Navier–Stokes equation or alternatively the Stokes equation.

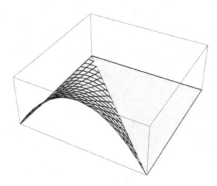

Figure 8.2.5.   Edge node basis function.

In both cases, you have two dependent variables, the velocity field φ and the pressure P. As developed in Section 6.3, we needed to model φ at a order higher than P. Should we use the degree 1 polynomials (three local degrees of freedom) for φ, then the pressure will be modeled as constant (one local degree of freedom) on each element. That can be and is often done. However, the six degrees of freedom model for the flow field and three degrees of freedom model for pressure will produce C^0 approximations for both dependent variables.

We next turn to a C^1 model with triangular elements. The model is analogous to the C^1model for the rectangle based on Hermite polynomials. We use a six nodal triangle (see Figure 8.2.2) with 21 degrees of freedom. In particular, we define six degrees of freedom at each vertex node associated to differential operators ∂^α, where α takes values $(0,0)$, $(1,0)$, $(0,1)$, $(2,0)$, $(1,1)$, or $(0,2)$. Hence, 18 of the basis functions for V_e^h will be determined as dual to these degrees of freedom. The remaining three basis functions will be determined by the value of the normal derivative at the edge midpoint nodes. When we developed DoF and the notation for partial derivatives, we did not include directional derivatives. We could have done this by using a vector to identify the direction of the differentiation. In this case, the partials would be the directional derivaties in the directions of the standard orthogonal, unit vectors. However, the notation we are using is more standard. Nevertheless, for the current discussion and theorem, we will suppose that α combines a direction vector and the identifier for a differential operator. Finally, we remark that Theorem 1.3.8 can be extended to include the Vandermonde matrix that arises from this case.

Theorem 8.2.6. *Suppose that U has two spatial dimensions and is partitioned into six nodal triangular elements with a total of m vertex nodes and q edge nodes in D. Then for each element E_e, there are two variable fifth degree polynomials $\varphi_{i,\alpha}^e$ associated to the 21 degrees of freedom discussed above. The local space, V_e^h, spanned by the $\varphi_{i,\alpha}^e$, is $21D$, which is also the dimension of the space of degree 5 polynomials in two variables. In turn, for each node p_i define $\psi_{i,\alpha}$ as the piecewise defined function arising from the $\varphi_{i,\alpha}^{e_j}$ associated to a given degree of freedom $F_{i,\alpha}$. If V^h is the linear span of $\{\psi_{i,\alpha} : p_i$ is a node and α identifies a differential operator as described above$\}$, then V^h is $5m + q$-dimensional. Further, the elements of V^h are C^1 continuous.*

Proof. With an argument similar to the one used in Section 1.3, the Vandermonde matrix that is used to derive the 21 polynomials is non-singular. In turn, it is immediate that the 21 polynomials that are dual to the 21 degrees of freedom described above are linearly independent. Hence, the space V_e^h has the required properties.

We define the global basis functions for each node by defining $\psi_{i,\alpha}$ piecewise. If p_i is a vertex node, then $\psi_{i,\alpha}$ is defined from the local functions associated to a given degree of freedom at this node. For a midpoint node, $\psi_{i,\alpha}$ is the joint of two local basis polynomials. It is immediate that these piecewise polynomial functions are linearly independent.

It remains to prove that the model is C^1. Consider an edge of a triangular element E_e with end points (a,b), (c,d) expressed in parametric form $\gamma(t) = (a,b) + t\tau$, $t \in [0,1]$ and $\tau = (c,d) - (a,b)$. In addition, set an element basis function $\varphi_{i,\alpha}^e$ associated to one of the nodes, $p_i = (a,b)$. The graph of the basis function at the edge may be expressed as a one variable quintic polynomial,

$$q(t) = \varphi_{i,\alpha}^e(\gamma(t)) = \sum_{i=0}^{5} a_i t^i.$$

Also, note that since the curve is a line segment, then the normal and tangential directions are constant. The tangential direction is given by $\tau = (c - a, d - b)$ and the normal is $(d - b, a - c)$.

Now, the six coefficients for q are determined by the six independent constraints given by

1. The value of $\varphi_{i,\alpha}^e$ at the two vertex nodes,
2. the derivative of q at the two vertex nodes,
3. the second derivative of q at the two vertex nodes.

Item 2 is known since we know the two partials of $\varphi_{i,\alpha}^e$ at the segment end points. Hence, we know the derivative in the direction of the edge, which is the derivative of q. Item 3 is known since we know the Hessian of $\varphi_{i,\alpha}^e$ at the two vertex nodes. The same applies to $\varphi_{i,\alpha}^{e'}$ associated to a second element $E_{e'}$ sharing the same degree of freedom. In particular, when the two elements share the edge γ, then we may conclude that $\psi_{i,\alpha}$ is continuous across this edge.

To prove that $\psi_{i,\alpha}$ is smooth across the bounding edge, we must verify that each pair of directional derivatives across the boundary are parallel. We argue that the gradient is the same as calculated from either side. The necessary smoothness follows.

It is immediate that each partial derivative is deterimined by five constraints. We are given values for the partials at the end points. We know the tangential derivative at the edge midpoint and we are given the normal derivative at the edge midpoint. Hence, with this information for two independent directions, we can solve for the partials at the midpoint. Altogether, this gives us three constraints for each partial. Finally, we have the second partials at the end points. Altogether, we have five independent contraints. Hence, we know the gradient at the common edges and it is the same, independent of which bounding element we consider.

The final step, the verification that the derivative is continuous, is left to the reader. $\qquad\square$

This last model is not easy to program but is nevertheless important and useful.

Exercises

1. For Theorem 8.2.2, prove that the basis functions of V^h are C^1 continuous.
2. Let $\mathcal{R}$ be a rotation about the origin in $\mathbb{R}^2$. Let f be a C^n continuous function also on $\mathbb{R}^2$. Prove that $f \circ \mathcal{R}$ is C^n continuous. Use this to extend the results of this section to rectangular partitions that are not oriented parallel to the coordinate axes.
3. Complete the proof of Theorem 8.2.4.

    a. Proceeding analogously to Theorem 8.2.3, define the space V^h.
    b. Prove that the basis functions of V^h are C^1 continuous.
    c. Verify that V^h is p-dimensional.
4. Suppose there is an FEM partition of $D = [-1, 1]$ with two elements $[-1, 0]$, $[0, 1]$. For the Hermite polynomial model in one dimension, plot the two global basis functions for the node at $(0, 0)$.
5. Repeat 4 for the 2D case. This time the domain is $[-1, 1] \times [-1, 1]$ with four elements $[-1, 0] \times [-1, 0]$, $[-1, 0] \times [0, 1]$, $[0, 1] \times [-1, 0]$, $[0, 1] \times [0, 1]$. Plot the four basis functions for the node $(0, 0)$.
6. Consider the case of a four-sided element where two opposite sides are formed by arcs of two concentric circles and two opposite sides are segments of lines which pass through the common center (see Figure 8.3.1). Using the conversion to polar coordinates, determine the associated rectangle. Find the basis for the local space of the rectangle based on first-order Lagrange polynomials. Determine the basis for the local space of the four sided element.
7. Continuing Exercise 6, determine four contiguous elements sharing a single node. Define a global basis function at that node and prove that it is continuous. Plot the resulting global basis functions.
8. Construct a case where global basis functions for the 6 nodal triangle model (six degrees of freedom) are not C^1.
9. Complete the following details for the proof of Theorem 8.2.6.

    a. Define the basis functions for V^h.
    b. Verify that the basis functions defined in (a) are linearly independent.
    c. Verify that the derivatives of the elements of V^h are continuous.

8.3. Serendipity

If there are elements with four edges that are C^1 curves, then your first response should be to parameterize the elements. In $\mathbb{R}^2$, you would look to find a two variable parametrization of the elements against a rectangle. For instance, an annular segment such as Figure 8.3.1 may be mapped to a rectangle via the polar coordinate transformation, $(x, y) \rightarrow (r, \theta)$ with $r^2 = x^2 + y^2$ and $\theta = \tan^{-1}(y/x)$. With this association, you may apply the techniques available for rectangles to define basis functions for the element. In this section, we are concerned with elements for which no such transformation is readily available.

There is an interesting procedure which generates the basis functions for quadrilateral elements with curved edges. The technique is named *serendipity* referencing a story from Sri Lankan mythology. The story's intent is to remind us that the best solution to a difficult problem may well lie with techniques that are well known, simple and close at hand. And the result may not be exactly what was expected.

The basic idea for serendipity, as we use the word, is to begin with a domain U and a finite element partition with quadrilateral

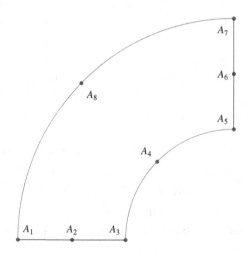

Figure 8.3.1. Annulus segment with 8 nodes.

elements. Let E_e be one element. We want to determine a two variable polynomial and a bijective mapping of a square to a planer figure F_e which approximates the original element E_e. Such a function will yield a parametrization for F_e. Moreover, when the process is repeated for each element then we will have a finite element partition with parameterized elements that approximates the original one. Since each element in the new partition is parameterized, then we will be able to define basis functions and proceed with the FEM. On the other hand, $\cup_e F_e \neq D$, rather it is an approximation of D. The process is successful when $\cup_e F_e$ is an acceptable approximation of D.

We begin our development within the context of an example. We start with a element which is a segment of an annulus, Figure 8.3.1. Note that we have identified eight nodes. Next, we consider an 8 nodded square centered at the origin as in Figure 8.3.2. This square will function as a reference element. For notational simplicity we set the lower left and upper right vertices at $(-1, -1)$ and $(1, 1)$ respectively. Now we associate each edge of the annulus segment to a corresponding edge for the square. For the following discussion, we identify the edges by their nodes and associate the edges

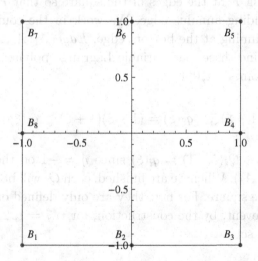

Figure 8.3.2.   Square with 8 nodes.

as follows:

$$Edge(A_1, A_2, A_3) \leftrightarrow Edge(B_1, B_2, B_3),$$
$$Edge(A_3, A_4, A_5) \leftrightarrow Edge(B_3, B_4, B_5),$$
$$Edge(A_5, A_6, A_7) \leftrightarrow Edge(B_5, B_6, B_7),$$
$$Edge(A_7, A_8, A_1) \leftrightarrow Edge(B_7, B_8, B_1).$$

The first task is to determine a reasonable approximation of the annulus segments. We do this by defining the three point polynomial interpolation of an edge in Figure 8.3.1 using the corresponding edge (line segment) in Figure 8.3.2. Since the edges in Figure 8.3.1 are lines and circular arcs, then we expect this degree 2 (three point) interpolation to provide a good approximation of each segment. Nevertheless, the interpolation process does not yield the original edges, and the resulting figure will not be an annulus segment. Note that for more complicated figures we can use a square with 12 nodes and cubic interpolating polynomials. We consider the cubic case below.

We let $P(\xi, \eta)$ denote a polynomial defined on the square. We start by determining P at the edges of the square so that P interpolates the corresponding annulus edge. We work in the counterclockwise direction beginning at the bottom edge, $Edge(A_1, A_2, A_3)$. The first step is to define three, one variable Lagrange polynomials that are dual to the points $-1, 0, 1$.

$$q_1(\xi) = -\frac{1}{2}(1-\xi)\xi, \quad q_2(\xi) = (1-\xi)(1+\xi), \quad q_3(\xi) = \frac{1}{2}(1+\xi)\xi.$$

In turn, we set $Q_i(\xi, -1) = q_i(\xi)$ since $\eta = -1$ on the lower edge, $Edge(A_1, A_2, A_3)$. When we are finished, each Q_i will be a polynomial defined on the square. For now they are only defined on the bottom edge. In any event, by the construction, for $i, j = 1, 2, 3$, $Q_i(B_j) = \delta_{i,j}$. Next, we set

$$P(\xi, -1) = Q_1(\xi, -1)A_1 + Q_2(\xi, -1)A_2 + Q_3(\xi, -1)A_3.$$

Since $P(\xi, -1) = p(\xi)$, this is exactly the polynomial interpolation of the edge by the interval $[-1, 1]$. Additionally, P maps -1 to A_1, 0 to A_2 and 1 to A_3.

Moving on to $Edge(A_3, A_4, A_5)$, we again use Lagrange polynomials to define

$$r_3(\eta) = -\frac{1}{2}(1 - \eta)\eta, \quad r_4(\eta) = (1 - \eta)(1 + \eta), \quad r_5(\eta) = \frac{1}{2}(1 + \eta)\eta$$

and set $Q_i(1, \eta) = r_i(\eta)$. This assignment assures us that for $i, j = 3, 4, 5$, $Q_i(B_j) = \delta_{i,j}$. As for the first edge, we set

$$P(1, \eta) = Q_3(1, \eta)A_3 + Q_4(1, \eta)A_4 + Q_5(1, \eta)A_5.$$

Note that Q_3 is repeated in the interpolation of both edges. Since $Q_3(1, -1) = 1$ in both cases, we have maintained consistency.

We complete the definition of P on the two remaining edges by setting $Q_5(\xi, 1) = q_3(\xi)$, $Q_6(\xi, 1) = q_2(\xi)$ and $Q_7(\xi, 1) = q_1(\xi)$ for the third edge and setting $Q_7(-1, \eta) = r_5(\eta)$, $Q_8(-1, \eta) = Q_4(\eta)$ and $Q_1(-1, \eta) = r_3(\eta)$ for the fourth. We then set

$$P(\xi, 1) = Q_5(\xi, 1)A_5 + Q_6(\xi, 1)A_6 + Q_7(\xi, 1)A_7,$$

$$P(1, \eta) = Q_7(-1, \eta)A_7 + Q_7(-1, \eta)A_7$$

$$+ Q_8(-1, \eta)(-1, \eta)A_8 + Q_1(-1, \eta)A_1.$$

Having defined P on the four edges, the next step is to extend each Q_i to a function of the entire square. We begin by deciding that we want Q_i to be dual to the nodes. In particular, $Q_i(B_j) = \delta_{i,j}$ for $i, j = 1, 2, \ldots, 8$. If we set $Q_2(\xi, \eta) = (1 - \xi)(1 + \xi)f(\eta)$ for some polynomial f, then $1 = Q_2(B_2) = Q_2(0, -1) = f(-1)$ and $0 = Q_2(B_6) = Q_2(0, 1) = f(1)$. Hence, we should take $f(\eta) = 1/2(1 - \eta)$. Using analogous reasoning for the other cases, we have the following assignments for Q_i, $i = 2, 4, 6, 8$:

$$Q_2(\xi, \eta) = \frac{1}{2}(1 - \eta)(1 - \xi)(1 + \xi), \tag{8.3.1}$$

$$Q_8(\xi, \eta) = \frac{1}{2}(1 - \xi)(1 - \eta)(1 + \eta), \tag{8.3.2}$$

$$Q_4(\xi, \eta) = \frac{1}{2}(1 + \xi)(1 - \eta)(1 + \eta), \qquad (8.3.3)$$

$$Q_6(\xi, \eta) = \frac{1}{2}(1 + \eta)(1 - \xi)(1 + \xi). \qquad (8.3.4)$$

It is now routine to verify that these functions satisfy $Q_i(B_j) = \delta_{i,j}$.

Turning to Q_1, we note that we require $Q_1(0, -1) = Q_1(1, -1) = 0$ and $Q_1(-1, -1) = 1$. If we take $Q_1(\xi, \eta)$ equal to $1/4(1 - \eta)\eta(1 - \xi)\xi$, then $Q_1(B_1) = 1$ and $Q_1(B_i) = 0$ for all other nodes on the square. But this yields a fourth degree polynomial while the other Q_i (that have been resolved to this point) are degree 3. On the other hand, $1/4(1 - \eta)(-\xi - \eta - 1)(1 - \xi)$ is degree 3 and also satisfies the requirement, $Q_1(B_i) = 0$, if $i \neq 1$. Hence, we make the (less intuitive) degree 3 selection.

Similarly, we make the remaining assignments.

$$Q_1(\xi, \eta) = \frac{1}{4}(1 - \eta)(-\xi - \eta - 1)(1 - \xi), \qquad (8.3.5)$$

$$Q_3(\xi, \eta) = \frac{1}{4}(1 - \eta)(\xi - \eta - 1)(1 + \xi), \qquad (8.3.6)$$

$$Q_5(\xi, \eta) = \frac{1}{4}(1 + \eta)(\xi + \eta - 1)(1 + \xi), \qquad (8.3.7)$$

$$Q_7(\xi, \eta) = \frac{1}{4}(1 + \eta)(-\xi + \eta - 1)(1 - \xi). \qquad (8.3.8)$$

Continuing the example, we plot the element produced by the interpolating functions (Figure 8.3.3). Visually this looks the same as the annulus segment. If we had overlaid the two images the error would be apparent.

We leave it as an exercise to prove that the function $P(\xi, \eta) = \sum_{i=1}^{8} Q_i(\xi, \eta) A_i$ maps the square bijectively to the interpolation of the annulus segment. We caution the reader that this is not always the case. It is possible that the image of two edges will intersect several times. Indeed, it is possible that the image of the rectangle will look like a figure 8.

We now define the basis functions for the element as $\varphi_i(x, y) = q_i P^{-1}(x, y)$, where the q_i form a polynomial model on the square.

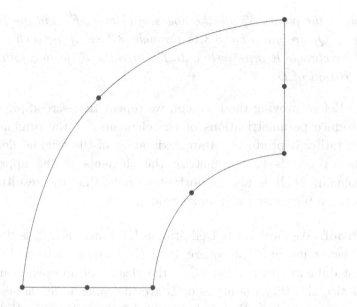

Figure 8.3.3.   Interpolation of the annulus segment.

This is just the analogous construction to the one used for reference elements. But, in this case P is not affine and the local basis functions are not polynomials.

We restate this for the general setting. Suppose that our domain D is partitioned into quadrilateral elements E_e with curved edges, γ_i^e, $i = 1, 2, 3, 4$. Suppose that each element has eight nodes, A_j^e, $j = 1, 2, \ldots, 8$. For each edge, γ_i^e, two of the nodes are endpoints and one node is interior to the edge. Define the eight functions Q_i using Eqs. (8.3.1)–(8.3.8) and set $P_e(\xi, \eta) = \sum_{i=1}^{8} Q_i(\xi, \eta) A_i^e$. If R denotes the eight nodal rectangle with lower left vertex at $(-1, -1)$ and upper right vertex $(1, 1)$, then set the image, $P_e(R) = F_E$ and set $\overline{D} = \cup_e F_e$. Further, let δ_i^e denote the second degree polynomial interpolation of γ_i^e based on the three nodes. With this notation and terminology we state formally,

Theorem 8.3.1. *Suppose that U is partitioned into eight nodal quadrilaterals with curved edges and define*

$$P_e(\xi, \eta) = \sum_{i=1}^{8} Q_i(\xi, \eta) A_i^e, \qquad (8.3.9)$$

where the points A_i^e are the nodes of element E_e and the eight func-
tions Q_i are given by (8.3.1) through (8.3.8). If for each e, P_e maps
the rectangle R injectively onto F_e, then the F_e form a finite element
partition of $\overline{D}$.

Before proving the theorem, we repeat that serendipity does not
produce parametrizations of the elements for the original domain
D, rather it produces an approximation of the original domain and
then proceeds to parametrize the elements of the approximated
domain, $\overline{D}$. It is also important to note that the resulting model
is not a piecewise polynomial model.

Proof. We must verify Definition 8.1.1. Since each F_e is the smooth,
bijective image of the square, then F_e is a quadrilateral. In turn it is
not difficult to show that $\overline{D}$ is the closure of an open bounded set.
Notice also that the nodes of $\overline{D}$ are the same as the nodes for U. It
is now immediate that Definition 8.1.1 items (i), (iii), (iv), and (v)
are all satisfied. Since the parametrization maps the interior to the
interior and boundary to boundary, then (ii) holds. Finally, we need
only verify that if two elements meet, and the intersection is not a
single point, then it is a side. Indeed, in this case they share a mid-
edge node. Now the result follows since they must share three nodes
along the intersection and in both cases the three common nodes
induce the same degree 2 Lagrange interpolation.   □

The assumption that each P_e is injective is essential. When using
the techniques of this section you must verify (with an automated
process using Jacobians) that these functions are indeed one to one.
The problem may arise when two edges of F_e intersect at a point that
is not an end point. You can minimize the possibility of this difficulty
by choosing the non-vertex nodes as close as possible to the center
of the edge as determined by arc length. In Exercise 3, we suggest
steps that lead to an example where P_e is not bijective.

If P_e is bijective, then define $\varphi_i^e(x, y) = q_i P_e^{-1}(x, y)$ on the element
F_e and extend φ_i^e to $\overline{D}$ so that the support of φ_i^e is contained in
the element. Since $\varphi_i^e(A_j^e) = \delta_{i,j}$, then it is immediate that these
eight local basis functions are linearly independent. As usual, we

designate their span as V_e^h. The following theorem states formally the properties of the FEM model for serendipity elements.

Theorem 8.3.2. *Given a set U contained in $\mathbb{R}^2$ with a decomposition into four sided, eight nodal elements and the interpolated domain $\overline{D}$ with elements, $F_e = P_e(R)$ derived from the eight serendipity polynomials Q_i. A polynomial model q_i defined on R and dual to the nodes induces a linearly independent set of functions $\varphi_i^e = q_i P_e^{-1}$ on each F_e. For a node p_k of $\overline{D}$, we define ψ_k as the piecewise join of the local functions φ_k^e with $\varphi_k^e(p_k) = 1$. The set, $B = \{\psi_k : p_k \in N\}$ where N denotes the set of nodes, is a set of linearly independent continuous functions that define an FEM model on $\overline{D}$.*

Proof. It remains to verify that the global basis functions are linearly independent and continuous. As stated, we associate each global basis function to a node p_k and define ψ_k piecewise from all φ_k^e with $\varphi_k^e(p_k)$ not zero. Keep in mind that for each e there is at most one φ_k^e is not zero at p_k. It is now immediate from Section 1.3 that B is linearly independent.

To verify the continuity of the basis functions ψ_k we consider a pair of adjacent elements F_e and $F_{e'}$. The question hinges on whether two components of ψ_k, φ_k^e, and $\varphi_k^{e'}$, are equal on the shared boundary of F_e and $F_{e'}$. If the intersection of two elements is a node, then both functions evaluate to 1 at the node. If the shared boundary is an edge, then since the two functions take the same values at the three nodes on the common edge and since the functions are derived from the second degree polynomial interpolation of the three points, then ϕ_k^e and $\varphi_k^{e'}$ are indeed equal on the edge. The result follows. $\square$

The functions ψ_k are sometimes referred to collectively as the *serendipity basis*.

The cubic serendipity model is a significantly more robust means of interpolating shapes. For this model, we will need 12 nodes with 4 on a side. The resulting interpolating Lagrange polynomials are degree 3. The four vertex polynomials are given by

$$Q_1(\xi,\eta) = \frac{1}{32}(1-\xi)(1-\eta)\left[9(\xi^2+\eta^2)-10\right], \quad (8.3.10)$$

$$Q_4(\xi,\eta) = \frac{1}{32}(1+\xi)(1-\eta)\left[9(\xi^2+\eta^2)-10\right], \quad (8.3.11)$$

$$Q_7(\xi,\eta) = \frac{1}{32}(1+\xi)(1+\eta)\left[9(\xi^2+\eta^2)-10\right], \quad (8.3.12)$$

$$Q_{10}(\xi,\eta) = \frac{1}{32}(1-\xi)(1+\eta)\left[9(\xi^2+\eta^2)-10\right]. \quad (8.3.13)$$

The four polynomials associated to the left and right edges are (respectively)

$$Q_{11}(\xi,\eta) = \frac{9}{32}(1-\xi)(1-\eta^2)(1+3\eta), \quad (8.3.14)$$

$$Q_{12}(\xi,\eta) = \frac{9}{32}(1-\xi)(1-\eta^2)(1-3\eta), \quad (8.3.15)$$

$$Q_5(\xi,\eta) = \frac{9}{32}(1+\xi)(1-\eta^2)(1+3\eta), \quad (8.3.16)$$

$$Q_6(\xi,\eta) = \frac{9}{32}(1+\xi)(1-\eta^2)(1-3\eta). \quad (8.3.17)$$

And finally, interchanging ξ and η yields the functions for the upper and lower edges.

$$Q_2(\xi,\eta) = \frac{9}{32}(1-\eta)(1-\xi^2)(1+3\xi), \quad (8.3.18)$$

$$Q_3(\xi,\eta) = \frac{9}{32}(1-\eta)(1-\xi^2)(1-3\xi), \quad (8.3.19)$$

$$Q_8(\xi,\eta) = \frac{9}{32}(1+\eta)(1-\xi^2)(1+3\xi), \quad (8.3.20)$$

$$Q_9(\xi,\eta) = \frac{9}{32}(1+\eta)(1-\xi^2)(1-3\xi). \quad (8.3.21)$$

The derivation of the 12 functions is based on the same insight as the eight nodal case. We begin with Lagrange polynomials dual to the nodes and interpolate the curved edges. These functions are subsequently extended to two variable polynomials with one variable constant. We then get vector valued functions as in the prior case. Finally, the same process is used to extend these basically one variable functions into true two variable polynomials. Only in this case, these details are more involved (see Heubner *et al.* (2001)).

We now state results corresponding to Theorems 8.3.1 and 8.3.2. As in the prior case, the proofs hinge on the following relations between opposite sides of the quadrilateral. These relations are used to bridge from a given element to adjacent elements.

$$Q_{10}(\xi,1)A_{10} + Q_9(\xi,1)A_9 + Q_8(\xi,1)A_8 + Q_7(\xi,1)A_7$$
$$= Q_1(\xi,-1)A_{10} + Q_2(\xi,-1)A_9 + Q_3(\xi,-1)A_8 + Q_4(\xi,-1)A_7,$$
$$(8.3.22)$$
$$Q_4(1,\eta)A_{10} + Q_5(1,\eta)A_9 + Q_6(1,\eta)A_8 + Q_7(1,\eta)A_7$$
$$= Q_1(-1,\eta)A_4 + Q_{12}(-1,\eta)A_5 + Q_{11}(-1,\eta)A_6 + Q_{10}(-1,\eta)A_7.$$
$$(8.3.23)$$

The notation is the same as in the previous theorems, the proofs are left as exercises for the reader.

Theorem 8.3.3. *Suppose that U is partitioned into twelve nodal elements with curved (smooth) edges and define*

$$P_e(\xi,\eta) = \sum_{i=1}^{12} Q_i(\xi,\eta)A_i^e, \qquad (8.3.24)$$

where the points A_i^e are the nodes of element E_e and the 12 functions Q_i are given by (8.3.6) through (8.6.11). If for each e, P_e maps R injectively onto F_e, then the F_e form a finite element partition of $\overline{D} = \cup_e F_e$.

Theorem 8.3.4. *Given a set U contained in $\mathbb{R}^2$ with a decomposition into four sided, 12 nodal elements and the interpolated domain $\overline{D}$ with elements, $F_e = P_e(R)$ where P_e is defined in (8.3.24). Then the functions $\varphi_i^e = q_i P_e^{-1}$ on F_e are extended to $\overline{D}$ so that they are supported by F_i and form a linearly independent set of functions defined*

on $\overline{D}$. For each node p_k, we set ψ_k equal to the function that is piecewise defined from the φ_i^e for all φ_i^e with $\varphi_i^e(p_i)$ not zero, then the set
$B = \{\psi_k : p_k \in N\}$ forms a set of linearly independent, continuous
functions defined on $\overline{D}$. Finally, we set V^h to the linear span of B.
Then, the elements and space V^h determine a C^0 model on $\overline{D}$.

Exercises

1. Referring to the quarter annulus, compute the Jacobian of the
   function P and prove using Mathematica that it has no roots
   inside the square. In this case the eight nodes are $A_1 = (-2, 0)$,
   $A_2 = (-1.5, 0)$, $A_3 = (-1, 0)$, $A_4 = (-\sqrt{2}/2, \sqrt{2}/2)$, $A_5 = (0, 1)$,
   $A_6 = (0, 1.5)$, $A_7 = (0, 2)$, $A_8 = (-\sqrt{2}), \sqrt{2}$. Conclude that P is
   injective.
2. Prove that for each point $A = (x, y)$ in the interpolated annulus,
   there is a unique ξ_0 in $[-1, 1]$, so that A lies on the curve $Q(\xi_0, \eta)$.
   Note that this is immediate for the annulus. It is also true for the
   interpolated figure. Conclude that P is surjective.
3. Let δ_1 and δ_2 be two adjacent sides of the eight nodal square
   centered at the origin. Denote the nodes by A_i, $1 \le i \le 8$. Show
   by example that it is possible for $P(\delta_1)$ and $P(\delta_2)$ to intersect at
   a point that is not an end point. (*Hint*: Take the edge node near
   to the common vertex.)
4. Verify (8.3.22) and (8.3.23).
5. Using an example, verify that the basis functions ψ_i defined in
   Theorem 8.3.2 are not necessarily differentiable.
6. Prove Theorem 8.3.3.
7. Prove Theorem 8.3.4.

8.4. The Linear System

In this section, we focus on the development the linear system of
equations that arises from the Galerkin weak form finite element
method, GFEM.

Continuing the notation from the previous sections, U denotes a
connected and open set in $\mathbb{R}^n$, D is its closure and Γ is the boundary.

Since we are concerned with FEM, then we may assume that D is bounded, hence compact. We also suppose that U has a finite element partition with elements denoted E_e. With the partition, we have the FEM model including local and global basis functions. We denote the global function space by V^h and local spaces by V_e^h. The latter are spaces of polynomial functions whose support is contained in the compact element E_e. The superscript h was defined in Section 8.1. It is the mesh parameter associated to the partition. Again, h plays no specific role at this time. Rather, it serves to remind the reader of the limit process supporting finite element method.

We fix a basis $\{\varphi_{i,\alpha}^e : F_{i,\alpha} \text{ is a DoF}\}$ for each V_e^h. The dimension of V_e^h is the number of local degrees of freedom. In turn, the piecewise polynomial functions $\psi_{i,\alpha}$ form a basis for V^h. These functions are formed by joining polynomials $\varphi_{i,\alpha}^e$ for each element that shares the degree of freedom $F_{i,\alpha}$. The dimension of V^h is the number of degrees of freedom for the model. We will see below that it is also the size of the GFEM linear system. This connection between the size of the linear system and the number of degrees of freedom is the source for the terminology. Finally, we remark that if the associated node is interior to D then $\psi_{i,\alpha}$ has compact support.

We consider a linear second-order partial differential equation as an affine mapping of the space of twice differentiable functions taking values in $L^2(U)$, the space of squared Lebesgue integrable functions on U. In particular, if L is a linear transformation with values in $L^2(U)$ and T_f denotes the translation of $L^2(U)$ by a function f, then the composition $T_f \circ L$ yields the affine operator. For instance, in the case of the Laplace equation, $L = \nabla^2$, $f = 0$ and the PDE is $\nabla^2 u = 0$. For a general element u of the domain of $T_f \circ L$, we call $T_f \circ L(u) = L(u) + f$ the *residual*.

Since elements of $L^2(U)$ are only known up to a set of measure zero and since the boundary of U is a set of measure zero, then it makes no sense to talk about boundary values as $g = u|_\Gamma$. Rather, we suppose that there is a linear operator $T : L^2(U) \to L^2(\Gamma)$ with $T(u) = g$. We see in Section 9.3 that such a T exists provided the boundary is Lipschitz and in that case, we may choose g so that $g(x) = u(x)$ for x in Γ. In particular, the PDE has *boundary value zero* means that u lies in the kernel of T. We also see later in this section and more formally in Chapter 9 that if $T_f \circ L$ is a PDE with nonzero boundary value, then there is a function h so that $T_h \circ L$ has boundary value zero.

We know that $L^2(U)$ is a Hilbert space with inner product σ. For f and g in $L^2(U)$, we write $\sigma(f,g) = \int_D fg$. The phrase orthogonal elements of $L^2(U)$ always refers to this inner product. Further, we say that u is a weak solution to $T_f \circ L$ with respect to V^h if the residual is orthogonal to V^h. Equivalently,

$$0 = \int_U T_f \circ L(u)\psi_{i,\alpha}$$

$$= \sum_e \int_{E_e} L(u)\psi_{i,\alpha} + \int_U f\psi_{i,\alpha}$$

$$= \sum_e \int_{U_e} L(u)\varphi^e_{i,\alpha} + \int_U f\psi_{i,\alpha}, \qquad (8.4.1)$$

since each $\psi_{i,\alpha}$ restricted to an element, is a local basis function, $\varphi^e_{i,\alpha}$ associated to that degree of freedom.

Using multi-indices, we may write a second-order derivative as ∂^γ where $\sum_i \gamma_i = 2$ or as $\partial^\alpha \partial^\beta$ with $|\alpha| = |\beta| = 1$. If L involves a second-order derivative denoted $\partial^\alpha \partial^\beta$, then we make the following assertion for any u with compact support in U:

$$\sum_e \int_{E_e} \partial^\alpha \partial^\beta(u)\varphi^e_{i,\gamma} = -\sum_e \int_{E_e} \partial^\beta u \partial^\alpha \varphi^e_{i,\gamma}. \qquad (8.4.2)$$

This identity in fact arises from integration by parts with the additional conditions that u vanishes on Γ and that the integrals along interior element edges sum to zero. We see (8.4.2) again in Chapter 9 when we define Sobolev spaces.

The condition that integrals along interior edges sum to zero is natural in this context. If u is in V^h, then it is a linear combination of piecewise polynomial functions $\psi_{i,\gamma}$ which in turn is the join of polynomial functions $\varphi^e_{i,\gamma}$ for the same degree of freedom but distinct adjacent elements. For instance, if two adjacent elements are E_e and

$E_{e'}$ with common edge $\Gamma_{e,e'}$, then the left-hand side of (8.4.2) will have a summand

$$\int_{E_e} \partial^\alpha \partial^\beta (\varphi_{i,\gamma}^e) \varphi_{j,\delta}^e = -\int_{E_e} \partial^\beta (\varphi_{i,\gamma}^e) \partial^\alpha (\varphi_{j,\delta}^e) + \int_{E_e} \partial^\alpha \left[\partial^\beta (\varphi_{i,\gamma}^e) \varphi_{j,\delta}^e \right]$$

$$= -\int_{E_e} \partial^\beta (\varphi_{i,\gamma}^e) \partial^\alpha (\varphi_{j,\delta}^e) + \int_{\Gamma_e} \overline{\partial^\beta (\varphi_{i,\gamma}^e)} \varphi_{j,\delta}^e,$$

by Green's Theorem. Moreover, the integral over the edge $\Gamma_{e,e'}$ is a summand of the last term on the right. On the other hand, the left hand side of (8.4.2) also has a corresponding summand for $E_{e'}$. Since the components of $\psi_{i,\gamma}$ are equal on common edges and edge traversal is oriented according to the element interior (see Figure 8.4.1) then

$$\int_{\Gamma_{e,e'}} \overline{\partial^\beta \varphi_{i,\gamma}^{e'}} \varphi_{j,\delta}^{e'} = -\int_{\Gamma_{e,e'}} \overline{\partial^\beta \varphi_{i,\gamma}^e} \varphi_{j,\delta}^e.$$

Equation (8.4.2) now follows.

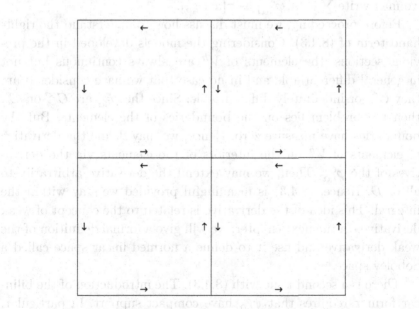

Figure 8.4.1. Four adjacent elements with positive traversal.

Going forward, we suppose that our PDE is not more than second order. Hence, using (8.4.2) we may write (8.4.1) using only first-order derivatives. Further, it is now convenient to change notation to reflect the fact that second-order derivatives have been reduced to two first-order applied separately to each factor of the integrand. In particular, the linear operator of the first argument in (8.4.1) is now a bilinear function of the two arguments. For instance $\int_U (Lu)\varphi = \int_U (\partial^{(2,0)} u)\varphi$ gives rise to $\tau[u, \varphi] = -\int_U \partial^{(1,0)} u \partial^{(1,0)} \varphi$. We write the modified (8.4.1) using τ to represent the bilinear form involving differential operators up to first order.

$$
0 = \int_U T_f L(u)\psi_{i,\alpha} = \sum_e \int_{E_e} L(u)\varphi_{i,\alpha}^e + \int_U f\psi_{i,\alpha}
$$

$$
= \sum_e \tau[u, \varphi_{i,\alpha}^e] + \int_U f\psi_{i,\alpha} = \tau[u, \psi_{i,\alpha}] + \int_U f\psi_{i,\alpha},
$$

$$(8.4.3)$$

for each basis function $\psi_{i,\alpha}$ with compact support in U. Note that we may write $\sum_e \tau[u, \varphi_{i,\alpha}^e]$ as $\tau[u, \psi_{i,\alpha}]$.

Before proceeding, we must discuss how to understand the right-hand term of (8.4.3). Considering the models developed in the previous sections, the elements of V^h are always continuous but not in general differentiable and in no case that we have considered are they C^∞ or indefinitely differentiable. Since the $\varphi_{i,\alpha}^e$ are C^∞ on E_e, then the problem lies on the boundaries of the elements. But the boundaries have measure zero. Hence, we may define the derivative of elements of V^h on the interiors of the elements via the derivatives of the $\varphi_{i,\alpha}^e$. Then, we may extend the derivative arbitrarily to all of D. Hence, (8.4.3) is meaningful provided we stay within the integral. This idea of the derivative is related to the concept of weak derivative. In the next chapter, we will give a formal definition of the weak derivative and use it to define a normed linear space called a Sobolev space.

There is a second issue with (8.4.3). The introduction of the bilinear form τ requires that $\psi_{i,\alpha}$ have compact support. In particular, $\psi_{i,\alpha}$ may not associate to a boundary node. Alternatively, $\psi_{i,\alpha}$ does not associate to a boundary node if the PDE has zero boundary condition. We have already noted that this assumption does not decrease generality.

To be more precise, if we permitted piecewise polynomial functions associated to boundary nodes, then the right-hand side of (8.4.3) would have an extra term that we can resolve as follows.

$$\tau\left[u,\psi_{i,\alpha}\right] + \int_U f\psi_{i,\alpha} + \sum_e \int_{\Gamma_e} \overline{\partial^\beta(\varphi_{i,\gamma}^e)}\varphi_{i,\alpha}^e = \tau\left[u,\psi_{i,\alpha}\right] + G(\psi_{i,\alpha}),$$

$$(8.4.4)$$

where G is a linear functional of V_h. With a boundedness assumption that we introduce in Chapter 9, the Reisz representation theorem implies there is a g with $G(\psi_{i,\alpha}) = \int_U g\psi_{i,\alpha}$ [Royden (2011)]. We merely replace f with g and proceed.

Now continuing with (8.4.3) for any u in V^h, we write $u = \sum_{j,\beta} a_{j,\beta}\psi_{j,\beta}$, and substitute into (8.4.3) to get

$$0 = \tau\left[u,\psi_{i\alpha}\right] + \int_U f\psi_{i,\alpha} = \sum_{j,\beta} a_{j,\beta}\tau\left[\psi_{j,\beta},\psi_{i,\alpha}\right] + \int_U f\psi_{i,\alpha}, \quad (8.4.5)$$

for each basis function $\psi_{i,\alpha}$. Note that (8.4.5) is a linear system of equations. Moreover, with the comment on zero boundary values, the system is $n \times n$ where n is the number of degrees of freedom in the model.

Definition 8.4.1. Let $T_f \circ L$ be a linear second-order PDE defined on D with an FEM partition and let τ denote the related bilinear form as determined above. Then, a solution to Eq. (8.4.5) is called a *weak form* solution to the PDE with respect to V^h. Alternatively, it is also referred to as the *Galerkin weak form* solution.

Now, we expand (8.4.5) on the elements using the linearity of τ and the integral to get

$$0 = \sum_{j,\beta} a_{j,\beta}\left(\sum_{e\in S_{i,\alpha}} \tau\left[\varphi_{j,b}^e, \varphi_{i,\alpha}^e\right]\right) + \sum_{i,\alpha}\sum_{e\in S_{i,\alpha}}\int_{U_e} f\varphi_{i,\alpha}^e, \quad (8.4.6)$$

where we have separated out the constituent components of the piece-wise polynomial functions and used $S_{i,\alpha}$ to designate the set of all e such that E_e supports $\psi_{i,\alpha}$. If, in turn we use (8.4.6) to write a separate system for each element we have

$$0 = \sum_{j,\beta} b_{j,\beta}^e \left(\tau \left[\varphi_{j,\beta}^e, \varphi_{i,\alpha}^e \right] \right) + \int_{U_e} f \varphi_{i,\alpha}^e, \qquad (8.4.7)$$

where the coefficient of the linear system in (8.4.6) are the sums over the elements of the corresponding coefficients of (8.4.7). The point here is that it is not easy to setup the piecewise polynomials. Indeed, even the number of components is indeterminant. Hence, it is not easy to compute the linear system (8.4.5) directly. But (8.4.7) provides an alternate approach. We need only compute the linear systems (8.4.7) and then accumulate the coefficients to form the system (8.4.5). In matrix form, the two systems are $K^e b = s^e$ and $Ka = s$.

The system $Ka = s$ is the restatement to system (8.4.5) in matrix format. It is the *global system*. When the global system is solved, then we have the weak GFEM solution to the PDE. The system $K^e b = s^e$ is the matrix form of (8.4.7) and is called the *local system*. The solution to this system is not of interest to us. Its sole purpose is to build the global system. The accumulation process has been described in Section 3.1. The entries of the local system depend on the particular PDE. There are seven specific examples in Chapters 3 and 6.

Before ending this section, we discuss interior nodes and a node removal process called *condensation*. There are situations where interior nodes arise naturally. For instance, if we wanted to use general two variable, degree 3 polynomials for the model, then we would need 10 nodes to have a full model. This would necessitate a quadrilateral with eight edge/vertex nodes and two interior nodes (Figure 8.4.2) or a triangle with nine edge/vertex nodes and one interior node at the centroid (see Figure 8.2.3).

We now describe the process. To start, look back at (8.4.7) which we wrote in matrix form, $K^e a = R^e$. Each coordinate of a associates to a global degree of freedom. We reorder the coordinates of a,

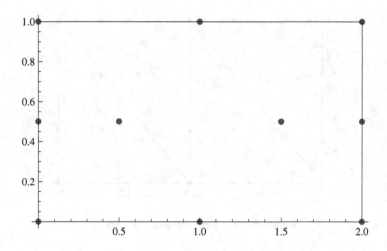

Figure 8.4.2. Rectangle with 10 nodes.

separating the variable vector a into coordinates that refer to vertex/ edge nodes, a_1, and an interior node part, a_2. We write this as (ignoring the element identifier in the superscript)

$$\begin{pmatrix} A_{11} & A_{12} \\ A_{21} & A_{22} \end{pmatrix} \begin{pmatrix} a_1 \\ a_2 \end{pmatrix} = \begin{pmatrix} s_1 \\ s_2 \end{pmatrix}. \qquad (8.4.8)$$

Both A_{11} and A_{22} are square matrices. If A_{22} is non-singular, then we can solve (8.4.8) for a_2 in terms of a_1 and substitute. In particular, (8.4.8) yields $A_{21}a_1 + A_{22}a_2 = s_2$. Hence, $a_2 = A_{22}^{-1}(s_2 - A_{21}a_1)$. We then insert this expression into $A_{11}a_1 + A_{12}a_2 = s_1$ to get

$$\left(A_{11} - A_{12}A_{22}^{-1}A_{21} \right) a_1 = s_1 - A_{12}A_{22}^{-1}s_2.$$

The idea behind condensation is that interior nodes may be useful for the initial definition of local basis functions. But as they do not play any role in continuity questions associated to V^h, they may be safely removed.

Theorem 8.4.1. *Consider a linear system in $n + m$ unknowns as given in (8.4.8), $a_1 \in \mathbb{R}^n$. If A_{22} is non-singular, then a_1 satisfies matrix relation $(A_{11} - A_{12}A_{22}^{-1}A_{21})a_1 = s_1 - A_{12}A_{22}^{-1}s_2$. In turn, $a_2 = (A_{22}^{-1})(s_2 + A_{21}a_1)$.*

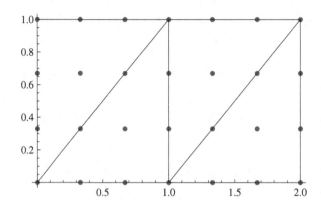

Figure 8.4.3.   Four 10 nodal triangles.

Condensation is done at the element level. When a variable is removed, then the node is removed from the model thereby decreasing the degrees of freedom and then the size of the linear system (8.4.5). If there are 1,000 elements and 1,000 interior nodes that may be removed, then we have decreased the size of the linear system by 1,000. This can make an observable difference in processing time.

Exercises

1. Figure 8.4.3 shows a configuration of four triangles and each triangle has 10 nodes. Model the Laplace equation on the four element domain using two variable cubics. For each element determine whether the matrix A_{22} is non-singular (see Theorem 8.4.1).
2. Consider a 1D model with first-order Lagrange polynomials and five elements. Write out the assembly for the global linear system. In this case, there are two degrees of freedom for each element.
3. Repeat Exercise 3 for Hermit polynomials. Now there are four degrees of freedom for each element.

8.5. Boundary Values, Neumann, Dirichlet, and Robin

In this section, we will determine the relationship between Neumann and Dirichlet boundary values. We begin by restricting attention to the steady-state case. Our results on boundary values do have extensions to transient PDE, however, the results depend on the particular time stepping implementation. We will consider a single transient case later in the section.

We start wtih the matrix formulation of Eq. (8.4.4), $Aa = R + F$, where the matrices are derived from a linear ordering of the degrees of freedom. In turn, the separate equations are written

$$r_{i,\alpha} + s_{i,\alpha} = \sum_{j,\beta} \delta_{i,\alpha;j,\beta} a_{j,\beta}, \qquad (8.5.1)$$

where each $r_{i,\alpha}$ is accumulated from local boundary integrals

$$\int_{\Gamma_e} \frac{\partial u}{\partial n} \varphi_{i,\alpha}^e. \qquad (8.5.2)$$

In the last section, we determined that the accumulated value of $r_{i,\alpha}$ is zero at each interior node. Hence, the only possible non-zero values are at boundary degrees of freedom. Since we do not know u, then we cannot know its normal derivative along the boundary of U. In the physical setting, these normal derivatives at the boundary often represent external forces on the system. As such, they should be measurable, the resulting functions inserted into (8.5.2), the integral computed and accumulated to yield $r_{i,\alpha}$.

Recall the channel flow problem of Section 3.1. In this case, we set the force at the inflow and outflow edges. These forces drive the flow through the channel. Along the channel boundaries, we set the force to zero. This prevents the fluid from penetrating these boundaries. Equivalently, these boundary values keep the flow within the channel.

Besides setting the normal derivative of u, we could also have set a value to the integral in (8.5.2) or to the accumulated $r_{i,\alpha}$. These are all equivalent choices. Indeed, depending on the setting, this latter alternative may be justified. No matter the case, when we give a value to $r_{i,\alpha}$, we say that we are setting a *Neumann boundary value*. Alternatively, these are sometimes called *natural boundary values*. It is common to designate these degrees of freedom by Γ_N.

There are times when we just want to have a particular value $a_{i,\alpha}$. Looking at the standing wave problem of Section 3.2, we selected a location and designated a particular wave height at the point. We accomplish this at the i, α^{th} degree of freedom by setting row (i, α) of A to the i, α^{th} row of the identity matrix and setting $r_{i,\alpha} + s_{i,\alpha} = \omega$, the desired value. Hence, (8.5.1) in this case becomes $a_{i,\alpha} = \omega$. When we do this we say we are setting a *Dirichlet boundary value*. We denote the set of these boundary degrees of freedom by Γ_D.

Since you cannot do the two processes to the same entry $r_{i,\alpha}$, then $\Gamma_D \cap \Gamma_N = \emptyset$. Since we must set some value to each boundary $r_{i,\alpha}$ then $\Gamma_D \cup \Gamma_N$ is the set of boundary degrees of freedom. Eventhough setting a Dirichlet boundary value seems to be a non-mathematical process, it can be resolved within the mathematical development. In fact, the two processes are equivalent. We will prove this in Theorem 8.5.1.

There is one particular point here that needs to be mentioned. First, the matrix A must be non-singular. Without this assertion, we do not have a well-posed problem at the discrete level (see Definition 7.1.1). Recall the Laplace equation, $\nabla^2 u = 0$. The function $u = 0$ is an obvious solution. Indeed, any degree 2, order 1 polynomial is a solution. Hence, the problem is inherently ill-posed. In Section 3.1, we resolved this by setting a stagnation point; that is, we set a Dirichlet value of zero to a point on the channel obstruction. On the one hand, this is a process that has physical validity. On the other hand, changing a row of A to an identity row is enough to make it non-singular. Since this was necessary, then Dirichlet boundary values are sometimes called *essential boundary values*.

Suppose that there is more than one degree of freedom at a given boundary node. Suppose further that one is the value of the function at the location and the others are values of the derivative there. If we were to set values for the function and its derivatives then we say that we are setting a *Robin condition* or *boundary value* at the node. Recall the Gaussian collocation treatment of the Black–Scholes equation in Section 5.2. The model used degree 3 Hermite polynomials. The two polynomials at the left-hand end point model the value and derivative of the solution. For the problem, we needed only two boundary values and we chose Dirichlet values at the two end points. Alternatively, we could have a set function value and the derivative at the left-hand end point. In that case, we would say we have set Robin boundary values at the left end point.

Recall that we first encountered this process of setting Dirichlet boundary values when doing FDM. In fact, FDM boundary values are nearly always Dirichlet. Nevertheless, it is still possible to set Neumann conditions (Thomas, 1995). On the other hand, Neumann conditions are an artifact of the weak form. Therefore, they are most often encountered in FEM. Collocation method with a polynomial model that includes DoF with derivatives allows for setting derivaitve values. This is analogous to Neumann boundary values.

Next, we consider the relationship between Dirichlet and Neumann boundary conditions.

Equation (8.5.1) is the discrete form of the differential equation, the mathematical representation of an equilibrium. Whether we set Neumann or Dirichlet boundary values, we must maintain the equilibrium. We have noted that the $r_{i,\alpha}$ often represent a force at the boundary that drives the system, while $a_{i,\alpha}$ is a value of the function that results from the external force.

At the level of the physical process or continuous mathematical model, we know that the process (function) and the external drivers (forces) are equivalent; $r_{i,\alpha}$ determines $a_{i,\alpha}$ and vice versa. For instance, in the case of a fluid flow, the $r_{i,\alpha}$ represent the external force driving the flow, while the $a_{i,\alpha}$ represent the flow potential. Since the velocity vector is derived from the potential function, then it is reasonable to suppose that the forces and the velocities are interchangeable. However, these continuous model statements do not directly apply to the discrete FEM format.

On the discrete side of the requirement, maintaining the equilibrium means that we must maintain equivalent linear systems. In other words, if (j, β) is a boundary degree of freedom, then we can choose to set $r_{j,\beta}$ or designate a value for $a_{j,\beta}$ provided the resulting linear systems are equivalent. Hence, we must determine the exact relationship.

First, we may suppose that we have a well-posed problem; that is, A is non-singular. Next, we rewrite the linear system as $R = Aa$, where the entries of R are $r_{i,\alpha} + s_{i,\alpha}$. We set a Neumann condition by providing a value for $r_{i,\alpha}$ or equivalently a value for the normal derivative in (8.5.2). We set a Dirichlet condition $\omega = a_{i,\alpha}$ by determining a system $R' = Ba$ where $r'_{i,\alpha} = \omega$ and the coefficient matrix is altered by setting j, βth row to the corresponding row of the identity matrix or equivalently the j, βth standard basis vector $e_{j,\beta}$. Furthermore, no other changes are made.

By non-singularity, each $e_{i,\alpha}$ is a linear combination of the rows of A. In particular, we write

$$e_{j,\beta} = \sum_{k,\gamma} b_{k,\gamma} A_{(k,\gamma)} = \sum_{k,\gamma \neq j,\beta} b_{k,\gamma} A_{(k,\gamma)} + b_{j,\beta} A_{(j,\beta)}, \qquad (8.5.3)$$

where $A_{(k,\gamma)}$ designates the row of A associated to the DoF k, γ. If $b_{j,\beta} \neq 0$, then we can recast (8.5.3) in terms of elementary row operations.

$$B = \left[\prod_{k,\gamma \neq j,\beta} E_{b_{k,\gamma}(k,\gamma)+j,\beta} \right] E_{b_{j,\beta}(j,\beta)} A = EA, \qquad (8.5.4)$$

where $E_{b_{j,\beta}(j,\beta)}$ indicates multiplying row (j, β) of A by $b_{j,\beta}$ and $E_{b_{k,\gamma}(k,\gamma)+j,\beta}$ is the operation of adding $b_{k,\gamma}$ times row k, γ to row j, β. It is immediate that for each $k, \gamma \neq j, \beta$, $B_{(k,\gamma)} = A_{(k,\gamma)}$, $B_{(j,\beta)} = e_{j,\beta}$ and $Ba = ER = R'$ is equivalent to $Aa = R$. Now, we are free to set $r_{j,\beta}$ as we choose so that the j, βth entry of ER is ω. That choice of $r_{j,\beta}$ is a Neumann condition that results in ω as the corresponding Dirichlet condition. Since E is non-singular, then the process is reversible; given a linear system with Dirichlet conditions set, the we can derive the corresponding Neumann values.

The question remains, what if $b_{j,\beta} = 0$? In this case, we must employ a row interchange operation to implement the Dirichlet boundary value. The linear system $Aa = r$ solves for the $a_{j,\beta}$ as coefficients in the FEM solution $\sum_{i,\beta} a_{i,\beta} \psi_{i,\beta}$. By interchanging rows, we are only changing the order in which we solve for the coefficients. Next, we can continue the procedure as necessary provided the total number of nonzero entries in the rows of A associated to DoF in Γ_N is no smaller than the number of DoF in Γ_D. But in this case, A is singular. Therefore, A non-singular implies that we can set

Dirichlet boundary values as necessary and maintain an equivalent linear system.

In particular, we have proved the first statement of the following result (Loustau and Bob-Egbe, 2010). Note that the theorem is stated as a linear algebra result. Hence, we may use a simpler notation to describe the method.

Theorem 8.5.1. *Let $Aa = R$ be a non-singular linear system, then the following hold.*

(i) *For any j there is an equivalent linear system with the value of a_j prescribed by the Dirichlet process.*

(ii) *For any given R there is a non-singular linear transformation E such that the Dirichlet value at a_j is given by $(ER)_{(j)}$.*

(iii) *If b is the vector with entries b_i, as in (8.5.3), then b is the jth row of A^{-1}.*

Proof. The first assertion was proved prior to the statement of the theorem. For the second, we need only set E to the product of elementary matrices used in (8.5.4).

It remains to prove the last statement. We write $A^{(i)}$ for the ith column of A and write the transpose of A as A^T. Then,

$$e_j = \sum_k b_k A_{(k)} = \left(\sum_k (A^T)^{(k)} b_k \right)^T.$$

If we denote the entries of A^T by $\hat{\delta}_{j,k}$, then the jth coordinate of the vector $\sum_k (A^T)^{(k)} b_k$ is $\sum_k \hat{\delta}_{j,k} b_k$. But this is the jth entry of $A^T b$. Therefore, $e_j = (A^T b)^T$. Solving for b, we have $b = (A^T)^{-1}(e_j)^T$. Since post multiplying any matrix by e_j^T merely extracts the jth column, then $(A^T)^{-1}(e_j)^T$ is the jth row of A^{-1}. $\square$

By part (iii) of the theorem, we need only interchange rows before implementing a Dirichlet boundary values when A^{-1} has zero entries on the diagonal. In Exercise 2, we lead the reader through a proof that if there is no first-order term in the PDE, then A^{-1} has no zero entries on the diagonal. Finally, note that the Dirichlet conditions corresponding to a particular Neumann setting are linear combinations of the Neumann values.

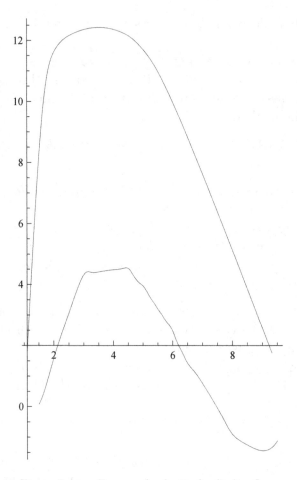

Figure 8.5.1.   Force and velocity for Stokes flow.

We consider a particular case encountered in Section 6.3, Stokes flow for a partially obstructed channel. The channel with element partition is shown in Figure 6.3.1. We previously executed the problem using an asymmetric inflow force. Figure 8.5.1 shows the inflow force plotted with the inflow velocity on the same graph. The upper curve is the force, the lower the x-coordinate of the velocity vector. Note that the force drops off to the right indicating negative velocities (back flow).

The situation for the transient case is more complicated. First of all, it is dependent on the particular FDM technique. For this discussion, we use the simple forward Euler. (For other time stepping procedures, see Loustau and Bob-Egbe (2010)). Suppose that the process starts from a steady state and then ramps up to a subsequent steady state. Since the driver of the system is the integrand in (8.5.2), then we should implement a time dependent function at certain boundary nodes. This function will provide the bridge between the states. In the literature, this is called a *start up*. For instance $\varphi(t) = 1 - e^{(-t/\lambda)}$, $\lambda > 0$, simulates a burst. Alternatively, we can simplify the mathematics and apply the start up function directly to the entries of R.

The transient version of (8.5.1) is $M \partial a(t) / \partial t = Aa(t) + R(t)$. We expand the right-hand side using the forward Euler formulation, $M(a^{n+1} - a^n) = \Delta t(Aa^n + R^n)$ and setup time stepping,

$$Ma^{n+1} = (M + \Delta t A)a^n + \Delta t R^n. \qquad (8.5.5)$$

We designate a Neumann boundary value determined by a start up function φ applied to node p_j at time t_n by setting $R_j^n = \varphi_j^n = \varphi(t_n, p_j)$. Now, M is the matrix representation of the positive definite L^2 inner product defined on V^h. Therefore, it is a positive definite symmetric matrix and the same is true of M^{-1}. In particular, M has no zeros on the diagonal. Hence, for each j, there is a non-singular N with $NM_{(j)} = e_j$, where N is the product of elementary matrices given in (8.5.4) as applied to M. In turn, the corresponding Dirichlet boundary value is given by the jth entry of $N(M + \Delta t A)a^n + N \Delta t R^n$. Since N is non-singular, then the process is reversible. Furthermore, repeating the process at several nodes determines a function $\psi(t_n, x_j) = \psi_j^n$ of Dirichlet boundary values, which corresponds to φ_j^n. It is apparent that the relationship is not simple. Nevertheless, we can write it down.

Theorem 8.5.2. *Suppose that a transient process is resolved spatially by GFEM and temporally by explicit FDM. Let φ be a start up function applied at a set of boundary degrees of freedom $\Gamma = \{F_{j,\beta} : p_j$ is a boundary node, β is a multi-index\}, then there exist non-singular* *linear transformations $N_{j,\beta}$, such that*

$$\prod_{j,\beta} N_{j,\beta} M a^{n+1} = \prod_{j,\beta} N_{j,\beta}(M + \Delta t A)a^n + \prod_{j,\beta} N_{j,\beta}\Delta t R^n \quad (8.5.6)$$

is equivalent to (8.5.5) with Dirichlet boundary conditions set at the nodes in Γ. The product $\prod_{j,\beta} N_{j,\beta} = N$ is dependent of the order in which the boundary degrees of freedom are listed in Γ, For each $n \geq 1$, the vector valued function $(\psi_{j,\beta})^n = N(M + \Delta t A)a^n + N\Delta t R^n$ is the corresponding Dirichlet start up function.

Proof. The proof of Theorem 8.5.2 is immediate from Equation (8.5.5), Theorem 8.5.1 and the discussion preceding the statement of the theorem. We note in passing that each $N_{j,\beta}$ is dependent on the order of nodes in Γ, whereas N is not. These matrices, as products of elementary matrices, are given by Theorem 8.5.1 applied to $\prod_{j,\beta} N_{j,\beta} M$. $\qquad\square$

Exercises

1. For the obstructed channel flow problem developed in Section 3.1, the boundary values at the inflow edge were set as Neumann boundary values. Use Theorem 8.5.1 to plot the corresponding Dirichlet boundary values of the flow potential at the inflow edge nodes.

2. Prove the assertion following Theorem 8.5.1. If the PDE has the form $\nabla^2 u + \lambda u = f$, then

   (i) the form τ (see Section 8.4) is a positive definite inner product.

   (ii) The symmetric matrix A is the matrix representation of τ.

   (iii) The eigenvalues of A are all positive.

   (iv) The symmetric matrix A^{-1} determines a positive definite inner product.

   (v) A^{-1} has no zeros on the diagonal.

3. Suppose a transient process is resolved via implicit time stepping. Derive the relationship between a start up function φ applied

as a Neumann boundary value function and the corresponding Dirichlet start up function.

4. Repeat Exercise 2 for Crank–Nicolson time stepping.

5. Consider the transient case of the obstructed channel flow problem of Section 3.1. The associated PDE is $\nabla^2 u = \partial u / \partial t$, where u is the flow potential.

  (i) Develop the GFEM model for the transient flow.

  (ii) Consider the burst start up function with $\lambda = 0.05$. Plot the values of ψ_j^n for each inflow node p_j. Do the plot for to time steps in the interval $[0.01, 0.1]$.

Chapter 9

Finite Element Method, the Theory

Introduction

In this chapter, we prove the convergence for Galerkin FEM, GFEM for a large class of elliptical PDE. Recall that elliptical PDE includes transient parabolic equations as the spatial part is elliptical and the temporal part is resolved with FDM or Runge–Kutta. Our convergence result will prove that given a PDE with solution u, then the GFEM solution u_h converges to u as $h \to 0$. We will see that in many respects, the road to the theorem is as interesting as the theorem itself.

The material in this chapter will bring together several ideas we have introduced but not yet developed. We began by formalizing the idea of a differential equation as an operator defined on a function space. At that time, we did not describe the domain space except to say that it should support the differential operators included in the PDE. Further, the Hilbert space L^2 arose in this context. We also considered several cases of a finite element decomposition and then developed sets of basis functions that are supported by element patches, sets of elements that share a given node. There we saw that higher levels of continuity and differentiability required more and more complicated models. In turn, these more complicated models would require more and more complicated programming and longer and longer execution times. On the other hand in Section 8.4, we also saw that the weak or Galerkin formulation allowed for slightly lower levels of differentiability than were present in the underlying differential equation. At that time, we alluded to the idea of weak differentiability. In this direction, we expected the global basis functions, defined as continuous piecewise polynomial functions, to be weakly differentiable.

In this chapter, we take this a step further. We formally develop the idea of a weak partial derivative. In particular, our C^0 basis functions from Section 8.2 are now weakly differentiable, and C^1 functions have two weak derivatives. Further, we will show that the space of weakly differentiable functions up to a given order form a complete normed linear space with a positive definite inner product, a Hilbert space. This space envelopes the spaces, V^h. And if we think of the differential operators occurring in the PDE as weak derivatives, then the Hilbert space becomes the natural domain for the PDE. This means that it contains both the FEM solutions and the actual solution. Hence, it is the space that supports the convergence theory. In the standard terminology, this space is called a Sobolev space.

The initial sections form the major steps toward the convergence theorem. First, we introduce the weak derivative and use it to define the Sobolev space, H. We see in subsequent sections that the Sobolev space contains both u, the actual solution and the GFEM approximations u_h. In this context, we will see that boundary values (Dirichlet or Neumann) may be stated in terms of linear mappings of the space H. Since H is a complete metric space, we might expect that we would proceed by showing u_h is Cauchy in h and then conclude convergence to u as well as the existence of u (see Section 6.5). Rather, the existence step precedes convergence. This step is concluded with the Lax–Milgram theorem. The convergence step relies on the theory of piecewise polynomial interpolation. Since you cannot interpolate a function without knowing that it exists, then convergence of GFEM without prior knowledge of the existence of a solution is impossible.

There are some general comments to be made about the following material. We have not stated the results at the level of generality as is usual in the literature. Rather, it is our intention to maintain proximity to the engineering/physics side of the topic. In other words, we maintain proximity to the examples presented in Part 1. In addition, our aim is not to present the mathematics for its own sake. Rather, we focus on what is necessary to reach our goal.

The context for the material in this chapter is the normed linear spaces L^1 and L^2, with norms defined via the Lebesgue integral. In this setting, functions are actually equivalence classes defined via the relation: f equivalent to g provided $f = g$ almost everywhere. As the material is presented in the literature, this fact is mostly ignored in the notation and usually in statements of the theorems. The resolution of this issue is the subject of the Sobolev embedding theorems. We include these results in Section 9.5.

The first three sections provide an introduction to Sobolev spaces. In Section 9.4, we prove the Lax–Milgram existence theorem. The steps toward the convergence theorem include the theory of piecewise polynomial interpolation (Section 9.6). This provides the major step toward our goal. We state and prove the existence theorem in Section 9.7.

9.1. The Weak Derivative

The motivation for the weak derivative lies with the several examples of FEM basis functions introduced in Chapter 8. Away from the element boundaries, these functions are well-known polynomial functions. But across element boundaries, they are often no more than continuous or once differentiable. However, the union of all element boundaries is only a set of measure zero. Therefore, as long as the functions occur within an integrand then we can largely ignore the problem area. Furthermore, we have seen that these piecewise polynomial functions do support integration by parts. We will see that the minimal requirement for a weakly differentiable function is that it should support a restricted form of integration by parts (the product rule with Green's theorem). Stated otherwise, it should be tailor made for FEM.

The second issue is generality. We want to make the definition of weak derivative general enough to apply to all the cases encountered in Chapter 8 as well as related cases not specifically mentioned there. Hence, we do not mention FEM.

Recall that we introduced the idea of *multi-index* in Section 8.2 as a compact means to denote partial differentiation operators and then used them to identify degrees of freedom. In particular, the multi-index $\alpha = (\alpha_1, \ldots, \alpha_n)$ is a finite sequence of non-negative integers so that if φ is defined on an open set U in $\mathbb{R}^n$ has at least $|\alpha| = \sum_{i=1}^{n} \alpha_i$ derivatives, we may write

$$\partial^\alpha \varphi = \frac{\partial^{\alpha_1}}{\partial x_1^{\alpha_1}} \frac{\partial^{\alpha_2}}{\partial x_2^{\alpha_2}} \cdots \frac{\partial^{\alpha_n}}{\partial x_n^{\alpha_n}} \varphi.$$

As before the *order* or *length* of α is the sum of the components of α. It is written $|\alpha|$.

Continuing with our usual setting, $\overline{U}$ denotes the compact set that is the closure of a connected, bounded open set and Γ denotes the boundary of U. We will suppose that Γ is piecewise smooth. Further, for O open we write $C_0^\infty(O)$ for the set of functions on O that are indefinitely differentiable with compact support. We have already used D to denote the closure of U. Now we want to use it for the weak gradient. There should be no confusion.

Definition 9.1.1. Let O be an open set contained in $\overline{U}$ and take real valued u, v in $L^1(O)$. We call v the α-*weak partial derivative* relative to O for u provided for any φ in $C_0^\infty(O)$,

$$\int_O u \partial^\alpha \varphi = (-1)^{|\alpha|} \int_O v\varphi. \qquad (9.1.1)$$

In this case, we write $D^\alpha u = v$. In addition, we write $Du = \sum_{|\alpha|=1} D^\alpha u$ and call this the *weak gradient* of u.

Our piecewise polynomial functions are easily seen to be absolutely integrable elements of L^1. We begin the presentation of the weak derivative with an elementary but necessary result. It follows from basic measure theory theorems that a function that is a.e. non-negative on a set of positive measure has positive integral.

Theorem 9.1.1. *If u has an α-weak partial derivative on O, then the weak derivative of u is unique up to a set of measure zero.*

Proof. If v and w are α-weak partial derivatives of u, then Eq. (9.1.1) implies that for any φ in $C_0^\infty(O)$, $\int_O (v-w)\varphi = 0$ If v is not almost everywhere equal to w, then we may select φ so that $(v-w)\varphi$ is non-negative and not zero on O. The result then follows. $\qquad\square$

The following result contains the information that the weak derivative generalizes the ordinary derivative. Note that we need continuous derivatives to ensure that the derivatives are integrable.

Corollary 9.1.1. *Let O be given as in the theorem and α be is a multi-index. If ψ is $C^{|\alpha|}$ on O, then D^α exists and is equal to ∂^α.*

Proof. Since $\partial^\alpha = \partial^\beta \partial^\gamma$, $|\beta| = 1$ and $|\gamma| = |\alpha - 1|$, then employing a standard induction argument, it suffices to prove the result for the case $|\alpha| = 1$. Noting that $\psi\varphi \in C_0^1$ on O, then $\int_O \psi\partial^\alpha \varphi = \int_O \partial^\alpha(\psi\varphi) - \int_O \partial^\alpha(\psi)\varphi$. Hence, the proof reduces to showing that $\int_O \partial^\alpha(\psi\varphi) = 0$. Since Γ is piecewise smooth, Green's theorem applies (Rudin, 1976). In particular, $\int_O \partial^\alpha(\psi\varphi) = \int_\Gamma(\psi\varphi) = 0$. $\qquad\square$

As an example of a function with a weak derivative but no derivative in the ordinary sense, we look at the first degree Lagrange polynomials, the basis functions that arise from a finite element problem in 1D. For notational convenience, we take $[-1,0]$ and $[0,1]$ as a pair if contiguous elements and consider the piecewise polynomial basis function ψ associated to the origin. Hence, $\psi(x) = 1 + x$ on $[-1,0]$, $\psi(x) = 1 - x$ on $[0,1]$ and $\psi(x) = 0$ otherwise. We claim that ψ is weakly differentiable on $O = (-1,1)$. To demonstrate this, we should expect to define v by means of the derivative of each component of ψ. Therefore, we set $v(x) = 1$, $x \in [-1,0)$, $v(x) = -1$, $x \in (0,1]$, $v(x) = 0$ otherwise. The reader will see that it does not matter how we define $v(0)$. Now, for any indefinitely differentiable φ with compact support, we must verify the definition of weak derivative. In particular, we separate the integral on segments and employ integration by parts to get

$$\int_O^1 \psi \frac{d\varphi}{dx} = \int_{-1}^0 \psi \frac{d\varphi}{dx} = \int_{-1}^0 (1+x)\frac{d\varphi}{dx} + \int_0^1 (1-x)\frac{d\varphi}{dx}$$

$$= \int_{-1}^0 \frac{d}{dx}[(1+x)\varphi] + \int_0^1 \frac{d}{dx}[(1-x)\varphi] - \int_{-1}^0 \varphi - \int_0^1 -\varphi$$

$$= \varphi(0) - \varphi(0) - \int_{-1}^0 \varphi + \int_0^1 \varphi = -\int_{-1}^0 \varphi + \int_0^1 \varphi,$$

since $\varphi(1) = \varphi(-1) = 0$. In turn,

$$\int_O v\varphi = \int_{-1}^0 v\varphi + \int_0^1 v\varphi = -\int_0^1 \varphi + \int_{-1}^0 \varphi.$$

Therefore, $\int_O v\varphi = -\int_O \psi\varphi'$. We conclude that ψ is weakly differentiable.

On the other hand, the jump discontinuity in v prevents it from being weakly differentiable. From the corollary, the weak derivative, if it exists should be $w(x) = 0$. Now, $\int_O w\varphi = 0$, whereas, $\int_O v d\varphi/dx = \int_{-1}^0 d\varphi/dx - \int_0^1 d\varphi/dx = \varphi(0) - \varphi(-1) - \varphi(1) + \varphi(0)$, for any indefinitely differentiable function φ with compact support in O. Therefore, any choice of φ with support contained in $(-1,1)$ and with $\varphi(0)$ not 0 will suffice to prove that v is not weakly differentiable.

We now prove a result which connects the FEM model to weak differentiability. We see that C^0 basis functions are weakly differentiable and C^1 basis functions are twice weakly differentiable.

Theorem 9.1.2. *Let $U \subset \mathbb{R}^n$ be a domain supporting a finite element partition with elements denoted E_e. Suppose that U is partitioned into intervals for $n = 1$, rectangular or triangular elements for $n = 2$ and hexahedron or tetrahedron for $n = 3$. If there exists a* *function ψ such that ψ is continuous on U and a polynomial function when restricted to each E_e, then ψ has first-order weak partial derivatives on U.*

Proof. We prove the result for rectangular or triangular elements in $\mathbb{R}^2$. The remaining cases are proved similarly.

It is necessary to demonstrate Definition 9.1.1 for any φ with support contained in the interior of $\cup_e E_e$. We need only consider differentiation with respect to any one of the independent variables, as the other case is similar. Letting $\partial/\partial x$ represent differentiation with respect to the first independent variable, then

$$\int_U \psi \frac{\partial}{\partial x}\varphi = \sum_e \int_{U_e} \psi \frac{\partial}{\partial x}\varphi = \sum_e \int_{U_e} \psi_e \frac{\partial}{\partial x}\varphi, \qquad (9.1.2)$$

where ψ_e is the polynomial which equals ψ when restricted to U_e. Now the results of ordinary calculus apply to each summand $\int_{U_e} \psi_e(\partial\varphi/\partial x)$. Hence,

$$\sum_e \int_{U_e} \psi_e \frac{\partial}{\partial x}\varphi = \sum_e \int_{U_e} \frac{\partial}{\partial x}(\psi_e\varphi) - \sum_e \int_{U_e} \varphi\left(\frac{\partial}{\partial x}\psi_e\right)$$

$$= \sum_e \int_{U_e} \frac{\partial}{\partial x}(\psi_e\varphi) - \int_U \varphi\left(\frac{\partial}{\partial x}\psi\right).$$

Therefore, it suffices to show that the first term on the right-hand side is zero. By Green's theorem,

$$\int_{U_e} \frac{\partial}{\partial x}(\psi_e\varphi) = \int_{\Gamma_e} (\psi_e\varphi)dn_{\Gamma_e},$$

where n_{Γ_e} denotes the outward pointing normal to the boundary of U_e.

We consider separately the segments of Γ_e. If a given segment is part of the boundary of U, then φ is zero there. Hence, the contribution to $\int_{\Gamma_e} \psi_e\varphi dn_{\Gamma_e}$ from the segment is zero. Otherwise, the segment is interior to U. In this case, the integral along this segment occurs twice in (9.1.2). And the two occurrences are in opposite directions. Suppose that the second occurrence arises from $\int_{U_{e'}} \partial/\partial x(\psi_{e'}\varphi)$.

Since ψ is continuous on U, then the functions $\psi_e\varphi$ and $\psi_{e'}\varphi$ are equal on $\Gamma_e \cap \Gamma_{e'} = \Lambda$. In particular, $\int_\Lambda (\psi_e\varphi)dn_{\Gamma_e} = -\int_\Lambda (\psi_{e'}\varphi dn_{\Gamma_{e'}}$. Therefore, the summation of the boundary integrals is zero as needed. $\square$

We end this section with some useful elementary properties of the weak derivative.

Theorem 9.1.3. *Let k be a positive integer. If u and v have weak derivatives of order α for $|\alpha| \le k$ with respect to an open set O, then*

(i) $D^\beta(D^\alpha u) = D^\alpha(D^\beta u) = D^{\alpha+\beta}(u)$, *provided* $|\alpha| + |\beta| \le k$.
(ii) *for real scalars a, b, $au+bv$ has weak derivatives of order $|\alpha| \le k$ and $D^\alpha(au + bv) = aD^\alpha u + bD^\alpha v$,*
(iii) *if O' is an open subset of O, then u has weak derivatives of order α with respect to O',*
(iv) *if $\xi \in C_0^\infty(O)$, $|\alpha| = 1$, then $D^\alpha(\xi u) = (D^\alpha\xi)u + \xi(D^\alpha u)$.*

Proof. The proof is left as an exercise. $\square$

We note that by (ii), the weak derivative is a linear operator. Item (iv) is the first-order product rule. In Exercise 2, we state the full product rule for weak differentiation.

Exercises

1. Prove Theorem 9.1.3.
2. Prove the following generalization of Theorem 9.1.3(iv). With the notation of Theorem 9.1.3:

$$D^\alpha(\xi u) = \sum_{|\beta|\le|\alpha|} C_\beta^\alpha D^\beta(\xi)D^{\alpha-\beta}(u),$$

   where C_β^α denotes the binomial coefficient.
3. Let u be a function on U with values in $\mathbb{R}$. We say that u is *differentiable a.e.* if u has a derivative in the usual sense on W contained in U where $U = W \cup X$ where X is measure zero. In

particular, for any x in W there exists $d \in \mathbb{R}^n$ with

$$\lim_{y \to x} \frac{|u(y) - u(x) - d(y - x)|}{\|y - x\|} = 0,$$

with d equal to the weak gradient, $(D^{\alpha_1}, \ldots, D^{\alpha_n})$, where $\alpha_1 = (1, 0, \ldots, 0)$, and so forth. Prove that the basis functions defined in Chapter 8 are differentiable a.e. Note that this concept is consistent with the definition of the derivative of a function of several variables (Rudin, 1976).

9.2. Sobolev Spaces

We now introduce the Sobolev space. The definition generally given in the literature is stated in terms of L^p spaces. Maintaining proximity to the GFEM, we restrict attention to the case for $p = 2$. Early in the section, we see that the Sobolev space is the natural domain for the PDE when considering GFEM. Indeed, we see that it is a Hilbert space that contains both the FEM approximate solution and the actual solution. Therefore, it is also the natural context for a convergence theorem.

Definition 9.2.1. Let U be an open set in $\mathbb{R}^n$, then the *Sobolev space* of order k on U is the set $H^k(U) = \{u \in L^1(U) : D^\alpha u \in L^2(U), for \ |\alpha| \leq k\}$.

In the more general setting, Sobolev spaces are denoted $W^{k,p}$ with the provision that $W^{k,2} = H^k$.

An immediate consequence of Theorem 8.1.4(ii) is that every H^k is a linear space. If we define

$$\sigma_{H^k}(u, v) = \sum_{|\alpha| \leq k} \int_U (D^\alpha u D^\alpha v),$$

then σ_H^k is a positive definite symmetric bilinear form on H^k. In turn, H^k is a normed linear space with norm induced from the inner product. We will see in the next theorem that it is a Hilbert space. The inner product is called the *Sobolev inner product* and the corresponding norm is the *Sobolev norm*.

Several comments are in order.

A. As we have already done, when U is otherwise known, then we will drop the reference and write H^k.

B. The case $|\alpha| = 0$ is included, H^0 is $L^1 \cap L^2$ with L^2 norm.

C. We are most interested in the cases $k = 1$, 2, and 3.

D. If for instance, we are looking at functions defined on $\mathbb{R}^2$ then the H^1 norm of u is given by

$$\left[\int_U u^2 + \int_U \left(D^{(1,0)} u \right)^2 + \left(D^{(0,1)} u \right)^2 \right]^{\frac{1}{2}} .$$

The first summand is necessary since otherwise constant functions would have zero norm.

E. For the case $n = 2$ and u, v in $H^1(U)$. The symmetric bilinear form

$$(u, v) = \int_U \left(D^{(1,0)} u \right) \left(D^{(1,0)} v \right) + \left(D^{(0,1)} u \right) \left(D^{(0,1)} v \right),$$

is positive semi-definite. The related form

$$|u|_{H^1} = \left[\int_U \left(D^{(1,0)} u \right)^2 + \left(D^{(0,1)} u \right)^2 \right]^{\frac{1}{2}}$$

is called a *semi-norm*. We will see this semi-norm in Section 9.6.

F. We will prove in Theorem 9.2.5 that integration by parts holds in Sobolev spaces.

G. We prove (Theorem 9.2.2) that the Sobolev space associated to piecewise defined functions determined by an FEM partition satisfies the following. $u \in H^k$, $k \geq 1$ if and only if $u \in C^{k-1}$ and has all weak derivatives of degree k. Hence, the inclusion of weakly differentiable functions only adds one order of differentiability.

H. As a consequence of G, every C^{k-1} model V^h is as subspace of H^k.

I. Given a PDE together with a C^{k-1} finite element model, then the spaces V^h are subspaces of $H^k(U)$. Hence, for each h, the GFEM solution u^h lies in $H^k(U)$. If the GFEM solutions converge to the actual solution, u, then by completeness (Theorem 9.2.1), u must also belong to $H^k(U)$.

We now prove the first theorem.

Theorem 9.2.1. *For any positive integer k, H^k is a Hilbert space.*

Proof. It suffices to prove that H^k is complete. Let $\{u_m : m = 1, 2, \ldots\}$ be a Cauchy sequence in H^k. Then for any $\epsilon > 0$, there is a positive integer N such that if $m, p \geq N$, then

$$\sigma_{H^k}(u_m - u_p, u_m - u_p) = \sum_{|\alpha| \leq k} \int_U (D^\alpha(u_m - u_p))^2 < \epsilon^2. \qquad (9.2.1)$$

In particular, for each α, $\int_U (D^\alpha(u_m - u_p))^2 < \epsilon^2$. Therefore, each sequence $\{D^\alpha u_m : m = 1, 2, \ldots\}$ is a Cauchy in L^2. Hence, $D^\alpha u_m$ converges to an element u_α in L^2. We claim that $u_\alpha = D^\alpha u$, where u is the L^2 limit of $u = D^0 u_m$. We verify for any indefinitely differentiable φ with compact support,

$$\int_U (D^\alpha u)\varphi = (-1)^{|\alpha|} \int_U u D^\alpha \varphi = (-1)^{|\alpha|} \lim_m \int_U u_m D^\alpha \varphi$$

$$= \lim_m (-1)^{2|\alpha|} \int_U (D^\alpha u_m)\varphi = \int_U u_\alpha \varphi, \qquad (9.2.2)$$

using convergence properties of the Lebesgue integral (Royden, 2011).

We set $\epsilon > 0$ and take q to be the number of multi-indices of length no greater than k. We calculate for each α and m sufficiently large, $\|D^\alpha u_m - D^\alpha u\|_2 = \|D^\alpha u_m - u_\alpha\|_2 < \epsilon/q$, where the subscript on the norm refers to L^2 space. Hence, $\|u_m - u\|_{H^k} < \epsilon$. $\square$

We actually have proved more than what is in the theorem. We state this result as a corollary.

Corollary 9.2.1. *If $\{u_m : m = 1, 2, \ldots\}$ is a sequence in H^k. Then the sequence converges to u in H^k if and only if for every α, $|\alpha| \leq k$, $D^\alpha u_m$ converges to $D^\alpha u$ in L^2.*

The following result provides an alternate characterization for the elements of a Sobolev space. This is done in terms of the underlying FEM context developed in Chapter 8. Recall Theorem 9.1.2, given a domain U and a finite element partition with a C^0 (respectively C^1) model, then the piecewise polynimial basis functions belong to $H^1(U)$ (respectively $H^2(U)$). We now consider the following converse.

Theorem 9.2.2. *Suppose that U is a domain partitioned into finite elements. If u belongs to $H^k(U)$ and every $C^k(E_e)$, then $u \in C^{k-1}(U)$.*

Proof. It suffices to prove the result for $k = 1$. In this case, we suppose that there is an element u in $H^1(U)$ which is C^1 on each E_e and discontinuous on U. Hence, the discontinuities must occur at the element boundaries. In particular, there must be a point $a \in \Gamma_e \cap \Gamma_{e'}$ with $\lim_{y \to a} u(y) \neq \lim_{z \to a} u(z)$ where $y \in E_e$ and $z \in E_{e'}$. Note that we may suppose that a is not an element vertex. We denote $u|_{E_e}$ by u_e and define $u_{e'}$ similarly. Further, take a neighborhood V of a, $V \subset E_e \cup E_{e'}$ and φ indefinitely differentiable with compact support contained in V. In addition, we may suppose that φ is non-negative (see for instance, the function η in Figure 9.2.1). Using standard results of calculus, we calculate, for $|\alpha| = 1$,

$$\int_U (D^\alpha u)\varphi = \int_{E_e} (\partial^\alpha u_e)\varphi + \int_{E_{e'}} (\partial^\alpha u_{e'})\varphi$$

$$= \int_{\Gamma_e} u_e \varphi dn - \int_{E_e} u_e(\partial^\alpha \varphi) + \int_{\Gamma_{e'}} u_{e'} \varphi dn' - \int_{E_{e'}} u_{e'}(\partial^\alpha \varphi),$$

where n and n' designate the outward pointing normals. Alternatively, as u is weakly differentiable,

$$\int_U (D^\alpha u)\varphi = -\int_U u(D^\alpha \varphi) = -\int_{E_e} u_e(D^\alpha \varphi) - \int_{E_{e'}} u_{e'}(D^\alpha \varphi).$$

Therefore, if $\Lambda = V \cap \Gamma_{E_e} \cap \Gamma_{E_{e'}}$, then

$$\int_\Lambda u_e \varphi dn = -\int_\Lambda u_{e'} \varphi dn'.$$

Since φ is positive on Λ and the normals from either side point in opposite directions, then u_e must equal $u_{e'}$ on Λ. This contradiction completes the proof. $\qquad\qquad\qquad\qquad\qquad\qquad\qquad\qquad\qquad\qquad\square$

Even though the weak derivative is a natural construct within this context, there are times when working with weakly differentiable functions is inconvenient. In Theorems 9.2.4 and 9.2.5, we show that weakly differentiable functions may be approximated by C^∞ functions. These results use the idea of mollification. We digress to introduce this procedure.

For any x in the open neighborhood of points with $\|x\| < 1$, we set $\eta(x) = C Exp[1/(\|x\|^2 - 1)]$ and extend η to be zero for each $\|x\| > 1$. The constant C is chosen so that $\int_{\mathbb{R}^n} \eta = 1$. It is easy to see that η is C^∞ since $\lim_{\|x\|\to 1} D^\alpha \eta = 0$ for each α. The function η has a bell-shaped graph. In Figure 9.2.1, we show the plot of η for $n = 1$.

Next, we set $\epsilon > 0$ and define the *standard mollifier*, $\eta_e(x) = (1/\epsilon^n)\eta(x/\epsilon)$. It is immediate that η_ϵ is C^∞, $\int_{\mathbb{R}^2} \eta_\epsilon = 1$ with support contained in the open neighborhood $N_\epsilon(0)$. In the literature, η and η_ϵ are also referred to as *cutoff functions*.

Given a locally integrable, real valued function f on U, we define the *mollification* f^e of f via the convolution filter of f with the standard mollifier. In particular, $f^\epsilon(x) = (\eta_\epsilon * f)(x) = \int_U \eta_\epsilon(x - y)f(y)dy = \int_{N_\epsilon(0)} \eta_\epsilon(y)f(s - y)dy$, where x lies in $U_\epsilon = \{x \in U :$

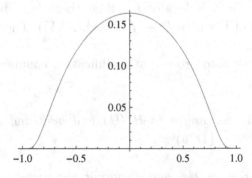

Figure 9.2.1. $\eta(x) = Exp\left(\frac{1}{\|x\|^2 - 1}\right)$ on $\mathbb{R}^1$.

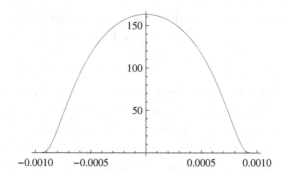

Figure 9.2.2.   Standard Mollifier η_ϵ, $\epsilon = 10^{-3}$.

$\|x - \gamma\| > \epsilon$ for every γ in Γ}. Usually, we say that U_ϵ is the
set of points of U with distance from Γ greater than ϵ. The following are the standard properties of f^e. For detailed proofs see
Evans (1998).

M(i) f^ϵ is indefinitely differentiable on U_ϵ,
M(ii) f^ϵ converges to f almost everywhere as $\epsilon \to 0$,
M(iii) if f is continuous on U, then the convergence in (ii) is uniform
on compact subsets of U,
M(iv) if f is locally L^2, then the convergence identified in (ii) is
locally L^2.

In the current setting, a proper open subset V of U has compact
closure. Hence, if f is locally L^2, then there is a finite cover V_i,
$i = 1, 2, \ldots, m$ of V where $f^\epsilon \to f$ in each $L^2(V_i)$. Therefore, $f^\epsilon \to f$
in $L^2(V)$ also.

Our first task is to prove that mollification commutes with weak
differentiation.

Lemma 9.2.1. *For any u in $H^k(U)$ and multi-index α, $|\alpha| \leq k$,*
$D^\alpha u^\epsilon = \eta_\epsilon * D^\alpha u = (D^\alpha u)^\epsilon$.

Proof. Taking x in U_e, and denoting the (weak) derivative of
$\eta_\epsilon(x - y)$ with respect to $x = (x_1, \ldots, x_n)$ as D_x^α, we compute

(using properties of the Lebesgue integral and properties of the weak derivative),

$$D^\alpha u^\epsilon(x) = D^\alpha \int_U \eta_\epsilon(x-y)u(y)dy$$

$$= \int_U D^\alpha\left(\eta_\epsilon(x-y)u(y)\right)dy$$

$$= \sum_{|\beta|\leq|\alpha|} \binom{\alpha}{\beta} \int_U D_x^\beta(\eta_\epsilon(x-y))D_x^{\alpha-\beta}(u(y))dy$$

$$= \int_U D_x^\alpha(\eta_\epsilon(x-y))u(y)dy$$

$$= (-1)^{|\alpha|} \int_U D_y^\alpha(\eta_\epsilon(x-y))u(y)dy, \qquad (9.2.3)$$

using the product rule, the fact that $u(y)$ is independent of x and finally using $D_x^\alpha \eta_\epsilon(x-y) = (-1)^{|\alpha|}D_y^\alpha \eta_\epsilon(x-y)$. Since the standard mollifier is C_0^∞, then we may apply weak differentiation to the right-hand expression in (9.2.3) to yield

$$D^\alpha u^\epsilon(x) = (-1)^{2|\alpha|} \int_U \eta_\epsilon(x-y)D^\alpha u(x) = (\eta_\epsilon * D^\alpha u)(x),$$

as was needed. □

We now prove that inside U, the mollifiers converge in H^k to the underlying function.

Theorem 9.2.3. *Suppose that u belongs to $H^k(U)$ and that V is an open set with closure contained in U. Then u^ϵ converges to u in $H^k(V)$ as $\epsilon \to 0$.*

Proof. Take V open with closure contained in U. By the comment following $M(iv)$, for any α, $|\alpha| \leq k$, $(D^\alpha u)^\epsilon \to D^\alpha u$, as $\epsilon \to 0$ with respect to the $L^2(V)$. By the lemma, $(D^\alpha u)^\epsilon = D^\alpha u^\epsilon$. Therefore, $u^\epsilon \to u$ in $H^k(V)$. □

In the next theorem we prove that the C^∞ functions in the Sobolev space $H^k(U)$ form a dense subset. We see later that C_0^∞ is not dense in H^k. Toward our immediate end, we use smooth partitions of unity (see Rudin (1986)).

Let U be an open set. Suppose that $\{V_i : i = 1, 2, \ldots\}$ is a locally finite family of open sets with $\cup_i V_i = U$. A family of functions $\xi_i : V_i \to \mathbb{R}$ forms a *smooth partition of unity* subordinate to the V_i provided

PU(i) for each i, $\xi_i \in C_0^\infty(V_i)$,
PU(ii) for any x in V_i, $\xi_i(x) \in [0, 1]$,
PU(iii) $\sum_i \xi_i = 1$.

By *locally finite*, we mean that for any x in U, there is a neighborhood of x that meets only finitely many V_i. Hence, the summation $\sum_i \xi_i(x)$ is finite. We can now prove

Theorem 9.2.4. $C^\infty(U) \cap H^k(U)$ *is dense in* $H^k(U)$.

Proof. For each positive integer i, let U_i be the points in U whose distance from Γ is greater then $1/i$. Hence, $U = \cup_i U_i$. We set $V_0 = U_2$, V_i equal to the complement of $\overline{U}_i$ relative to U_{i+3}. Next, we determine a smooth partition of unity $\{\xi_i : i = 1, 2, \ldots\}$ subordinate to the V_i. By Exercise 2 in Section 9.1, for any u in $H^k(U)$ and any i, $\xi_i u$ is also in $H^k(U)$. Moreover, $\sum_i \xi_i u = u$.

The next step is to mollify the weakly differentiable functions $\xi_i u$. We set $u^i = \eta_{\epsilon_i} * (\xi_i u)$ where the parameters $\epsilon_i > 0$ are to be determined. By Theorem 9.2.3, $u^i \to \xi_i u$ in $H^k(V_i)$. Since the support of $\xi_i u$ is inside V_i, then u^i is zero in U outside of $\overline{V}_i$. Hence, the convergence is in $H^k(U)$. Therefore, given $\delta > 0$ we may select the $\epsilon_i > 0$ with

$$\|u^i - \xi_i u\|_{H^k(U)} \leq \frac{\delta}{2^{i+1}}.$$

And the support of u^i is contained in $\overline{U}_{i+3} \subset U_{i+4}$.

Now set $v = \sum_i u^i$. For any open V with $\overline{V}$ contained in U, there is an m with $\overline{V}$ contained in U_m. Hence, v restricted to V, is actually a finite sum $\sum_{i=1}^n u^i$. Hence, as a finite sum of functions in $C^\infty(V)$,

v is in $C^\infty(V)$. In turn, since the V_i cover U, we have $v \in C^\infty(U)$. Now, we compute

$$\|v - u\|_{H^k(V)} = \left\| \sum_i u^i - \sum_i \xi_i u \right\|_{H^k(V)} \leq \sum_i \|u^i - \xi_i u\|_{H^k(V)}$$

$$\leq \sum_i \|u^i - \xi_i u\|_{H^k(U)} \leq \delta \sum_i \frac{1}{2^{i+1}} = \delta.$$

Since this inequality holds for every V, it follows that $\|v - u\|_{H^k(U)} \leq \delta$ and the proof is complete. $\qquad\qquad\qquad\qquad\qquad\qquad\qquad\qquad$ $\square$

The next theorem shows that $C^\infty(D) \cap H^k(U)$ is dense in $H^k(U)$ provided Γ is Lipschitz. We introduced Lipschitz boundary in Section 8.1 and noted then the piecewise continuously differentiable polyhedral boundaries are included. Indeed, for the cases that interest us, the boundary is formed from points, line segments or planar patches.

We state the result now without proof. Even with the simplification to the expected boundaries the proof is long and uses techniques similar to those already encountered in Theorem 9.2.4.

Theorem 9.2.5. *If Γ is a Lipschitz boundary, then the $C^\infty(D) \cap H^k(U)$ is dense in $H^k(U)$.*

Proof. See Evans (1998). $\qquad\qquad\qquad\qquad\qquad\qquad\qquad\qquad\qquad\qquad$ $\square$

Our final theorem in this section is important not only for its content but also the proof technique. The result is well known as Green's theorem. It holds for H^1 functions since these functions may be expressed as the limit of C^∞ functions. Indeed, it is a standard proof technique for Sobolev spaces to prove a result for C^∞ functions and then extend to result to the Sobolev space by appealing to one of the prior two theorems. In the literature, the argument is often referred to as the *density proof*.

Theorem 9.2.6. *(Integration by Parts) Suppose that U is a domain with Lipschitz boundary, then for u, v in $H^k(U)$, continuous on D*

and $|\alpha| = 1$,

$$\int_U (D^\alpha u)v = \int_\Gamma uv - \int_U uD^\alpha v.$$

Proof. First note that $H^k \subset H^1$. Hence, we may take u and v in $H^1(U)$. We begin by applying Theorem 9.2.5. Take u in $H^1(U)$ and let u_n be a sequence of elements of $C^\infty(D)$ which converges to u. Define v_n similarly. Since the theorem holds for pairs u_n and v_n, it suffices to prove that

$$\int_U (D^\alpha u_n)v_n$$

$$\rightarrow \int_U (D^\alpha u)v, \quad \int_U u_n(D^\alpha v_n) \rightarrow \int_U u(D^\alpha v), \quad \int_\Gamma u_n v_n \rightarrow \int_\Gamma uv.$$

We prove the first now, leave the second as an exercise. The third is a consequence of Theorem 9.3.1 from the next section. We hold this verification to after the proof of Theorem 9.3.1. We compute

$$\left| \int_U (D^\alpha u_n)v_n - \int_U (D^\alpha u)v \right|$$

$$= \left| \int_U (D^\alpha u_n)v_n - \int_U (D^\alpha u_n)v + \int_U (D^\alpha u_n)v - \int_U (D^\alpha u)v \right|$$

$$\leq \left| \int_U (D^\alpha u_n)v_n - \int_U (D^\alpha u_n)v \right| + \left| \int_U (D^\alpha u_n)v - \int_U (D^\alpha u)v \right|$$

$$\leq \|u_n\|_{H^1(U)}\|v_n - v\|_{L^2(U)} + \|u_n - u\|_{H^1(U)}\|v\|_{L^2(U)},$$

by Cauchy–Schwartz applied to $L^2(U)$ and the inequality $\|Dv\|_{L^2(U)} \leq \|v\|_{H^1(U)}$.

Since u_n is an H^1 convergent sequence, then $\|u_n\|_{H^1(U)}$ is bounded. Also since the $L^2(U)$ norm is no larger than the $H^1(U)$ norm, it follows that this last expression converges to zero.

The proof is complete pending the exercise and the deferred step in the next section. $\qquad\square$

Exercises

1. For the proof of Theorem 9.2.4, Prove that M(iii) implies that for any α, $|\alpha| \leq k$, $(D^\alpha u)^\epsilon \rightarrow D^\alpha u$, as $\epsilon \rightarrow 0$ with respect to the $L^2(V)$ for any V open in U.
2. With the notation of Theorem 9.2.4, prove that the support of each u^i is contained in U_{i+4}.
3. In the proof of theorem 9.2.7, verify the second assertion,

$$\int_U u_n D^\alpha v \rightarrow \int_U u D^\alpha v.$$

4. Prove the following converse to Theorem 9.2.2. Suppose that U has a finite element partition with a C^k finite element piecewise polynomial model, V^h. Prove that V^h is a subspace of H^{k+1}.

9.3. Boundary Values

In Section 8.4, we determined that there are logical problems with stating the boundary values for a PDE in weak form. An element of a Sobolev space actually is not a function but an equivalence class of functions where the equivalence relation is $f \doteq g$ provided $f = g$ a.e. Therefore, we cannot merely state that a solution is equal to some designated function k on the boundary since the boundary is a set of measure zero. In particular, the statement $u|_\Gamma = k$ is meaningless. The usual way around this problem is to define a linear transformation, $T : H^k(U) \rightarrow L^2(\Gamma)$. Then given u in $H^k(U)$, we can think of u on Γ via $T(u)$.

In Section 1.1, we proved that any linear transformation T of a finite dimensional normed linear space is bounded. In Section 2.3, we saw an example of a linear transformation of a Banach space that was not bounded. Also in Section 1.1, we saw that the set of bounded linear operators of a normed linear space is itself a normed linear space. Finally, bounded linear operators are always uniformly continuous (see Exercise 1). It is also important to note that the bounded linear transformations between Banach spaces is itself a Banach space. This latter fact came up in conjunction with the Lax Equivalence theorem. In particular, a bounded element of $Hom[H^k(U), L^2(\Gamma)]$ is called a *trace.* The following theorem provides sufficient conditions for the existence of a trace. It is important to notice that the result requires a Lipschitz boundary.

Before embarking on the theorem, we make a brief digression. The proof of the theorem uses a technique called *edge flattening*. This procedure is valid for sets with Lipshitz boundary. We will develop this idea for the specific cases that concerns us. See [Evans (1998)] for the general case. In particular, we consider only the 2D case and we suppose that the boundary is a polygon. Given any boundary point q, there are two possibilities, the point is on an edge or is a vertex.

In the first case, the edge is locally flat near the point q. We may rotate and translate the plane so that the edge is on the x-axis and q is mapped to $\hat{q}$. We will refer to this affine mapping by γ. In addition, there is a neighborhood N of q so that $\gamma(N)$ contains $\hat{q}$ and $\gamma(N) \cap \gamma(\Gamma)$ lies on the x-axis. Moreover, we may view γ as a change of variable transformation that leaves integrals unchanged.

If q is a vertex, then rotate and translate so that $\hat{q}$ is the origin and the edges adjacent to q are in the first and second quadrants. Take N so that $N \cap \Gamma$ only contains q and points of the two adjacent sides. Next, consider a point $s = (a, b)$ inside $\gamma(N)$. It is immediate that we can identify a point (a, b_0) in $\gamma(N) \cap \gamma(\Gamma)$. Now map $s = (a, b - b_0)$. (Note that b_0 is a function of a.) This function maps $\hat{q}$ to itself, maps $\gamma(N)$ to a neighborhood of $\hat{q}$, and the two adjacent sides are now mapped to the x-axis. In addition, it may also be used as a change of variables transformation provided we separate the integral along the y-axis. For notational simplicity, we refer to is also as γ.

Theorem 9.3.1. *If Γ is a Lipschitz boundary, then there exists a trace T in $Hom[H^1(U), L^2(\Gamma)]$. Moreover, if $u \in H^1(U)$ and is continuous on D, then for any x in Γ, $T(u)(x) = u(x)$.*

Proof. We begin with the case for u continuously differentiable on $D \subset \mathbb{R}^n$. We will complete the proof by making a density argument that appeals to Theorem 9.2.7. By our choice of u, it makes sense to talk about u restricted to Γ. Furthermore, restricting a function to a subset of the domain is a linear process. Hence, there is a linear transformation of the subspace of continuously differentiable functions on D which takes values in the real valued functions on Γ. Our immediate goal is to prove that this transformation is bounded.

For the case at hand, we will see that

$$\|u\|_{L^2(\Gamma)} \leq C\|u\|_{H^1(U)}$$

for some constant C. Note that since u is continuous on D and Γ is compact, then u is square integrable on Γ. Also, there is no problem with identifying u on U and u restricted to Γ.

We start with x_0 in Γ and suppose that Γ is flat near x_0. We denote the associated open neighborhood as a disc $N_r(x_0)$ of radius r. Set $N_r^+ = N_r(x_0) \cap U$ and select φ in $C_0^\infty(N_r(x_0))$. We may also take φ to be non-negative on $N_r(x_0)$ and identically 1 on $N_{r/2}(x_0)$. We now set $\Gamma_0 = \Gamma \cap N_{r/2}(x_0)$. Employing the change of variables, we may suppose that $\gamma(\Gamma_0)$ is an open interval on the x-axis. Take $x' \in \gamma(\Gamma_0)$ and take α to be the multi-index $(0,1)$. Then we calculate,

$$\int_{\gamma(\Gamma_0)} |u(x')|^2 dx' = \int_{\gamma(\Gamma_0)} \varphi(x')|u(x')|^2 dx'$$

$$= \int_{\gamma(N_r^+)} D^\alpha(\varphi(x)|u(x)|^2) dx,$$

since φ is identically 1 on Γ_0, the integrand is zero on the boundary except for $\gamma(\Gamma_0)$ and Green's theorem. Next, we use properties of the weak derivative proved in Section 9.1,

$$\int_{\gamma(\Gamma_0)} |u(x')|^2 dx'$$

$$= \int_{\gamma(N_r^+)} |u(x)|^2 D^\alpha(\varphi(x)) + 2|u(x)||D^\alpha(u(x))|\varphi(x) dx$$

$$\leq C_1 \int_{\gamma(N_r^+)} |u(x)|^2 + 2 \int_{\gamma(N_r^+)} |u(x)||D^\alpha(u(x)) dx,$$

since φ takes values in the unit interval.

For any real a and b, $0 \leq (a-b)^2$, or $2ab \leq a^2 + b^2$. In particular, $2|u(x)||D^\alpha(u(x))| \leq |u(x)|^2 + |D^\alpha(u(x))|^2$. Hence, there exists a postitve constant C_1 with

$$\int_{\gamma(\Gamma_0)} |u(x')|^2 dx' \leq C_1 \int_{\gamma(N_r^+)} |u(x)|^2 + \int_{\gamma(N_r^+)} |u(x)|^2$$

$$+ \int_{\gamma(N_r^+)} \sum_{|\alpha|=1} |D^\alpha(u(x))|^2 dx.$$

Now, it follows immediately that for $M > 0$,

$$\int_{\gamma(\Gamma_0)} |u(x')|^2 dx' \leq M \int_{\gamma(U)} |u(x)|^2 + |D^\alpha(u(x))|^2 dx.$$

If Γ is not flat near x_0 then, as in the discussion prior to the proof, we can flatten the boundary near x_0 by using the flattening change of variables, to derive the same inequality.

Restating the inequality without the change of variables, we have

$$\int_{\Gamma_0} |u(x)|^2 \leq M \int_U |u(x)|^2 + \sum_{|\alpha|=1} |D^\alpha(u(x))|^2 dx. \qquad (9.3.1)$$

And (9.3.1) holds for any $x_0 \in \Gamma$ and an associated open neighborhood Γ_0 in Γ of x_0. Furthermore, taking square roots of both sides of (9.3.1) it follows that the $L^2(\Gamma_0)$ norm of u is bounded by M times the $H^1(U)$ norm.

Since D is compact, its boundary is also. Therefore, there is a finite open cover of Γ so that (9.3.1) holds on each open set in the cover. If we denote these sets by Γ_i, $i = 1, 2, \ldots, p$, then $\|u\|_{L^2(\Gamma_i)} \leq M\|u\|_{H^1(U)}$, for each i. In turn,

$$\|u\|_{L^2(\Gamma)} \leq \sum_i \|u\|_{L^2(\Gamma_i)} \leq pM\|u\|_{H^1(U)}. \qquad (9.3.2)$$

Next we extend T to all of $H^1(U)$. By Theorem 9.2.6, we know that the elements of $C^\infty(D)$ are dense in $H^1(U)$. Therefore, for any u in $H^1(U)$, there is a sequence u_k of C^1 functions converging to u in $H^1(U)$. By (9.3.2), T is continuous on $C^1(D) \cap H^1(U)$. Therefore, $\{u_k\}$ is Cauchy in $L^2(\Gamma)$. Since $L^2(\Gamma)$ is complete, then the sequence is convergent. We set this limit to $T(u)$. Now we have for $\epsilon > 0$ and large k,

$$\|Tu\|_{L^2(\Gamma)} = \|Tu - Tu_k + Tu_k\|_{L^2(\Gamma)}$$
$$\leq \|Tu - Tu_k\|_{L^2(\Gamma)} + \|Tu_k\|_{L^2(\Gamma)} \leq \epsilon + pM\|u\|_{H^1(U)}.$$

$$(9.3.3)$$

The right-hand side of (9.3.3) is independent of k. Hence, T is bounded.

Finally, if u is continuous on D, then by M(iii), the convergence of the mollifiers is uniform. Since we know that the u_k are derived via mollifications of u, the result follows. □

Completion of the proof of Theorem 9.2.8

Proof. To demonstrate the third assertion, we rely on Theorem 9.3.1. Namely, if U has Lipschitz boundary and u is continuous on D and lies in $H^1(U)$, then there is a bounded linear transformation $T : H^1(U) \to L^2(\Gamma)$ so that $T(u)$ determines u on Γ. We now introduce the trace as something that was intended but omitted in Theorem 9.2.8. We compute

$$\left| \int_\Gamma u_n v_n n_\Gamma - \int_\Gamma uv n_\Gamma \right|$$

$$= \left| \int_\Gamma T(u_n)T(v_n)n_\Gamma - \int_\Gamma T(u)T(v)n_\Gamma \right|$$

$$\left| \int_\Gamma T(u_n)T(v_n)n_\Gamma - \int_\Gamma T(u_n)T(v)n_\Gamma \right.$$

$$\left. + \int_\Gamma T(u_n)T(v)n_\Gamma - \int_\Gamma T(u)T(v)n_\Gamma \right|$$

$$= \left| \int_\Gamma T(u_n)\,(T(v_n) - T(v))\,n_\Gamma + \int_\Gamma (T(u_n) - T(u))\,T(v)n_\Gamma \right|$$

$$\leq \left| \int_\Gamma T(u_n)\,(T(v_n) - T(v))\,n_\Gamma \right| + \left| \int_\Gamma (T(u_n) - T(u))\,T(v)n_\Gamma \right|$$

$$\leq \|T(v_n) - T(v)\|_{L^2(\Gamma)} \left| \int_\Gamma T(u_n)n_\Gamma \right|$$

$$+ \|T(u_n) - T(u)\|_{L^2(\Gamma)} \left| \int_\Gamma T(v)n_\Gamma \right|$$

$$\leq \|T\| \left(\|v_n - v\|_{L^2(\Gamma)} \left| \int_\Gamma T(u_n)n_\Gamma \right| \right.$$

$$+ \|u_n - u\|_{L^2(\Gamma)} \left| \int_\Gamma T(v)n_\Gamma \right| \Bigg).$$

This last expression converges to zero since T is continuous and all other terms either go to zero or are bounded. $\square$

Note that in the terminology developed for the FEM, a trace determines Dirichlet boundary values. Further, the set of all u with $T(u) = 0$, the kernel of T, is a linear subspace of $H^1(U)$. Any other boundary value configuration must determine a coset of the kernel. In the next result, we characterize the kernel of T. For this purpose, we set $H_0^1(U)$ equal to the closure in $H^1(U)$ of the set of indefinitely differentiable functions on U with compact support, $C_0^\infty(U)$. We mentioned earlier that $C_0^\infty(U)$ is not dense in $H^1(U)$. This is apparent from the next theorem.

Theorem 9.3.2. *If Γ is a Lipschitz boundary, then the kernel of T is $H_0^1(U)$.*

Proof. See Evans (1998). $\square$

We next look at the dual or the set of bounded linear functionals on $H_0^1(U)$. By the Riesz representation theorem, the elements of the dual space of a Hilbert space can be identified with the space itself. In particular, if (H, σ) is a Hilbert space with positive definite inner product σ, then for any bounded linear functional, f, on H, there is an element u in H with $f(v) = \sigma(u, v)$ for all v in H. Hence, the dual spaces of $L^2(U)$ and $H^1(U)$ are known. The next result gives us a handle on the dual to H_0^1. The standard notation for this space is $H^{-1}(U)$.

Theorem 9.3.3. *If F is in $H^{-1}(U)$, then there exist elements f^0, f^1, ..., f^n in $L^2(U)$ so that for any v in $H_0^1(U)$*

$$F(v) = \int_U f^0(v) + \sum_i f^i(D^{\alpha_i}v), \qquad (9.3.4)$$

where for each i, $\alpha_i = (0, \ldots, 1, \ldots, 0)$ with 1 occurring in the ith entry. The norm of F is determined by the $L^2(U)$ norm of the f^i via $\|F\| = \inf[\int_U \sum_{i=0}^n (f^i)^2]^{1/2}$ where the inf is taken over all $f^0, f^1, \ldots, f^n$ in $L^2(U)$ which together satisfy (9.3.4).

Proof. The inner product σ_{H^1} induces a positive definite inner product on the subspace $H_0^1(U)$. Since $H_0^1(U)$ is a closed subspace of a Hilbert space, then $H^1(U)$ is the direct sum of $H_0^1(U)$ and its orthogonal complement (see [Rudin (1986)]). Hence, there is a natural projection of $H^1(U)$ onto $H_0^1(U)$. In turn, any F in $H^{-1}(U)$ induces a bounded linear functional on $H^1(U)$ by composing F with the projection. By the Riesz theorem, there is an element u in $H^1(U)$ such that for any for any v in $H_0^1(U)$,

$$F(v) = \sigma_{H^1}(u, v) = \int_U \left(uv + \sum_i D^{\alpha_i}u D^{\alpha_i}v\right)$$

$$= \sigma_{L^2}(u, v) + \sum_i \sigma_{L^2}(D^{\alpha_i}u, D^{\alpha_i}v). \qquad (9.3.5)$$

Looking at the Riesz theorem in the other direction, there are bounded functional f^0, f^i in the dual space of $H^1(U)$, so that $f^0(v) = \sigma_{L^2}(u, v)$ and $f^i(D^{\alpha_i}v) = \sigma_{L^2}(D^{\alpha_i}u, D^{\alpha_i}v)$ for $i > 0$. Hence, (9.3.4) is established.

It remains to prove the final assertion. If $F(v) = \int_U g^0 v + \sum_i g^i D^{\alpha_i} v$, g^i in $L^2(U)$, then for $v = u$,

$$F(u) = \sigma_{H^1}(u, u) = \int_U \sum_{|\alpha| \leq 1} |D^\alpha u|^2 = \int_U \sum_{i=0}^n |f^i|^2 = \int_U \sum_{i=0}^n g^i D^{\alpha_i} u.$$

In turn, as u may be taken as an H^1 unit vector, then by Cauchy Schwarz applied to $L^2(U)$

$$F(u) \leq \sum_i \|g^i\|_{L_2} \leq \sum_i \int_U |g^i|^2.$$

Therefore,

$$\sum_i \int_U |f^i|^2 \leq \sum_i \int_U |g^i|^2.$$

Since $\|F\|$ must be $[\int_U \sum_{i=0}^n (f^i)^2]^{1/2}$, then the final assertion is proved. $\square$

The standard notation for the image of T is $H^{(1/2)}(\Gamma)$. This is because it supports an inner product

$$\sigma_{1/2}(u, v) = \sigma_2(u, v) + \int_{\Gamma \times \Gamma} \frac{[Du(x) - Du(y)]^2}{\|x - y\|^{1+n}} dx dy,$$

where n is the dimension of the underlying vector space. Hence, it is more than $L^2(\Gamma)$ but less than $H^1(\Gamma)$. Continuing the notation, the closure of the elements of compact support is denoted $H_0^{1/2}$. Both $H^{1/2}$ and $H_0^{1/2}$ are Hilbert spaces. The dual space of $H_0^{1/2}$ is denoted $H^{-1/2}(\Gamma)$.

In the next section, we will see that (9.3.5) in central to understanding boundary values.

Exercises

1. Let $T : X \to Y$ be a linear operator defined for normed linear spaces X and Y. Prove that T bounded implies that T is uniformly continuous.

2. Consider the set of all real sequences $x = (x_i : i = 1, 2, \ldots)$ such that $x_i = 0$ except for a finite number of i. It is straight forward to verify that X is a vector space. By defining $\|x\| = (\sum_{i=1}^{\infty} x^2)^{1/2}$ we make X into a normed linear space. Define $T : X \to \mathbb{R}$ by $Tx = \sum_{i=1}^{\infty} i x_i$.

   a. Prove that T is linear.
   b. Prove that T is not bounded.
   c. Prove that X has no finite spanning set.

3. For the flattening funciton determine b_0 as a function of a. Write out the associated change of variables transformation.

4. Explain the statement, $H^{-1}(U)$ is the set of solutions for the second-order PDE with boundary value zero.

9.4. Weak Solutions of Elliptical Equations

Our purpose in this section is to make precise statement of the weak boundary value problem and to prove the Lax–Milgram theorem on the existence of weak solutions to elliptical PDE. In terms of our basic goal, the convergence of GFEM, the existence of a solution in the Sobolev space H^k is a critical step. Indeed, the convergence theory is based on the theory of polynomial interpolation. In order to apply these theorems, we must know that there is something to interpolate. Secondly, with the Lax–Milgram theorem, we will be able to conclude that the GFEM solution exists for a large class of elliptical PDE. The GFEM solution is a piecewise polynomial in V^h. This function is derived from a linear system of equations. In this context, the Lax–Milgram theorem states that the linear system has unique solution. The theorem is stated as the answer to a dual space question for an abstract Hilbert space. With this generality, we have both applications from the same theorem.

Throughout this section and the following ones we will suppose that U has Lipschitz boundary. Often we state this specifically for emphasis.

With the structures introduced in Sections 9.1 and 9.2 along with the notation of Section 8.4, we can now make a precise statement of a linear boundary value problem. In Chapter 8, we presented a linear PDE in terms of a geometric mapping (linear transformation followed by a translation) between function spaces. In Section 8.4, we introduced the idea of the weak form of a PDE. In this context, we saw the possibility of applying the PDE as a mapping to functions which were less than differentiable. We now recognize this idea as the weak derivative. In particular, the domain of the PDE is a Sobolev space $H^k(U)$ and the range is $L^2(U)$. The translation part is a function in $L^2(U)$ and the linear part is a linear combination of weak partial derivatives D^α where the multi-index satisfies $|\alpha| \le k$. The coefficients may be constants or functions of the independent variable. The smallest k is the order of the largest derivative that appears.

If α and β are multi-indices with $|\alpha + \beta| \le k$, then for any u in $H^k(U)$, there is no ambiguity in writing $D^{\alpha+\beta}u$ as the iteration of differntialble operators. Furthermore, if a PDE coefficient is sufficiently differentiable (see Exercises 1 and 2), then $(D^\beta)(aD^\alpha u)$ may be written as a linear combination of terms of the form $bD^\gamma u$. We now have the notation necessary to write the expression for the PDE linear part L in the case of a second-order PDE,

$$L(u) = \sum_{|\alpha|,|\beta|=1} D^\beta \left[a_{(\alpha,\beta)} D^\alpha u \right] + \sum_{|\alpha|=1} b_\alpha D^\alpha u + cu, \qquad (9.4.1)$$

where the $a_{(\alpha,\beta)}$ are C^1. Moreover, we will assume symmetry; that is, $a_{(\alpha,\beta)}(x) = a_{(\beta,\alpha)}(x)$ for any $\alpha \ne \beta$.

Recall that in Section 5.3, we encountered a PDE of the form given in 9.4.1. The context was the river pollution problem. In that setting, the second-order term represented *diffusion*, the first-order term was the *transport* and the zero-order term was the decay or more generally, *reaction*. Finally, the pollution source occurred as the translation term. For this reason the translation part is often called the *source* or *forcing* term. Since it is natural to suppose that the diffusion is positive or negative in all directions, it is reasonable to suppose that the symmetric bilinear second-order term is positive or negative definite. For instance, combustion is negative, heat is positive.

Finally, we set boundary values via a trace operator (Section 9.3).

Definition 9.4.1. The function u in $H^2(U)$ is a *solution to a linear boundary value problem of order* 2, provided there exists L in $Hom[H^2(U), L^2(U)]$ given by (9.4.1), an element f in $L^2(U)$ and g in $L^2(\Gamma)$ with $L(u) + f = 0$ and $T(u) = g$, where T is a trace operator.

We have previously stated that a PDE is elliptical provided for any x, the eigenvalues of the matrix $[a_{\alpha,\beta}(x)]$ are all positive or all negative. Note that we define the matrix using lexicographic ordering applied to the multi-indices. Hence, $a_{(1,0),(0,1)}$ is $a_{1,2}$. The following definition introduces a concept that is extended in equation (9.4.5).

Definition 9.4.2. A linear second-order symmetric PDE is *uniformly elliptical* if there exists a positive θ such that for every ξ and $x \in \mathbb{R}^n$,

$$\xi^T [a_{(i,j)}(x)]\xi \geq \theta \|\xi\|^2, \tag{9.4.2}$$

holds almost everywhere.

Note that (9.4.2) is equivalent to $\xi^T [a_{(i,j)}(x)]\xi \geq \theta$ for any ξ of unit length. Moreover, if ξ is a unit eigenvector of eigenvalue λ, then $\xi^T [a_{(i,j)}(x)]\xi = \xi^T \lambda \xi = \lambda$. Hence, $\lambda \geq \theta$, independent of x. This latter statement is an alternate definition (see Exercise 3.)

We define a (not necessarily symmetric) bilinear form on $H_0^1(U)$, the closure of $C_0^\infty(U)$ in $H^1(U)$ and the kernel of the trace opeartor,

$$\tau[u, v] = (-1)^{|\beta|} \int_U \sum_{|\alpha|, |\beta| = 1} a_{\alpha, \beta} D^\alpha u D^\beta v + \sum_{|\alpha| = 1} b_\alpha (D^\alpha u) v + cuv. \tag{9.4.3}$$

In Section 8.4, we considered the bilinear operator $\int_U L(u)v$ on element spaces V_e^h and then accumulated to V^h. In this context (9.4.3) is equivalent to this opeartor for the case of a PDE with zero boundary values. In the more general context, the two are equivalent on $H^2 \cap H_0^1$.

If we fix u and consider the v as the independent variable in (9.4.3), then we have an element in the dual space of $H_0^1(U)$ associated to u and L. In Theorem 9.3.3, we characterized elements of the dual to $H_0^1(U)$ by means of functions f^i in $L^2(U)$. In particular, there is an element w in $H^1(U)$ such that for each $i = 0, 1, \ldots, n$, $f^0 = w$, $f^i = -D^{\alpha_i}w$, $i > 0$. Upon substitution, we get $\tau[u, v] = \int_U (f^0 - \sum_{i=1}^n f^i)v$. Since σ_2 is non-degenerate and $\tau[u, \circ] - (f^0 - \sum_{i=1}^n f^i) = 0$, then $-f^0 + \sum_{i=1}^n f^i$ is the translation part of the PDE.

Now with this notation, we can precisely define the weak solution of a boundary value problem. Before we write the definition, we comment on the notation. Considering $\tau[u, v] + \int_U fv = 0$, we see that the assignment $v \to \int_U fv$ defines a linear functional on $H_0^1(U)$. By Cauchy–Schwartz, the functional is bounded. It is standard in the literature to identify this linear functional with f itself. This identification then yields expressions like $f(v) = \int_U fv$. This is justified by appealing to the Riesz representation. However, for clarity we choose to write this functional with F.

Definition 9.4.3. Suppose that τ is the bilinear function defined on $H_0^1(U)$ given by (9.4.3) and F is a bounded linear functional on H_0^1 with $F(v) = \int_U (f^0 - \sum_i^s)v$ for $f^i \in L^2(U)$. An element u in H_0^1 is a *weak solution to the boundary value problem with boundary value zero* provided, $\tau[u, v] = F(v)$ for every v.

More generally, if Γ is Lipschitz, and $u \in H^1$ satisfies $\tau[u, v] = F(v)$, but $T(u) = g \neq 0$ on Γ, then we may take w with $w|_\Gamma = T(w) = g$ (see Exercise 4). Then $u - w$ is in $H_0^1(U)$ and for any v, $\tau[u - w, v] = F(v) - \tau[w, v]$. We may then designate the right-hand side as $G(v)$ with $G \in H^{-1}$. Now, by the Riesz theorem, there is a g in $L^2(U)$ with $G(v) = \int_U gv$. Hence, the translation part of the associated differential equation is changed. In short, in the weak form, the general boundary value problem is related to the boundary value zero problem via a change in the translation term. Alternatively, u and w lie in the same $H_0^1(U)$ coset.

Because of the importance of the preceding discussion we stop here to summarize. Given a second-order PDE in weak form with given boundary values, there is a related second-order weak form PDE with boundary value zero. The two equations are related by a change in the translation term of the PDE. Alternatively, the solution to the original PDE is unique modulo $H_0^1(U)$. And restated for the context of the PDE, the solution to the original PDE is unique up to the forcing term.

The solution question for elliptical PDE is largely resolved by the following theorem. The statement is very general. When applied to H^1 it states sufficient conditions for the existence of a solution of an elliptical PDE. Applied to V^h, it provides conditions for GFEM to be well-posed. In the case of a transient parabolic PDE, the spatial part is elliptical. Therefore, the spatial part may be resolved by means of GFEM. The following theorem concerns the elliptical part while the Lax Equivalence speaks to the time stepping part. Formal statements of these results follow the proof. The condition pair, Eqs. (9.4.4) and (9.4.5), are often referred to as the *elliptical hypothesis* or *elliptical regularity*. Condition 9.4.5 alone is called *coercive*.

Theorem 9.4.1. (*Lax–Milgram*) *Suppose that τ is bilinear form on a real Hilbert space H. Suppose that there exist positive constants a and b such that for any u and v in H,*

$$|\tau[u, v]| \le a\|u\|\|v\|, \tag{9.4.4}$$

$$b\|u\|^2 \le \tau[u, u]. \tag{9.4.5}$$

In addition, suppose that F is a bounded linear functional on H. Then there exists a unique element u in H with $\tau[u, v] = F(v)$ for every v in H.

Proof. If we fix u in H, then by (9.4.4) $g_u(v) = \tau[u, v]$ defines a bounded linear functional on H. Hence, by the Riesz representation theorem, there exists a w in H with $g_u(v) = \sigma(w, v)$ where σ denotes the inner product on H. We use the association $u \to w$ to define a function M mapping H to H given by $M(u) = w$. We claim that M determines a bounded, bijective linear transformation of H.

First, since τ is linear in its first argument, then M is linear. Now, $\sigma(M(u), v) = \tau[u, v]$ for every v, so taking $v = M(u)$, it follows that $\sigma(M(u), M(u)) = \tau[u, M(u)]$. Therefore, $\|M(u)\|^2 = \tau[u, M(u)] \le a\|u\|\|M(u)\|$ by (9.4.4). It then follows that $\|M(u)\| \le a\|u\|$. Hence, M is also bounded.

Using (9.4.5) we have $b\|u\|^2 \leq \tau[u,u] = \sigma(M(u),u) \leq \|M(u)\|\|u\|$, by Cauchy–Schwarz. It follows that $M(u) = 0$ only if $u = 0$. Therefore, M is injective.

To prove that M is surjective, we first prove that the image of M is closed in H. Indeed, if v is in the closure of the image of M, then there is a sequence u_m in H with $Mu_m \to v$. Since $b\|u_n - u_m\| \leq \|M(u_n - u_m)\|$ by (9.4.5) and Cauchy–Schwarz, then u_m is Cauchy. It is therefore convergent. Let the limit be u. As M is uniformly continuous, $M(u) = \lim_m M(u_m) = v$. Therefore, v is in the image of M verifying closure.

Now, if M is not surjective, then the image of M is not equal to H and we know that there exists a nonzero w, which is orthogonal to the image. To see this, recall that any closed subspace Hilbert space has orthogonal compliment (Rudin, 1986). Now, if w is orthogonal to the image of M, then by (9.4.5) we have $b\|w\|^2 \leq \tau[w,w] = \sigma(M(w),w) = 0$. But this is impossible unless $w = 0$, Hence, the image of M is H and M is surjective.

Finally, if F is a bounded functional on H, then from the Riesz representation theorem, there is a vector w with $F(v) = \sigma(w,v)$ for every v. Taking u with $M(u) = w$, we have $F(v) = \tau[u,v]$. Now it remains to prove that u is uniquely determined by τ and F. Indeed, if in addition $F(v) = \tau[\hat{u},v]$, then we get $0 = \tau[u-\hat{u},v] = \sigma(M(u-\hat{u},v)$ for any v in H. Hence, $M(u - \hat{u}) = 0$. Since M is bijective, it follows that $u = \hat{u}$. $\square$

We now state and prove two corollaries to Theorem 9.4.1. Each concerns a specific setting that is central to this development. Keep in mind that we may always assume that the elliptical PDE has boundary value zero.

Corollary 9.4.1. *Consider a second-order PDE with domain U and boundary value zero. If the bilinear form (9.4.3) satisfies the elliptical hypothesis, then the PDE has solution in H_0^1.*

It is not difficult to see that the GFEM will generate (in V^h) the discrete form of the solution to the weak boundary value problem. Since V^h is finite dimensional and σ_{H^1} is positive definite, then we can write H^1 as the orthogonal sum of V^h and a direct compliment W. In this context, the GFEM generated solution is the weak form solution modulo W.

Corollary 9.4.2. *Consider a second-order PDE with domain U and boundary value zero. If the bilinear form (9.4.3) applied to the space of piecewise polynomial functions V^h satisfies the elliptical hypothesis, then the linear system of Eq. (8.4.5) has unique solution.*

Proof. With the hypothesis, V^h is a closed subspace of $H_0^1(U)$. By Theorem 9.4.1, there is a unique u in V^h with $\tau[u,v] = F(v)$, v in V^h. Replacing v by basis elements $\psi_{j,\beta}$ and writing $u = \sum_{j,\beta} a_{j,\beta} \psi_{j,\beta}$, we have the familiar linear system

$$\sum_{j,\beta} a_{j,\beta} \tau[\psi_{j,\beta}, \psi_{i,\alpha}] = F\psi_{i,\alpha}.$$

The existence of a unique u implies that the existence of a unique solution to the linear system. □

We remark in passing that there are several important applications of Theorem 9.4.1. We will not include any others here. See for instance (Evans, 1998). However, we will review some of the result regarding the inequalities (9.4.4) and (9.4.5). The first is usually credited to Poincarè and Friedrichs. For any u in $H_0^1(U)$,

$$\|u\|_2 \leq diam(U) \left(\sum_{|\alpha|=1} \|D^\alpha u\|_2^2 \right)^{\frac{1}{2}}. \qquad (9.4.6)$$

Notice that the restriction to trace zero functions eliminates the constants. For the Poisson equation

$$\tau[u,u] = \int_U \sum_{|\alpha|=1} D^\alpha u D^\alpha u = \sum_{|\alpha|=1} \|D^\alpha u\|_2^2.$$

Therefore, (9.4.5) holds in this case. Inequality 9.4.4 is more like Cauchy–Schwartz. In this case, we calculate from (9.4.3) using D to denote the weak gradient.

$$|\tau[u,v]| = |\int_U \sum_{|\alpha|,|\beta|=1} a_{(\alpha,\beta)} D^\alpha u D^\beta v + \sum_{|\alpha|=1} b_\alpha (D^\alpha u)v + cuv|$$

$$\leq \|a_{(\alpha,\beta)}\|_\infty \int_U |Du||Du| + \|b_\alpha\|_\infty \int_U |Du||v| + \|c\|_\infty \int_U |u||v|$$

$$\leq \alpha \left[\int_U |Du||Dv| + \int_U |Du||v| + \int_U |uv| \right],$$

for an appropriate choice of α. Since each summand is an L^2 inner product we may apply Cauchy–Schwarz,

$$|\tau[u,v]| \leq \alpha \left[\|Du\|_2\|Dv\|_2 + \|Du\|_2\|v\|_2 + \|u\|_2\|v\|_2 \right].$$

Since the L^2 norm of u and Du are no larger than the H^1 norm of u, then

$$|\tau[u,v]| \leq 3\alpha\|u\|_{H^1}\|v\|_{H^1}.$$

Hence, (9.4.4) is satisfied by our basic case. This is not a demanding assumption. For Theorem 9.4.1, inequality (9.4.5) is the serious issue.

Exercises

1. Prove that if $u \in H^1(U)$ and $a \in C^1(U)$, then $au \in H^1(U)$ and $D^\alpha(au) = D^\alpha(a)u + aD^\alpha(u)$, where $|\alpha| = 1$.
2. State and prove a result analogous to Exercise 1 for $H^k(U)$ and $|\alpha| = k$.
3. Prove that a linear, symmetric second-order PDE is uniformly bounded provided there is a positive constant θ so that for each x the eigenvalues of $[a_{(i,j)}(x)] > \theta$.
4. Let $g \in H^{(1/2)}(\Gamma)$. Prove that there an element $w \in C^1(D)$ with $T(w) = g$, where T is the trace given in Theorem 9.3.1. (*Hint:* Use a density argument.)

5. Prove that the Helmholtz equation (see Section 3.1) with respect to the Hilbert space H_0^1 satisfies the elliptical hypothesis. Then consider the elliptical equation without first-order term

$$L(u) + f = \sum_{|\alpha|,|\beta|=1} D^\beta a_{(\alpha,\beta)} D^\alpha u + \mu u + f = Au + \mu u + f.$$

(9.4.7)

   If there exists $\lambda > 0$ such that λ is a lower bound for the eigenvalues of $A\mu \geq \lambda$ then for any f, (9.4.7) has a weak solution in H_0^1.
6. Consider a linear second-order PDE $f \circ L$ defined on the function space of all polynomial functions of a domain D. Prove that a weak form solution is also a strong form solution.

9.5. Sobolev Embedding

The stage is now set for the FEM convergence theorem. The theorem we prove will be for elliptical PDE that satisfy the hypothesis of the Lax–Milgram theorem. Towards this goal, we have FEM partitions and associated piecewise polynomial interpolations. We will want to know when the interpolations of a function u converge to u. In this section, we lay the foundation for piecewise polynomial interpolation associated to a finite element partition of a compact domain.

Elements of a Sobolev space belong to L^2 and elements of this space are only known up to a set of measure zero. In short, elements of a Sobolev space are actually equivalence classes of functions. But the polynomial interpolation of a function requires knowledge of the function at a fixed finite set of points. Hence, it makes no sense to interpolate an element of a Sobolev space. There is a work-around.

It is possible to embed some of the Sobolev spaces $H^k(U)$ into the spaces of continuous functions $C^0(\bar{U})$. Conceptually, we are identifying a continuous element in each equivalence class. The process is called *Sobolev embedding*.

Up to this point, we have avoided mentioning the L^p spaces. But now we need these Banach spaes. In a sense, the purpose of the L^p spaces is to generalize the L^2. In particular, we start with the set of functions f that are p-power integrable ($p \geq 1$) and then define the norm as

$$\|f\|_p = \left[\int_U |f|^p \right]^{1/p}.$$

The resulting normed linear space is a Banach space (see Rudin (1986) or Royden (2011)). Of course, for $p = 2$, this is the usual Hilbert space of square integrable functions. As with L^2, there are Sobolev spaces associated to the various L^p spaces. These are denoted $W^{k,p}(U)$. As expected, $W^{k,p}$ is the space of all k-times weakly differentiable functions in $L^p(U)$. The norm is given by

$$\|f\|_{W^{(k,p)}} = \left[\sum_{|\alpha| \leq k} \int_U (D^\alpha u)^p \right]^{1/p}.$$

The proof that the $W^{(k,p)}$ are Banach spaces is analogous to the corrsponding completeness proof of H^k.

We begin with the statement of a well-known result. It is beyond the scope of this text to prove every necesary theorem, the following included. As it plays a central role in this developement, we include the statement. For a proof, see Evans (1998).

Theorem 9.5.1. (*Extension Theorem*) *Take U with boundary Γ and suppose that Γ has a C^1 parameterization. For any open V containing $\overline{U}$ there is a bounded operator $T: W^{1,p}(U) \to W^{1,p}(\mathbb{R}^n)$ $(1 \leq p \leq \infty)$, such that*

i. *$Tu = u$ a.e.,*
ii. *The support of $Tu \subset V$.*

The condition that the boundary has a C^1 parameterization is often referred to more simply as having a C^1 *boundary*.

For $1 \leq p < n$, we define the *Sobolev conjugate* of p as $p^* = np/(n-p)$. We now state the first result toward Sobolev embedding. Notice the similarity with the Poincaré–Friedrich's Inequality quoted in the last section.

Theorem 9.5.2. (*Gagliardo, Nirenberg and Sobolev*) *Given* $1 \leq p < n$ *and* u *in* $C_0^1(\mathbb{R}^n)$,

$$\|u\|_{L^{p^*}(\mathbb{R}^n)} \leq C\|Du\|_{L^p(\mathbb{R}^n)}, \qquad (9.5.1)$$

where C *depends only on* n *and* p.

Proof. Since u has compact support then for each $x = (x_1, \ldots, x_i, \ldots, x_n)$, we write

$$u(x) = \int\limits_{-\infty}^{x_i} \frac{\partial}{\partial i} u(x_1, \ldots, y_i, \ldots, x_n) dy_i,$$

where the differentiation is taken with respect to the ith component. Hence, for the absolute value

$$|u(x)| \leq \int\limits_{-\infty}^{\infty} |Du(x_1, \ldots, y_i, \ldots, x_n)| dy_i$$

and

$$|u(x)|^{n/(n-1)} \leq \prod_{i=1}^{n} \left[\int\limits_{-\infty}^{\infty} |Du(x_1, \ldots, y_i, \ldots, x_n)| dy_i \right]^{1/(n-1)}.$$

Next, we integrate both sides.

$$\int\limits_{-\infty}^{\infty} |u(x)|^{n/(n-1)} dx_1$$

$$= \int\limits_{-\infty}^{\infty} \prod_{i=1}^{n} \left[\int\limits_{-\infty}^{\infty} |Du| dy_i \right]^{1/(n-1)} dx_1$$

$$= \left[\int\limits_{-\infty}^{\infty} |Du| dy_1 \right]^{1/(n-1)} \int\limits_{-\infty}^{\infty} \prod_{i=2}^{n} \left[\int\limits_{-\infty}^{\infty} |Du| dy_i \right]^{1/(n-1)} dx_1$$

$$\leq \left[\int\limits_{-\infty}^{\infty} |Du| dy_1 \right]^{1/(n-1)} \left[\prod_{i=2}^{n} \int\limits_{-\infty}^{\infty} \int\limits_{-\infty}^{\infty} |Du| dy_i dx_1 \right]^{1/(n-1)},$$

by the Hölder identity with $n-1$ factors (Royden, 2011). We now repeat the process for x_2, x_3 and so forth to get

$$\int\limits_{-\infty}^{\infty} \cdots \int\limits_{-\infty}^{\infty} |u(x)|^{n(n-1)} dx_1 \ldots dx_n$$

$$\leq C \left[\prod_{i=1}^{n} \int\limits_{-\infty}^{\infty} \cdots \int\limits_{-\infty}^{\infty} |Du| dx_1 \ldots dx_n \right]^{1/(n-1)}$$

$$= C \left[\int\limits_{\mathbb{R}^n} |Du| \right]^{n(n-1)}.$$

In particular,

$$\int\limits_{\mathbb{R}^n} |u|^{n/(n-1)} \leq C \left[\int\limits_{\mathbb{R}^n} |Du| \right]^{n/(n-1)}$$

and we now have verified (9.5.1) for the case $p = 1$, $p^* = n/(n-1)$. To complete the proof we set $a = p(n-1)/(n-p) > 1$. Then

$$p^* = \frac{np}{n-p} = \frac{np}{n-p}\frac{p-1}{p-1} = \frac{np-n}{n-p}\frac{p}{p-1}$$

$$= \left[\frac{p(n-1)}{n-p} - \frac{n-p}{n-p} \right] \frac{p}{p-1} = (a-1)\frac{p}{p-1} = a\frac{n}{n-1}.$$

Therefore,

$$\left[\int\limits_{\mathbb{R}^n} |u|^{p^*} \right]^{(n-1)/n} = \left[\int\limits_{\mathbb{R}^n} |u|^{an/(n-1)} \right]^{(n-1)/n}$$

$$= \left[\int\limits_{\mathbb{R}^n} (|u|^a)^{n/(n-1)} \right]^{(n-1)/n}$$

$$\leq C \int\limits_{\mathbb{R}^n} |Du^a| = aC \int\limits_{\mathbb{R}^n} |u|^{a-1}|Du|$$

$$\leq aC \left[\int_{\mathbb{R}^n} |u|^{(a-1)p/(p-1)} \right]^{(p-1)/p} \left[\int_{\mathbb{R}^n} |Du|^p \right]^{1/p}$$

$$= aC \left[\int_{\mathbb{R}^n} |u|^{p^*} \right]^{(p-1)/p} \|Du\|_p,$$

using the Hölder identity one more time. Hence,

$$\left[\int_{\mathbb{R}^n} |u|^{p^*} \right]^{1/p^*} \leq a\|Du\|_p,$$

which completes the proof. □

The following result states the particular formulation of Theorem 9.5.2 that we need later.

Theorem 9.5.3. *Suppose $1 \leq p < n$. If U is a bounded open set with C^1 boundary, then there is a constant C depending on p, n and U such that $\|u\|_{L^{p^*}(U)} \leq C\|u\|_{W^{1,p}(U)}$ for any u in $W^{1,p}(U)$.*

Proof. By Theroem 9.5.1, there is a bounded operator T mapping $W^{1,p}(U)$ to $W^{1,p}(\mathbb{R}^n)$. Furthermore, if $v = T(u)$, then $v|_U = u$ and v has compact support. We denote $\|T\|$ by C so that

$$\|v\|_{W^{1,p}(\mathbb{R}^n)} \leq C\|u\|_{W^{1,p}(U)}. \tag{9.5.2}$$

Since v has compact support in a Sobolev space, then there is a sequence of $C_0^\infty(\mathbb{R}^n)$ functions, v_m that converges to v in $W^{1,p}(\mathbb{R}^n)$. Now, by the G–N–S theorem,

$$\|v_m - v_n\|_{L^{p^*}(\mathbb{R}^n)} \leq C\|Dv_m - Dv_n\|_{L^p(\mathbb{R}^n)} \leq C\|v_m - v_n\|_{W^{1,p}(\mathbb{R}^n)}.$$

Hence, the sequence is Cauchy in L^{p^*}. Therefore, it is convergent in L^{p^*} and the limit must be v. We apply (9.5.1) to v_m and then use the triangle inequality to conclude inequality (9.5.2) for the function v that is not necessarily C^1.

$$\|v\|_{L^{p^*}(\mathbb{R}^n)} \leq C\|Dv\|_{L^p(\mathbb{R}^n)}.$$

Since v equals u on U, then we apply Theorem 9.5.1 to yield

$$\|u\|_{L^{p^*}(U)} = \|v\|_{L^{p^*}(U)} \leq \|v\|_{L^{p^*}(\mathbb{R}^n)}$$

$$\leq C\|Dv\|_{L^p(\mathbb{R}^n)} \leq C\|v\|_{W^{1,p}(\mathbb{R}^n)} \leq C_1\|u\|_{W^{1,p}(U)},$$

which verifies the inequality stated in the theorem.   □

We state the next theorem without proof. But first we need to develop the notation and terminology. A function $u : U \subset \mathbb{R}^n \to \mathbb{R}$ is called *Hölder continuous* provided $|u(x) - u(y)| \leq C\|x - y\|^a$ where C is a constant independent of x and y and $a \geq 0$ is called the *exponent* of continuity. If $a = 1$, then u is Lipschitz continuous. If $a = 0$, then u is bounded. We denote the set of Hölder continuous functions of exponent a on U by $C^{0,a}$. In general, for $a > 0$, Hölder continuous implies continuous.

Theorem 9.5.4. (*Morrey*) *Given u in $C^1(\mathbb{R}^n)$ and $n \leq p < \infty$, then there is a constant C independent of u so that*

$$\|u\|_{C^{0,a}(\mathbb{R}^n)} \leq C\|u\|_{W^{1,p}(\mathbb{R}^n)}, \tag{9.5.3}$$

where $a = 1 - n/p$.

Proof. See Evans (1998).   □

The following is a corollary to the result of Morrey. It is the actual result that we need. In the statement of the theorem u is an element of an L^p space, an equivalence class of functions, whereas u^* is a continuous function defined on a compact set, an element of C^0 with sup norm. Moreover, as an element of L^p, it lies in the class of u. In the proof of the theorem, we derive u^* from u.

Theorem 9.5.5. *Suppose that U is a bounded open set in $\mathbb{R}^n$ with C^1 boundary. Given u in $W^{1,p}$, there is a continuous function u^* on $\bar{U}$ with*

a. $u = u^*$ *a.e.,*
b. $\|u^*\|_{C^0(\mathbb{R}^n)} \leq C\|u\|_{W^{1,p}(\mathbb{R}^n)},$

where C depends on p, n and U, and $0 < p \leq \infty$.

Proof. We begin by applying Theorem 9.5.1 to induce $\overline{u} = Tu$ in $W^{1,p}(\mathbb{R}^n)$ with compact support satisfying $\overline{u}|_U = u$ and $\|\overline{u}\|_{W^{1,p}(\mathbb{R}^n)} \leq C\|u\|_{W^{1,p}(U)}$. As in the proof of Theorem 9.5.3, $\overline{u}$ is the $W^{1,p}$ limit of C_0^∞ functions.

By Morrey's Theorem, this sequence of C_0^∞ functions is $C^{0,a}$-space Cauchy $(a = 1 - n/p)$. Therefore, the sequence has a $C^{0,a}$ sup norm limit, a function in the ordinary sense, that we designate as u^*. Furthermore, u^* must equal $\overline{u}$ as an element of $W^{1,p}$. The final inequality holds for elements of the sequence, hence, it applies to u as well. $\square$

We can now prove the basic Sobolev embedding theoem. The particular cases that interest us are included in the corollary. In the following statement $\lceil a \rceil$ denotes the greatest integer function.

Theorem 9.5.6. *Suppose that U in $\mathbb{R}^n$ has Lipschitz boundary, and $k > n/2$, then there is a bounded, linear injection $H^k(U) \to C^r(U)$ where $r = k - \lceil n/2 \rceil - 1$.*

Proof. Take k with $2k > n$ and suppose that $n/2$ is not an integer. For u in $H^k(U)$, then we can see the inequality in Theorem 9.5.3 as a relationship between a kth Sobolev space and a $(k-1)^{st}$. In particular,

$$\|u\|_{L^2(U)} \leq C\|u\|_{W^{1,2}(U)} = C\|u\|_{H^1(U)},$$

may be recast as

$$\|u\|_{W^{k-1,2^*}(U)} \leq C\|u\|_{H^k(U)}.$$

In particular, the identity transformation defines a bounded linear operator from $H^k(U)$ to $W^{k-1,p^*}(U)$. Repeating the process, u lies in $W^{k-2,q}$ where $q = 2^{**}$. Now, we compute

$$\frac{1}{2^*} = \frac{n-2}{2n} = \frac{1}{2} - \frac{1}{n}, \quad \frac{1}{q} = \frac{1}{2} - \frac{2}{n}, \ldots, \frac{1}{r} = \frac{1}{2} - \frac{l}{n},$$

where the right hand identity is after l iterations, $l \leq k$. Therefore, taking l so that $l < n/2 < l+1$, we have $l = \lceil n/2 \rceil$ and $r = 2n/(n-2l) > n$.

Next we apply the Morrey inequality from Theorem 9.5.5 item b.

$$\|u^*\|_{C^0(\mathbb{R}^n)} \leq C\|u\|_{W^{1,p}(\mathbb{R}^n)}.$$

In particular, for u in $W^{k-l,r}$ and $|\alpha| \leq k - l - 1$, then $D^\alpha u \in W^{k-l-|\alpha|,r}$ contained in $W^{1,r}$. Hence, by Morrey, $D^\alpha u$ may be considered an element of $C^0(U)$. Therefore, u belongs to C^{k-l}. Now $k - l - 1 = k - \lceil n/2 \rceil - 1$. This completes the proof for n odd.

It remains to prove the result for n even. By (9.5.1), since $k > n/2$, then with $q = 2^*$, we have for each $|\alpha| \leq k - n/2$, $D^\alpha u \in L^q$, $q > n$. Hence, u lies in $W^{1,q}$, $q > n$. Therefore, Morrey applies to $D^\alpha u$ and it follows that u belongs to C^γ, $\gamma = k - (n/2) - 1$. $\qquad\square$

Corollary 9.5.1. *The following hold.*

a. *If $n = 1$, then $r = k - 1$ and $H^1(U)$ is continuously embedded in the continuous functions C^0.*

b. *If $n = 2$ or 3, then $r = k - 2$ and $H^2(U)$ is continuously embedded in the continuous functions C^0.*

c. *If $n > 3$, then $H^k(U)$ is embedded in the continuous functions C^0 for k at least $1 + \lceil n/2 \rceil$.*

Before ending this section we have an additonal result that will be useful in the next section. It is an applicaiton of the Extension and Theorem 9.5.3 (see [Evans (1998)]) that may be referred to as *Sobolev compact embidding.*

Theorem 9.5.7. *If U has C^1 bondary then $H^k(U)$ is a compact subspace of $L^2(U)$. In particular, $H^{k+1}(U)$ is a compact subset of $H^k(U)$.*

Finally, a comment on hypotheses. Several of the results of this section required that the domain U have C^1 boundary. Clearly, in the context of a domain with FEM partition the boundary is polygonal, not C^1. We understand this criteria as follows. If the original (pre-partitioned) domain for the PDE has C^1 boundary and the partitioned domain is contained within the original, then all of the results in this section apply to the FEM case. If the original domain bondary is not smooth, then we can approximate it closely with a C^1

curve inscribed in the domain. We then partition that domain and proceed.

Exercises

1. For the proof of Theorem 9.5.3 complete the following details.
    a. The sequence that converges to v in $W^{1,p}(\mathbb{R}^n)$ and is Cauchy in L^{p^*} also converges to v in L^{p^*}.
    b. Prove that

    $$\|u_m\|_{L^{p^*}(\mathbb{R}^n)} \leq C\|Du_m\|_{L^p(\mathbb{R}^n)},$$

    and $u_m \to v$ implies that

    $$\|v\|_{L^{p^*}(\mathbb{R}^n)} \leq C\|Dv\|_{L^p(\mathbb{R}^n)}.$$

2. Complete the proof of the final details of theorem Theorem 9.5.5.

9.6. Polynomial Interpolation on a Sobolev Space

Since Section 3.1, we have been working with piecewise polynomial interpolation associated to a finite element partition. In Section 5.1, we introduced the associated operator in order to faciliate the definition of collocation method. We now consider this concept, as defined on elements of a Sobolev space. With Corollary 9.5.1, we know that we can do polynomial interpolation on a Sobolev space H^s provided s is large enough relative to the dimension of the underlying real vector space. With this result in place, we move on to considerations associated to the piecewise polynomial spaces defined in Section 8.2. We now consider a function u with a sequence of interpolations u_n in a Sobolev space and ask when does the sequence of interpolations converge to u. It is well known that if we let the degree of the interpolations increase to ∞, then the sequence may not converge to u (see Exercise 4). On the other hand, if we fix the degree and take $h \to 0$, then convergence is assured, provided the associated family of partitons is regular (see Section 8.1). The purpose of this section is to prove this theorem.

We begin with a finite element partition of our compact domain $D = \cup_e E_e \subset \mathbb{R}^n$ with mesh parameter h equal to the maximal element diameter relative to the diameter of D. Next, we select the degree of the polynomials, k and the interpolation points Z_e in E_e. In this setting, the order of Z_e is no greater than the binomial coefficient C_k^{n+k-1}, the dimension of the space V_e^h of degree k polynomials in n variables. The basis for the polynomial space φ_i^e may be taken dual to the points Z_e; that is, $\varphi_i^e(z_{e,j}) = \delta_{i,j}$. Now, given u in $H^s(E_e) \subset C^0(E_e)$, we set the interpolation operator as $\mathcal{P}_{Z_e}^{h,k}(u) = \sum_i u(z_{e,i})\varphi_i^e$. Each $\mathcal{P}_{Z_e}^{h,k}$ is a bounded element of $Hom[H^s(E_e), V_e^h]$, $s \geq 1 + \lceil n/s \rceil$ (see Exercise 1). Since $\mathcal{P}_{Z_e}^{h,k}$ is the identity on V_e^h, then $\mathcal{P}_{Z_e}^{h,k}$ is a projection operator.

It is a simple matter to extend the idea of polynomials on an element to piecewise polynomials on D. In particular, we set $\mathcal{P}_Z^{h,k}(u) = \sum_e \sum_i u(z_{e,i})\varphi_i^e$, where $Z = \cup_e Z_e$. Furthermore, we suppose that there is sufficient element boundary information to ensure that $\mathcal{P}_Z^{h,k}(u)$ is continuous. In Section 8.2, we considered FEM partitions and associated piecewise polynomial interpolation. In this case, we write $\mathcal{P}_Z^{h,k}(u) = \sum_i u(P_i)\psi_i$, summing over all nodes P_i. Again $\mathcal{P}_Z^{h,k}$ is a bounded projection now defined on $H^s(U)$ with values in V^h, the space of continuous piecewise polynomial functions on D. We also need to extended the notation to include a reference element. In addition, we want the notation to carry an identifier for the particular partition.

We designate a reference interval (triangle, rectangle, tetrahedron and so forth) R. For each element E_e, there is a unique affine transformation $A_e^h : R \to E_e$ with linear part T_e^h and translation part t_e^h so that $A_e^h(x) = T_e^h(x) + t_e^h$. Further, we set B_e^h to be the inverse of A_e^h. Now, A_e^h determines a transformation of the polynomials φ_i on R to the polynomials φ_i^e given by $\varphi_i^e = \varphi_i \circ B_e^h = \mathcal{A}_e^h(\varphi_i)$. It is immediate that $\mathcal{A}_e^h$ is linear, non-singular.

We begin with a lemma. In this result, the norm is the Euclidean norm on $\mathbb{R}^n$.

Lemma 9.6.1. *The linear transformation T_e^h satisfies, $\|T_e^h\| \leq h_e/\rho_R$ and $\|(T_e^h)^{-1}\| \leq h_R/\rho_e$ where ρ_R (respectively ρ_e) is the* *diameter of the maximal in-sphere of R (respectively E_e) and h_R (h_e) is the diameter of R (E_e).*

Proof. To simplify notation we write A for A_e^h and T for T_e^h. We begin with $\|T\| = \sup_{x \neq 0} \|T\|/\|x\|$ and set $z = \rho_R x/\|x\|$. $\|z\| = \rho_R$ and

$$\|Tx\| = \frac{\|x\|}{\rho_R}\|Tz\| = \rho_R^{-1}\|Tz\|\|x\|$$

or equivalently,

$$\|T\| = \frac{1}{\rho_R} \sup_{\|z\|=\rho_R} \|Tz\|.$$

Next, we select a' and b' satisfying $z = b' - a'$. Hence, $\rho_R = \|b' - a'\|$. Taking c so that $a = a' + c$ and $b = b' + c$ lie in S_R, the in-sphere of R, we have $z = b - a$, $a, b \in S_R$. Therefore, we may consider the sup of $\|T(b - a)\|$ taken over all $a, b \in S_R$. Since $T(b - a) = T(b) - T(a) = A(b) - A(a) \in E_e$, then $\|T(b - a)\| \leq h_e$, and it follows that $\|T\|$ is bounded by h_e/ρ_R.

The proof of the bound for $\|T^{-1}\|$ is similar. $\qquad\square$

The affine map induces a bijection between $H^s(R)$ and $H^s(E_e)$. For v in $H^s(E_e)$ we denote the corresponding function by $\hat{v}$. In Section 9.2 we briefly introduced the semi-norm

$$|v|_m = \left[\sum_{|\alpha|=m} \int |D^\alpha v|^2 \right]^{\frac{1}{2}},$$

for $m \leq s$. The next theorem relates the value of the semi-norm over an element to the value over the reference element.

Theorem 9.6.1. *With the notation given above, there are constants C_1 and C_2 independent of R and E_e so that for each $m \leq s$, the semi-norms at R and E_e satisfy*

$$|\hat{v}|_{R,m} \leq C_1 \|T\|^m |\det T|^{-1/2} |v|_{e,m}, \tag{9.6.1}$$

$$|v|_{e,m} \leq C_2 \|T^{-1}\|^m |\det T|^{\frac{1}{2}} |\hat{v}|_{R,m}. \tag{9.6.2}$$

Proof. As in Lemma 9.6.1, we simplify the notation as no ambiguity will occur. We calculate for $\hat{x} \in R$,

$$|\hat{v}|_{R,m}^2 = \sum_{|\alpha|=m} \int_R |D^\alpha \hat{v}(\hat{x})|^2 d\hat{x}$$

$$= \int_{U_e} \sum_{|\alpha|=m} |D^\alpha \hat{v}(Bx)|^2 |\det T|^{-1} dx.$$

Now, v and $\hat{v}$ are functions from $\mathbb{R}^n$ to $\mathbb{R}$. Therefore, the derivative is a linear functoinal on $\mathbb{R}^n$. In particular, for v, the matrix representation is the gradient Dv and entries $D^\alpha v$, $|\alpha| = 1$. Furthermore, $Dv(x) \circ T = D\hat{v}(\hat{x})$. Hence, the entries of $D\hat{v}$ are the entries of Dv modified as linear combinations with the entries of T as coefficients. Continuing, the mth derivative is an m-linear functional on $\mathbb{R}^n$ which we can represent as the tensor product of linear functionals. In particular,

$$D^m \hat{v}(\hat{x})(\hat{y}_1, \ldots, \hat{y}_m) = D^m v(x)(T\hat{y}_1, \ldots, T\hat{y}_m)$$

$$= f_1 \otimes \cdots \otimes f_m(T\hat{y}_1, \ldots, T\hat{y}_m)$$

$$= \prod_i f_i(T\hat{y}_i) = \prod_i f_i \circ T(\hat{y}_i).$$

Therefore, if we take the absolute value and apply the basic properties of the linear transformation norm and multilinear transformations, we get an expression bounded by a multiple of $\|T\|^m$. Hence,

$$|\hat{v}|_{R,m}^2 = \int_{U_e} \sum_{|\alpha|=m} [D^\alpha \hat{v}(Bx)]^2 |det T|^{-1} dx$$

$$\leq C_1 \|T\|^{2m} |\det T|^{-1} \int_{U_e} \sum_{|\alpha|=m} [D^\alpha v(x)]^2 dx$$

$$= C_1 \|T\|^{2m} |\det T|^{-1} |v|_{e,m}^2.$$

Now, Eq. (9.6.1) follows. The verification of (9.6.2) is similar. $\qquad \square$

In the following result, we combine the lemma and the theorem.

Corollary 9.6.1. *Consider a finite element partition $D = \cup_e E_e$ for a mesh parameter h, associated to a reference element R and affine mappings $A_e^h = t_e^h \circ T_e^h$. If $v \in H^k(U_e)$ ($\hat{v} \in H^k(R)$ with $\hat{v} = v \circ A_e^h$), then for any $m \le k$, there are constants C_1 and C_2 so that the seminorms satisfy*

$$|\hat{v}|_{R,m} \le C_1 \left(\frac{h_e}{\rho_R}\right)^m |\det T_e^h|^{-\frac{1}{2}} |v|_{e,m}, \tag{9.6.3}$$

$$|v|_{e,m} \le C_2 \left(\frac{h_R}{\rho_e}\right)^m |\det T_e^h|^{\frac{1}{2}} |\hat{v}|_{R,m}. \tag{9.6.4}$$

There is one last technical result that we need before proving convergence of the interpolation operator. In this result, we connect the degree the polynomial interpolation k to the order of the Sobolev space $k + 1$. For convenience, we denote the polynomials on U of degree no larger than k by $\mathbb{P}_k$.

Lemma 9.6.2. *For v in $H^{k+1}(U)$ and p in $\mathbb{P}_k$, the following inequality holds,*

$$\inf_p \|v + p\|_{H^{k+1}(U)} \le C|v|_{k+1} \tag{9.6.5}$$

for a constant C that depends only on U.

Proof. Let $p_1, \ldots, p_m$ be a basis for the space of polynomials of degree no greater than k. Considering this space as a subspace of $H^{k+1}(U)$, select a dual basis of linear functionals on the polynomial space. By the Hahn–Banach theorem (Rudin, 1986), these bounded functionals may be extended to bounded functionals f_i of $H^{k+1}(U)$.

For any element of $H^{k+1}(U)$,

$$\|v\|_{H^{k+1}(U)} \le C|v|_{k+1} + \sum_i |f_i(v)|, \tag{9.6.6}$$

where C depends only on U (see Exercise 2). Select $p = -\sum_i f_i(v)p_i$. Then for each i, $f_i(v + p) = 0$. Therefore,

$$\|v + p\|_{H^{k+1}(U)} \le C|v|_{k+1},$$

since each $D^\alpha(p) = 0$, $|\alpha| = k + 1$, and therefore, $|v + p|_{k+1} \le C|v|_{k+1}$. $\square$

We now prove that in the case of a regular family of FEM partitions (see Definition 8.1.5) piecewise polynomial interpolation is convergent, $\mathcal{P}_Z^{h,k}(u) \to u$ in the Sobolev space norm. We begin with the reference element.

Lemma 9.6.3. *With the current notation, $|\hat{v} - \mathcal{P}_Y^{h,k}\hat{v}|_{R,m} \leq C|\hat{v}|_{R,k+1}$ in $H^{k+1}(R)$, where $n \leq k+1$ and $\mathcal{P}_Y^{h,k}$ denotes the degree k polynomial interpolation operator on R for a given point set Y.*

Proof. There are constants C_1 and C_2, the latter associated to the Sobolev embedding theorem, with

$$\|\mathcal{P}_Y^k \hat{v}\|_{H^m(R)} \leq \sum_i |\hat{v}(y_i)| \|\varphi_i\|_{H^m(R)}$$

$$\leq C_1 \|\hat{v}\|_\infty \leq C_1 C_2 \|\hat{v}\|_{H^{k+1}(R)}, \qquad (9.6.7)$$

where $y_i \in Y$ and the φ_i form a basis of the space of degree up to k polynomials on R. Hence, for any polynomial p of degree less than or equal to k,

$$|\hat{v} - \mathcal{P}_Y^k \hat{v}|_{R,m} \leq \|\hat{v} - \mathcal{P}_Y^k \hat{v}\|_{H^m(R)} = \|\hat{v} + p - p - \mathcal{P}_Y^k \hat{v}\|_{H^m(R)}$$

$$\leq \|\hat{v} + p\|_{H^m(R)} + \|p + \mathcal{P}_Y^k \hat{v}\|_{H^m(R)}$$

$$\leq \|\hat{v} + p\|_{H^m(R)} + \|\mathcal{P}_Y^k(p + \hat{v})\|_{H^m(R)}$$

$$\leq (1 + C_1 C_2)\|\hat{v} + p\|_{H^{k+1}(R)}, \qquad (9.6.8)$$

where the final inequality follows from (9.6.7). In particular, we have proved that there is a constant C_3 with

$$|\hat{v} - \mathcal{P}_Y^k \hat{v}|_{R,m} \leq C_3 \inf_p \|\hat{v} + p\|_{H^{k+1}(R)} \leq C|\hat{v}|_{R,k+1},$$

by Lemma 9.6.2. $\qquad\qquad\qquad\qquad\qquad\qquad\qquad\qquad\qquad\square$

The next step looks at the individual element. It is credited to two American mathematicians and dated in 1970. The theorem gives us an estimate for the polynomial interpolation error measured with the semi-norm.

Lemma 9.6.4. (*Bramble–Hilbert*) *Given the current notation, there is a constant C with*

$$|v - \mathcal{P}_{Z_e}^{h,k} v|_{e,m} \leq C \frac{h_e^{k+1}}{\rho_e^m} |v|_{e,k+1}$$

for v in $H^{k+1}(E_e)$.

Proof. We begin by setting the scene. To apply Lemma 9.6.3, start with v in $H^{k+1}(E_e)$ and set $\hat{v} = v \circ A_e^h$. Since A_e^h is affine (mapping polynomials to polynomials)

$$\mathcal{P}_Y^k \hat{v} = \mathcal{P}_Y^k (v \circ A_e^h) = \sum_i v \circ A_e^h(P_i) \varphi_i$$

$$= \sum_i v(P_{e,i}) \varphi_i = \sum_i v(P_{e,i}) \varphi_{e,i} \circ A_e^h = \mathcal{P}_{Z_e}^{h,k}(v) \circ A_e^h.$$

Hence, by Lemma 9.6.3 and Corollary 9.6.1 there are constants so that

$$|v - \mathcal{P}_{Z_e}^{h,k} v|_m = |\hat{v} B_e^h - (\mathcal{P}_Y^k \hat{v}) \circ B_e^h|_m$$

$$\leq C_1 \frac{h_R^m}{\rho_e^m} |\hat{v} - \mathcal{P}_Y^k \hat{v}|_m C_2 \frac{h_R^m}{\rho_e^m} |\hat{v}|_{k+1}$$

$$\leq C_3 \frac{h_e^{k+1}}{\rho_R^{k+1}} \frac{h_R^m}{\rho_e^m} |v|_{e,k+1} \leq C \frac{h_e^{k+1}}{\rho_e^m} |v|_{e,k+1}. \qquad \square$$

We can now prove convergence for piecewise polynomial interpolation with a regular sequence of refinements.

Theorem 9.6.2. *Suppose we have a regular sequence of finite element partitions of a domain D, then for any v in $H^{k+1}(D)$, there is a constant C with $\|v - \mathcal{P}_Z^{h,k}(v)\|_{H^m} \leq Ch^{k+1-m} \|v\|_{H^{k+1}}$, for any $m \leq k$.*

Proof. We calculate using the Bramble–Hilbert lemma,

$$\|v - \mathcal{P}_Z^{h,k} v\|_{H^m}^2$$

$$= \sum_e \|v - \mathcal{P}_{Z_e}^{h,k} v\|_{H^m(E_e)}^2 = \sum_e \sum_{s \leq m} |v - \mathcal{P}_{Z_e}^{h,k} v|_{e,s}^2$$

$$\leq \sum_e \sum_{s\leq m} C_s \left(\frac{h_e^{k+1}}{\rho_e^s}\right)^2 |v|_{e,k+1}^2 \leq \sum_e C_e \sum_{s\leq m} \left(\frac{h_e^{k+1}}{\rho_e^s}\right)^2 |v|_{e,k+1}^2$$

$$= \sum_e C_e \sum_{s\leq m} \left(\frac{\rho_e^m}{\rho_e^s}\right)^2 \left(\frac{h_e^{k+1}}{\rho_e^m}\right)^2 |v|_{e,k+1}^2$$

$$\leq \sum_e C_e K_e \left(\frac{h_e^{k+1}}{\rho_e^m}\right)^2 |v|_{e,k+1}^2 \leq \sum_e C_e K_e \left(\frac{h_e^{k+1}}{\rho_e^m}\right)^2 \|v\|_{H^{k+1}(E_e)}^2$$

$$\leq C_1 \sum_e \left(\frac{h_e^{k+1}}{\rho_e^m}\right)^2 \|v\|_{H^{k+1}(E_e)}^2,$$

where $K_e = \sum_{s\leq m}(\rho_e^m/\rho_e^s)^2$. By regularity, there is a constant κ with $h_e/\rho_e \leq \kappa$ or $1/\rho_e \leq \kappa/h_e$. Hence,

$$\frac{h_e^{k+1}}{\rho_e^m} \leq \frac{h_e^{k+1}}{h_e^m}\kappa^m = h_e^{k+1-m}\kappa^m.$$

We now are able to calculate

$$\|v - \mathcal{P}_Z^{h,k}v\|_{H^m}^2 \leq C_1 \sum_e \left(\frac{h_e^{k+1}}{\rho_e^m}\right)^2 \|v\|_{H^{k+1}(E_e)}^2$$

$$\leq \kappa^m C_1 \sum_e \left(h_e^{k+1-m}\right)^2 \|v\|_{H^{k+1}(E_e)}^2$$

$$\leq \kappa^m C_1 \left(h^{k+1-m}\right)^2 \sum_e \|v\|_{H^{k+1}(E_e)}^2.$$

If $\kappa \geq 1$, then

$$\|v - \mathcal{P}_Z^{h,k}v\|_{H^m}^2 \leq \kappa^{k+1} C_1 \left(h^{k+1-m}\right)^2 \|v\|_{H^{k+1}}^2.$$

if $\kappa < 1$, then

$$\|v - \mathcal{P}_Z^{h,k}v\|_{H^m}^2 \leq C_1 \left(h^{k+1-m}\right)^2 \|v\|_{H^{k+1}}^2.$$

The result now follows. $\qquad\qquad\qquad\qquad\qquad\qquad\qquad\qquad$ □

Exercises

1. Prove that $\mathcal{P}_{Z_e^{h,k}}$ defined on $C^0(E_e)$ with sup norm is a bounded operator. Conclude that $\mathcal{P}_{Z_e^{h,k}}$ is bounded on $H^s(E_e)$, $s \geq 1 + \lceil n/2 \rceil$.

2. Prove inequality (9.6.6) For convenience write $N(v) = |v|_{k+1} + \sum_i |f_i(v)|$. To begin, by Theorem 9.5.7, we know that $H^{k+1}(U)$ is compact in $H^k(U)$. Verify each of the following.

   a. If (9.6.6) fails then there is a sequence v_j in $H^{k+1}(U)$ with $\|v_j\|_{H^{k+1}(U)} = 1$ while $N(v_j) \to 0$.

   b. By compactness, conclude that v_j has an $H^{k+1}(U)$ convergent subsequence. Denote the limit by v. For notational convenience we may suppose that v_j is convergent.

   c. For the limit v, $|v|_{k+1} = 0$.

   d. The function v must be a polynomial of degree no greater than k.

   e. Prove that each $|f_i(v_j)| \to 0$, and conclude that $v = 0$. Equation (9.6.6) now follows.

3. Prove that ordinary polynomial interpolation is not convergent. What are the implications for spectral method?

4. Consider $u(x) = 1/(1 + x^2)$ on $[-3, 3]$.

   a. Consider the Taylor expansion of u at $x = 0$. Rewrite the expansion in terms of Lagrange polynomials to derive a sequence of polynomials of nth degree polynomials so that $\lim_{n \to \infty} u_n$ does not exist.

   b. The Weierstrass theroem states that the polynomials are dense in the $C^0[a, b]$. In this context the Bernstein polynomials b_n converge associated to u converge to u as $n \to \infty$ (Rudin, 1976). using a program, determine n so that $\sup_x \|u - u_n\|_\infty \leq 10^{-5}$ for the given case. Convergence for this sequence is generally considered to be too slow to be useful.

   c. Consdier the sequecne of interpolation points interoduced in Section 5.3. In this case, use Lagrange polynomials to interpolate u of successive sets of interpolations points. Determine the number of points (degree of the interpolation) necessary to ensure that $\sup_x \|u - u_n\|_\infty \leq 10^{-5}$.

5. Take a domain $D \subset \mathbb{R}^2$ with triangular FEM partition and a reference element R. Select an integer k and a set of interpolation points Y in R so the polynomials dual to the points in Y for a basis for the degree k polynomials in two variables. Distribute the points of Y to include the three vertices and $k - 1$ additional points evenly distributed points along each edge. The remaining points will be interior to R. Finally, we employ the notation used in this section.

   a. Prove that the given distribution of the points in R is possible.

      Distribute the polynomials φ_i to the elements via the usual affine transformations, do the same with the Y. We will denote the polynomials as φ_i^e. We will refer to the points as nodes. Let Z_e be the set of nodes for an element and set and $Z = \cup_e Z_e$.

   b. Prove that for each node there is continuous piecewiese polynomial function ψ_k determined as the join of polynomials φ_i^e. Furthermore, if p_i is a node, then $\psi_j(p_i) = \delta_{i,j}$.

      As usual we designate the space spaned by the ψ_i as V^h. and the piecewise interpolation operator by $\mathcal{P}_Z^{h,k} : C^0(D) \to V^h$.

   c. Using sup norm, prove that $\mathcal{P}_Z^{h,k}$ is bounded for each h with $\|\mathcal{P}_Z^{h,k}\|$ bounded by $\sum_i \|\varphi_i\|_\infty$.

   d. Prove that if u is C^2 on D, then $\mathcal{P}_Z^{h,k}(u) \to u$ as $h \to 0$ on sup norm. In this case convergence is $\mathcal{O}(h^2)$.

9.7. Convergence for Finite Element Method

The heavy work for convergence was done in Sections 5 and 6. At this stage, we need to only set the scene and then prove the result.

As above, U is a bounded open set with compact closed in $\mathbb{R}^n$. In addition, we have a linear elliptical PDE $Lu - f = 0$, where we modify f if necessary to ensure that boundary value zero. We suppose that the PDE satisfies the elliptical hypothesis (Eqs. (9.4.4) and (9.4.5)). Hence, the PDE has unique weak solution in $L^2(U)$. In addition, we suppose the function u is sufficiently smooth so that $H^k(U)$, the domain of the PDE, supports the Sobolev embedding theorem.

In addition we have a finite element partition of D with elements E_e associated to a partition with mesh parameter h. In order to ensure that piecewise polynomial interpolation associated to the FEM partition converges, we suppose that this partition is part of a regular family of partitions. Associated to the setting are the continuous, piecewise polynomial functions V^h of degree no greater than some predetermined integer k. As usual, functions in V^h are polynomials when restricted to an element.

Suppose that $u_{h,k}$ is the finite element solution to the PDE associated the partition. In particular, $u_{h,k}$ is a element of V^h which satisfies

$$0 = \int_U Lu_{h,k}v = \int_U fv$$

for every v in V^h. The following result refers to the bilinear form τ introduced in Sections 8.4 and 9.4. The result effectively impletes the proof of the convergence theroem. It dates from 1964.

Theorem 9.7.1. (*Céa*) *Let $\tau[w,v]$ be the bilinear form associated to the weak form of PDE $Lu - f = 0$. Suppose that τ satisfies the conditions (9.4.4) and (9.4.5), then there is a constant c so that the finite element solution satisfies*

$$\|u - u_{h,k}\|_2 \le c \inf_{v \in V^h} \|u - v\|_2,$$

where u is the weak solution determined in Theorem 9.4.1.

Proof. By the definition of the finite element method solution, $\tau[u_{h,k}, w] = 0$ for every w in V^h. Of course $\tau[u,w] = 0$ as well. Therefore, $\tau[u - u_{h,k}, u_{h,k} - w] = 0$.

We apply condition 9.4.5 to get for any w in V^h,

$$b\|u - u_{h,k}\|_2^2 \leq \tau[u - u_{h,k}, u - u_{h,k}]$$

$$= \tau[u - u_{h,k}, u - u_{h,k}] + \tau[u - u_{h,k}, u_{h,k} - w]$$

$$\leq \tau[u - u_{h,k}, u - w].$$

Next, we apply (9.4.4) to get

$$b\|u - u_{h,k}\|_2^2 \leq a\|u - u_{h,k}\|_2 \|u - w\|_2.$$

Hence,

$$\|u - u_{h,k}\|_2 \leq \sqrt{\frac{a}{b}} \|u - w\|_2$$

and the theorem is proved.   □

We now prove the convergence theorem. At this point there is little to do other than to reiterate accumulated hpotheses.

Theorem 9.7.2. (*FEM Convergence*) *Suppose the following.*

a. $Lu - f$ *is a linear elliptical PDE satisfying the elliptical hypothesis.*
b. *The domain* $D \subset \mathbb{R}^n$ *has a regular family of FEM partitions with mesh parameter* h.
c. *The FEM model is based on degree* k *polynomials.*
d. *The solution* u *belongs to the Sobolev space* $H^k(U)$, $\lceil n/2 \rceil + 1 \leq k$.

Then FEM solution $u_{h,k} \to u$ *in* L^2 *of order* h^{k+1}.

Proof. The proof is almost immediate. By the hypothesis, polynomial interpolation is well defined on u and $\mathcal{P}_Z^{h,k} u \to u$ order h^{k+1}. Taking $\mathcal{P}_Z^{h,k} u \in V^h$ then Céa's theorem implies

$$\|u - u_{h,k}\|_2 \leq \|u - \mathcal{P}_Z^{h,k} u\|_2 \leq Ch^{k+1} \|u\|_{H^{k+1}}$$

by Theorem 9.6.7 with $m = 0$.   □

Compare this theorem to Theorem 6.5.1. For the earlier result, the need to interpolate elements of the Sobolev space required that the approximate solutions lie in H^2. In Section 10.4, we extend Theorem 9.7.2 in an alternate direction.

Chapter 10

Collocation Method

Introduction

We defined collocation method (CM) in Chapter 5. At that time, we introduced two versions of the method. Spectral collocation using Cheybshev interpolation points is perhaps the oldest numerical PDE method (Hildebrand, 1974). In this chapter, we develop the theory supporting OSC or Gaussian collocation. For an overview of this method in one and several dimensions, see Bialecki and Fairweather (2001). Our development will follow the one laid out in Douglas and Dupont (1974) and Fairweather (2008). We introduce Legendre interpolation, we develop the collocation weak form and then we achieve convergence in a manner analogous to the corresponding result for FEM. In particular, in both cases, the result is a consequence of piecewise polynomial interpolation. Hence, arguments of this chapter rely on the material of Chapters 8 and 9. In turn, FEM and CM refer back to FDM (Chapters 2 and 7) or Runge–Kutta (Chapter 4) to resolve time stepping.

We conclude the chapter with a discussion of related concepts. As OSC collocation is well known to be superconvergent, we use this opportunity to discuss superconvergence for FEM (Fairweather, 2008; Wahlbin, 1995). Next, we discuss Gaussian collocation in 2D (Bialecki, 1998) and collocation based on triangular, tetrahedral partitions (Loustau *et al.*, 2013; Li *et al.*, 2008). There is an error estimated developed by Zienkiewicz (Zienkiewicz *et al.*, 2005). It uses superconvergence applied to Theorem 4.2.1. We will present this technique in this section.

Note that this not a complete list of collocation techniques. For instance, see Gupta and Kadalbajoom (2011), where the authors develop B-spline based collocation. B-spline curves and surfaces arise naturally when the data is generated by a *CAD* system. In this setting, there are B-spline-based FEM and collocation applications.

Finally, as we are primarily concerned with the spatial part of a PDE with one spatial dimension, then some of the material of the following sections is stated for ODE.

10.1. Interpolation with Legendre Polynomials

The Legendre polynomials are a sequence of orthogonal elements of $L^2[a, b]$. They arise naturally from a process that begins with Hermite interpolation. In Section 5.1, we introduced the cubic Hermite polynomials. These four polynomials are dual to the four degrees of freedom associated to the derivatives D^α, $\alpha = 0, 1$ at the two interval end points. Using these functions to interpolate over an interval ensures that the interpolating polynomial models both the value and slope of the function at the interpolation points. Hermite interpolation generalizes this idea to larger interpolation sets.

Definition 10.1.1. Let $a \le x_0 < x_1 < \cdots < x_{n-1} < x_n \le b$ and let f be a differentiable function on $[a, b]$. The expression,

$$h(x) = \sum_{i=0}^{n} f(x_i) H_i(x) + \sum_{i=0}^{n} f'(x_i) S_i(x)$$

is a *Hermite interpolation* of f provided the polynomials H_i and S_i satisfy the following:

(i) They are degree $2n + 1$.
(ii) They are dual to the DoF, $H_i(x_j) = \delta_{i,j}$, $S_i(x_j) = 0$, and $DH_i(x_j) = 0$, $\partial S_i(x_j) = \delta_{i,j}$.

Our first task is to show that the Hermite interpolation exists, that is, polynomials exist that satisfy the conditions of Definition 10.1.1.

We start with the $n + 1$, nth degree Lagrange polynomials,

$$N_i(x) = \frac{\prod_{j \neq i}(x - x_i)}{\prod_{j \neq i}(x_j - x_i)}.$$

Since $N_i(x_j) = \delta_{i,j}$, then the same is true for $h_i = N_i^2$. Furthermore, if we set $p(x) = \prod_{i=0}^{n}(x - x_i)$, then we may write $N_i(x) = p(x)/(x - x_i)p'(x_i)$ and compute

$$\frac{d}{dx}h_i = 2\frac{1}{x - x_i}\frac{p(x)}{p'(x_i)}\left(\frac{-1}{(x - x_i)^2}\frac{p(x)}{p'(x_i)} + \frac{1}{x - x_i}\frac{p'(x)}{p'(x_i)}\right).$$

Therefore, for $j \neq i$, $h_i'(x_j) = 0$.

At this point, we have $n + 1$ functions h_i, which are polynomials of degree $2n$. Next, we seek polynomials u_i and v_i satisfying the following,

a. Degree $u_i = 1$, $u_i(x_i) = 1$, $u_i' = -h_i'(x_i)$;
b. Degree $v_i = 1$, $v_i(x_i) = 0$, $v_i' = 1$.

Indeed, in this case, $H_i = u_i h_i$ and $S_i = v_i h_i$ satisfy the conditions of Definition 10.1.1. It is immediate that $v_i(x) = x - x_i$ is consistent with item b above. In turn, we set $u_i(x) = \alpha(x - x_i) + 1$ where $\alpha = -h_i'(x_i)$. Altogether, the existence of a Hermite interpolation is assured.

With existence in hand, the next task is to estimate the error.

Theorem 10.1.1. *Suppose that f is $2n + 2$ times continuously differentiable on (a, b) and that h is the Hermite interpolation of f with respect to a partition of $[a, b]$. If we denote the error, $e(x) = f(x) - h(x)$, then for each $\hat{x}$, there is $\xi_{\hat{x}}$ in the interval with*

$$e(\hat{x}) = \frac{f^{(2n+2)}(\xi_{\hat{x}})}{(2n + 2)!}p(\hat{x})^2. \tag{10.1.1}$$

If $M = \|f^{(2n+2)}\|_\infty$, then

$$|e(\hat{x})| \leq \frac{M}{(2n + 2)!}p(\hat{x})^2. \tag{10.1.2}$$

Furthermore, if f is a polynomial of degree less than or equal to 2n+1, then f = h.

Proof. The proof of (10.1.1) is analogous to the derivation of the error estimate for ordinary one variable polynomial interpolation. As in that case, it is an application of Rolle's theorem. Equation (10.1.2) is an easy consequence of (10.1.1). The final assertion is immediate.   □

The next step is to use the Hermite interpolation to estimate the integral of f. The Legendre polynomials arise from a special case.

Definition 10.1.2. Let f be a $C^{2n+2}[a,b]$ function. Then, the *Hermite quadrature* for the integral of f relative to a partition $a \le x_0 < x_1 < \cdots < x_n \le b$ is

$$\sum_{i=0}^{n} \gamma_i f(x_i) + \sum_{i=0}^{n} \delta_i f'(x_i), \qquad (10.1.3)$$

where the coefficients are given by

$$\gamma_i = \int_a^b H_i = \int_a^b [1 - 2h_i'(x_i)(x - x_i)]h_i, \quad \delta_i = \int_a^b S_i = \int_a^b (x - x_i)h_i.$$

As a consequence of Theorem 10.1.1, we estimate the Hermite quadrature error.

Theorem 10.1.2. *The error E for the Hermite quadrature is bounded by*

$$|E| \le \frac{M}{(2n+2)!} \int_a^b p^2 \le \frac{M}{(2n+2)!}(b-a)^{2n+3}.$$

Furthermore, if f is a polynomial of degree less than or equal to 2n+1, then the estimate for the integral of f given by (10.1.3) is exact.

Proof. It is only necessary to integrate the error estimate for Hermite interpolation and then note that $\|p^2\|_\infty \le (b-a)^{2n+2}$.   □

Next, we want to look at the case of (10.1.3) with $\delta_i = 0$ for each i. We see that this requirement corresponds to a particular choice of the x_i. For this case, the polynomials H_i are called Legendre polynomials.

In particular,

$$0 = \delta_i = \int_a^b (x - x_i) h_i = \int_a^b (x - x_i) N_i^2$$

$$= \int_a^b (x - x_i) \left(\frac{1}{x - x_i} \frac{p(x)}{p'(x)} \right)^2 = \frac{1}{p'(x)} \int_a^b p(x) N_i(x).$$

$$(10.1.4)$$

Hence, the condition $\delta_i = 0$ implies that p is L^2 orthogonal to the space of all polynomials of degree no larger than n. We now formally state this in the following theorem.

Theorem 10.1.3. *Suppose that $f \in C^{2n+2}$. Then the Hermite quadrature for f is given by weighted sum of function values,*

$$\int_a^b f = \sum_{i=0} \gamma_i f(x_i); \quad \gamma_i = \int_a^b H_i; \quad |E| \leq \frac{\|f^{(2n+2)}\|_\infty}{(2n+2)!} \int_a^b p^2,$$

$$(10.1.5)$$

if and only if p is L^2 orthogonal to the polynomials of degree no larger then n.

Proof. This result is a simple consequence of Theorem 10.1.2, Eqs. (10.1.3) and (10.1.4). $\qquad\square$

The following version of Theorem 10.1.3 will be useful.

Corollary 10.1.1. *For f in C^{2n+2}, the Hermite quadrature error is bounded by*

$$|E| \leq C\|f\|_{H^{2n+2}}, \qquad (10.1.6)$$

where C is a constant independent of f.

Proof. This result is an immedate consequence of Sobolev imbedding (see Section 9.5). $\qquad\square$

The next step is to determine when p is orthogonal to the space $\mathbb{P}_n$ of polynomials of degree no greater than n. As the Lagrange polynomials form a basis for this space, then we need only consider the integrals $\int N_i p = 0$.

Given $0 = \int_a^b S_i = \int_a^b (x - x_i) N_i$, we compute

$$\int_a^b N_j = \sum_{i=0}^n N_j(x_i) \int_a^b H_i = \int_a^b H_j = \int_a^b u_j h_j$$

$$= \int_a^b \left[1 - \frac{d}{dx} h_j(x_j)(x - x_j) \right] N_j^2$$

$$= \int_a^b N_j^2 - \frac{d}{dx} h_j(x_j) \int_a^b S_j = \int_a^b N_j^2.$$

We have now proved the following corollary. In the statement of the corollary, $\mathbb{P}_n$ denotes the space of polynomials of degree no larger than n.

Corollary 10.1.2. *The polynomial* $p(x) = \prod_{i=0}^n (x - x_i)$ *is orthogonal to* $\mathbb{P}_n$ *provided the Lagrange polynomials satisfy*

$$\int_a^b N_i = \int_a^b N_i^2. \tag{10.1.7}$$

In this case, the weights are given by $\int_a^b N_i > 0$

The special case of Hermite quadrature associated to (10.1.5) is called *Gaussian quadrature*. Gaussian quadrature with n points has the property that it is exact for polynomials of degree $2n + 1$. There is no numerical integration technique that is better. It all depends on the choice of the points $x_0, x_1, \ldots$. These points are commonly called *Gaussian quadrature points*.

We can solve for the sequence of functions $p_{n-1} = \prod_{i=0}^{n-1} (x - x_i)$ by starting with $p_0 = 1$ and then deriving succeeding functions using the Gram–Schmidt process. The roots of the polynomials are the points necessary to execute the process. The weights are given in Theorem 10.1.3. Additionally, the points and weights for the standard interval $[-1, 1]$ are given in most numerical analysis references, for instance, Hildebrand (1974).

The sequence of orthogonal polynomials $p_0, p_1, \ldots, p_{n-1}$ is referred to as the *Legendre polynomials*. They form a basis for $\mathbb{P}_{n-1}$. They are commonly written L_i.

Exercises

1. Verify that when n is 1, $a = x_0$ and $b = x_1$, then the Hermite cubics provide a Hermite interpolation of $[a, b]$.
2. Complete the verification that the functions $u_i h_i$ and $v_i h_i$ satisfy the requirements for a Hermite interpolation.
3. Prove Theorem 10.1.1.
4. With the notation developed in the section, suppose that w is a non-negative function on the interval $[a, b]$ and define quadrature for the weighted integral as

$$\int_a^b wf = \sum_{i=0}^{n} \zeta_i f(x_i) + \sum_{i=0}^{n} \eta f'(x_i) + E.$$

   Derive the expression for the weight parameters and error bound analogous to the corresponding results of this section.
5. Compute $p_0, \ldots, p_5$ for the interval $[0, 1]$.
6. Determine the Gaussian quadrature points for $n = 1$, 2, and 3 for the interval $[0, 1]$.
7. Extend the Gaussian quadrature to functions defined on rectangles. For a rectangle $[a, b] \times [c, d]$ and quadrature points x_i on the x-axis, y_i on the y-axis, the quadrature points for the rectangle will be the set of all pairs (x_i, y_j).
8. Extend the Gaussian quadrature to higher dimensional spaces.

10.2. The Collocation Inner Product

The *collocation inner product* is a symmetric bilinear form defined from the Gaussian quadrature expression for the integral. It provides the opportunity to define a collocation weak form. In this section, we define the symmetric bilinear form, prove some elementary properties and verify the claim that it supports the idea of a collocation weak form. We conclude the section by proving that the OSC or Gaussian collocation solution exists.

We begin with an interval $[a, b]$ that is partitioned into subintervals or elements $[\alpha_i, \alpha_{i+1}] = E_i$. We set $h_i = \alpha_{i+1} - \alpha_i$ and the mesh parameter $h = \max_i h_i$. The polynomial model will be the cubic Hermite polynomials $H_{i,j}, S_{i,j}$ where i identifies the element that supports the polynomial and $j = 1, 2$ identifies the element node. We complete the setup for Gaussian collocation introduced in Section 5.1 by identifying the collocation point set $Z = \cup_i Z_i$, for $Z_i = \{z_{i,1}, z_{i,2}\}$ the 2 point quadrature abscissa for the subinterval. We now define for $u, v \in C^0[a, b]$,

$$\sigma_C(u, v) = \sum_i h_i \sum_{j=1}^{2} \gamma_{i,j} u(z_{i,j}) v(z_{i,j}), \qquad (10.2.1)$$

where $\gamma_{i,j}$ is given in (10.1.5). Analogous to Section 9.6, we denote the piecewise Hermitian interpolation operator by $\mathcal{P}_Z^{h,3}$ where h identifies the partition by means of the mesh parameter h, Z is the interpolation point set and $\mathbb{P}_3$ is the polynomial space. With the notation established, we state the elementary properties of σ_C.

Lemma 10.2.1. *For σ_C defined on C^4, the following hold:*

a. *σ_C is symmetric, bilinear and semi-definite.*
b. *The kernel of σ_C contains the kernel of $\mathcal{P}_Z^{h,3}$.*
c. *$\sigma_C(f, f)$ defines a semi-norm.*
d. *$|\sigma_2(u, v) - \sigma_C(u, v)|$ is the numerical integration error for Gaussian quadrature. Hence, it is bounded above by the expressions (10.1.5) and (10.1.6).*
e. *If $u \in H_0^1 \cap V^h$ and $u(z) = 0$ for each z in Z, then $u = 0$.*

Proof. Parts a–d are immediate. To prove e, take u with $u(\alpha_0) = 0$. If $u'(\alpha_0) = 0$, then u on $[\alpha_0, \alpha_1]$ is a degree 3 polynomial with four roots. Hence, u is zero on the first element. But then the same must be true of the second element and so forth. Hence, $u = 0$.

On the other hand, if $u'(\alpha_0) \neq 0$, then we may repeat the prior argument and conclude that $u = 0$ or $u(\alpha_1)u'(\alpha_1) \neq 0$. Now by

Rolle's Theorem, there is a root of u in $[\alpha_1, \alpha_2]$. Therefore, we may suppose that $u(\alpha_2)u'(\alpha_2) \neq 0$ and so forth. But this argument leads to a contradiction as $u(b) = 0$. $\square$

We now state and prove the fundamental weak form result for collocation. We consider a linear PDE in one spatial variable,

$$u_t = a(x)u'' + b(x)u' + c(x)u - f(x) \quad \text{or} \quad L(u) = f, \qquad (10.2.2)$$

on the spatial interval $[a, b]$ with continuous coefficient functions. Since we are concerned here with the spatial rendering, then we may take the left-hand side to be zero and the equation to be an ODE. We will suppose that the ODE is elliptical satisfying the elliptical hypothesis (see Section 9.4). Therefore, the ODE is well-posed with unique weak form solution u in $H_0^1[a, b] \cup H^2[a, b]$.

Theorem 10.2.1. *The following holds in the Hilbert space $H_0^1[a, b] \cup H^2[a, b]$. The function u as a solution to $Lu = f$ implies that $\sigma_C(Lu, v) = \sigma_C(f, v)$. Conversely, any solution to $\sigma_C(Lu, v) = \sigma_C(f, v)$ is a collocation solution to $Lu = f$.*

Proof. The first part follows immediately from Lemma 10.2.1. In Definition 5.1.1, we identified the collocation solution as a piecewise polynomial function u_C with the property that $\mathcal{P}_Z^{h,3} L(u_C) = \mathcal{P}_Z^{h,3} f$. If we were to expand this idea to include elements of $H_0^1[a, b] \cup H^2[a, b]$, then we set $v = 1$ and the converse follows. $\square$

The purpose of the next result is to estimate the value of σ_C at the residual in terms of the L^2 inner product. In the statement of the following theorem, we denote the rth Legendre polynomial for the ith partition element by $L_{i,r}$.

Theorem 10.2.2. *Suppose that u and v are $C^2[a, b]$ piecewise polynomial functions formed with cubic Hermite polynomials associated to the partition. Then, there are constants, denoted k_i, such that*

a. $-\sigma_C(u'', v) = \sigma_2(u', v') + [u'v]_a^b + \sum_i k_i h_i^5 \int_{\alpha_{i-1}}^{\alpha_i} (L_{i,4}'')^2$,

b. If $u(a) = u(b) = 0$, $u \neq 0$, then $0 < \sigma_2(u', v') \leq -\sigma_C(u'', u)$.

Proof. For the proof of a, we use the continuity of the Hermite piecewise polynomial function. We compute using the elementary properties of Gaussian quadrature,

$$-\sigma_C(u'', v) = -\int_a^b (u''v) + \left(\sigma_2(u'', v) - \sigma_C(u'', v)\right)$$

$$= \sum_i \int_{\alpha_{i-1}}^{\alpha_i} (u'v') - [u'v]_{\alpha_{i-1}}^{\alpha_i} + \left(\sigma_2(u'', v) - \sigma_C(u'', v)\right)$$

$$= \int_a^b (u'v') - [u'v]_a^b$$

$$+ \sum_i \left(\int_{\alpha_{i-1}}^{\alpha_i} [k_i(L_{i,4}'')^2 + p_i] - \sigma_C(k_i L_{i,4}'')^2 + p_i) \right),$$

where $L_{i,4}$ is a degree 4 Legendre polynomial, p_i has degree less than four and $\sum_i k_i(L_{i,4}'')^2 + p_i$ interpolates $u''v$.

The additional hypothesis implies that $[u'v]_a^b = 0$ and that $\sigma_2(u', v')$ is a true norm. Further, if $u = v$ in (a), then the right-hand term is positive. We have verified b. $\qquad\square$

In the next result, we prove existence for the collocation solution.

Theorem 10.2.3. *The form $\tau_C[n, v]$ associated to $\sigma_C(Lu, v)$ is coercive, provided $a(x)$ is bounded above by $\alpha < 0$ and $c \geq 0$. Hence, existence for the collocation solution to the given model follows.*

Proof. We verify $\sigma_C(Lu, u) \geq 0$ by (10.2.1). By Theorem 10.2.2c, $a(x)\sigma_C(u'', u) > -\alpha\|u\|_{H_0^1}^2 > 0$ for any $u \neq 0$. It remains to consider $\sigma_C(u', u)$. Now, by the zero boundary value assumption,

$$\int_a^b uu' = \frac{1}{2}[u^2]_a^1 = 0,$$

since $u \in H_0^1[a, b]$. Therefore, $\sigma(u, u') \to 0$ as $h \to 0$. Hence, for h sufficiently small

$$\tau_C[u, u] > -\alpha \|u\|_{H_0^1}^2 + \left[\inf_x c(x)\right] \sigma_C(u, u) > 0.$$

The final assertion follows from the Lax–Milgram Theorem (Section 9.4). □

Exercises

1. Referring to Theorem 10.2.2a, derive an expression for k_i.
2. With the hypothesis of Theorem 10.2.3, prove that L is non-singular.

10.3. Convergence for Gaussian Collocation

As in Section 10.2, we consider the linear ODE (10.2.2) given by $a(x)u'' + b(x)u' + c(x)u = f(x)$ or $L(u) = f$ on the interval $[a, b]$ with continuous coefficient functions. We will suppose that the ODE is elliptical satisfying the elliptical hypothesis (see Section 9.4). Therefore, the ODE is well-posed with unique weak form solution in $H_0^1[a, b]$. In addition, we suppose that $a(x)$ is strictly negative, bounded and bounded away from zero (see Definition 9.4.2, uniformly elliptical). Also, we require that c is not negative (see Definition 9.4.2, uniformly elliptical). We take $0 > \alpha \geq a(x)$ for all x. Additionally, we can normalize 10.2.2 so that $a(x) = -1$, a negative constant. In order to apply (10.1.5) and (10.1.6), we will suppose that u is sufficiently smooth. By Theorem 10.2.3, the collocation solution exists.

We proceed directly to the convergence theorem.

Theorem 10.3.1. *Suppose that the solution to the ODE belongs to $H_0^1 \cap H^6$ for the interval $[a, b]$, then the collocation solution u_C converges to u order 4 in L^2 norm.*

Proof. For convenience, we write $w = u - u_C$. By the elliptical hypothesis, L is non-singular. We write L^σ for the adjoint of L with regard to the L^2 inner product σ_2 and write q for the L^σ pre-image of w. In particular,

$$\|q\|_{H^2} \leq C\|w\|_2. \tag{10.3.1}$$

Since u_C is the collocation solution, then $0 = \sigma_C(Lw, v)$ for any piecewise polynomial $v \in V^h$. We compute

$$\|w\|_2^2 = \sigma_2(w, L^\sigma q) = \sigma_2(Lw, q)$$
$$= \sigma_2(Lw, q - v_q) + \big[\sigma_2(Lw, v_q) - \sigma_C(Lw, v_q)\big]$$
$$\leq \|Lw\|_2 \|q - v_q\|_2 + E[(Lw)v_q],$$

where the last term denotes the Gaussian quadrature error for the product integrand. If we take v_q to be the piecewise polynomial interpolation of q, then by Theorem 9.6.2

$$\|w\|_2^2 \leq C\|w\|_{H^2}\|q\|_{H^2}h^2 + E[(Lw)v_q]. \tag{10.3.2}$$

We state the following inequality without proof. It is a consequence of (10.1.5) for the particular functions at hand.

$$E[(Lw)v_q] \leq C\big[h\|w\|_2 + h^4\|u\|_{H^6}\big]\|v_q\|_{H^1}. \tag{10.3.3}$$

Now we proceed. Since v_q piecewise interpolates q, then by Theorem 9.6.2,

$$\|v_q\|_{H^1} \leq \|q - v_q\|_{H^1} + \|q\|_{H^1} \leq Ch\|q\|_{H^2} + \|q\|_{H^1} \leq \kappa\|q\|_{H^2}.$$

We apply (10.3.3) to (10.3.2) and this last inequality to get

$$\|w\|_2^2 \leq C\big[h^2\|w\|_{H^2} + h\|w\|_2 + h^4\|u\|_{H^6}\big]\|q\|_{H^2}$$

and by (10.3.1)

$$\|w\|_2^2 \leq C\big[h^2\|w\|_{H^2} + h\|w\|_2 + h^4\|u\|_{H^6}\big]\|w\|_2.$$

Therefore, for small h

$$\|w\|_2 \leq C\big[h^2\|w\|_{H^2} + h^4\|u\|_{H^6}\big]. \tag{10.3.4}$$

The next step is to find a bound for $\|w\|_{H^2}$ as a linear combination of $\|w\|_2$ and $\|u\|_{H^s}$ for several values of s. Since the piecewise polynomial space V^h is finite dimensional, it is closed in $L^2[a, b]$. Hence, it has an orthogonal complement. Hence, given v in L^2 and $\bar{v}$ in V^h, there is a unique $\hat{v}$ with $\sigma_2(v - \hat{v}, \bar{v}) = 0$. Furthermore, we may write $\hat{v} = Qv$ for a linear projection Q. In other words, Qv is the V^h component of v with orthogonal complement $v - Qv$.

We compute

$$\sigma_2(Qu'' - u_C'', \bar{v}) = \sigma_2(Qu'' - u'' + u'' - u_C'', \bar{v}) = \sigma_2(w'', \bar{v})$$

$$= \sigma_2(Lw - b(x)w' - c(x)w, \bar{v})$$

$$= \sigma_2(Lw, \bar{v}) - \sigma_2(b(x)w' + c(x)w, \bar{v})$$

$$= \sigma_2(Lw, \bar{v}) - \sigma_C(b(x)w' + c(x)w, \bar{v}) = E[(Lw)\hat{v}].$$

Setting $\bar{v} = Qu'' - u_C''$ and applying the following inequality

$$E[(Lw)\hat{v}] \leq C\big[\|w\|_2 + h^3\|u\|_{H^5} + h^4\|u\|_{H^6}\big]\|\hat{v}\|_2, \qquad (10.3.5)$$

yields

$$\|Qu'' - u_C''\|_2 \leq C\big[\|w\|_2 + h^3\|u\|_{H^5} + h^4\|u\|_{H^6}\big].$$

Note that (10.3.5) is related to (10.3.3). Next we have

$$\|w''\|_2 = \|u'' - Qu''\|_2 + \|Qu'' - u_C''\|_2$$

$$\leq \|u'' - Qu''\|_2 + C\big[\|w\|_2 + h^3\|u\|_{H^5} + h^4\|u\|_{H^6}\big]$$

$$\leq C\big[\|w\|_2 + h^2\|u\|_{H^4} + h^3\|u\|_{H^5} + h^4\|u\|_{H^6}\big].$$

For the next step, we apply the following two inequalities.

$$\|w\|_{H^2} \leq \|w\|_{H^1} + \|w''\|_2; \quad \|w\|_{H^1} \leq C\big[h^{-1}\|w\|_2 + h\|w\|_{H^2}\big].$$

Applying these to the prior expression we have

$$\|w\|_{H^2} \leq C\big[h^{-1}\|w\|_2 + h\|w\|_{H^2} + h^2\|u\|_{H^4} + h^3\|u\|_{H^5} + h^4\|u\|_{H^6}\big].$$

For h small,

$$\|w\|_{H^2} \leq C\big[h^{-1}\|w\|_2 + h^2\|u\|_{H^4} + h^3\|u\|_{H^5} + h^4\|u\|_{H^6}\big].$$

Inserting this expresssion for $\|w\|_{H^2}$ into (10.3.4) completes the proof.

$$\|w\|_2 \leq C\big[\|w\|_2 + h^4\|u\|_{H^4} + h^5\|u\|_{H^5} + h^6\|u\|_{H^6} + h^4\|u\|_{H^6}\big]$$

or

$$\|w\|_2 \leq Ch^4\|u\|_{H^6}. \qquad \square$$

10.4. Remarks on Collocation and Superconvergence

In this section, we take a second look at Theorems 9.7.2 and 10.3.1. We begin with the obvious question of spatial dimension.

In Bialecki (1998), the author extends Gaussian collocation to higher spatial dimensions. As in the 1D case, the necessary level of smoothness for u is higher than that needed for FEM. And in turn the order of convergence is also larger than for FEM (Theorem 9.7.2). This result is restricted to rectangular or hexahedral partitions. In Li *et al.* (2008), the authors introduce collocation in 2 and 3 spatial dimensions for the Poisson equation. Their approach is similar in some respects to the discontinuous Galerkin method (Hesthaven and Warburton, 2008). The result holds for general partitions. In Loustau *et al.* (2013), we extended the work of Li *et al.* to general elliptical order 2 PDE. Subsequently, we have improved on the theory presented in that reference. The technique is significantly faster compared to FEM with numerically evaluated integrals (see Section 1.4). But, it must be considered as an area of ongoing research.

Spectral collocation techniques remain popular even though they are not supported mathematically. They are also remarkably reliable. We expect that one day there will be a theory to justify spectral collocation. Of course, we have the Weierstrass theorem [Rudin (1976)]. But given a continuous function on a compact set, the only sequence of polynomials we know of are the Bernstein polynomials. But they converge so slowly that this procedure cannot be used effectively.

The order of convergence in Theorem 10.3.1 is greater than the order of convergence given in Theorem 9.7.2. This gives rise to the following concept.

Definition 10.4.1. Suppose that $\hat{u}_\lambda$, $\lambda \in \Lambda$ is a family of approximate solutions to a PDE with solution u. Further, suppose that the approximate solutions converge to u order $r + \rho$, where r is the order of convergence for FEM. Then we say that $\hat{u}_\lambda$ is *superconvergent* order ρ.

The best known result on superconvergence states that FEM in one spatial dimension is superconvergent order 1 at the element nodes. The proof uses the Green's function. In his monograph, Wahlbin (1995) develops the theory for superconvergence.

There is an interesting error estimation procedure that uses nodal superconvergence (Zienkiewicz *et al.*, 2005). In this case, there is a PDE with solution u and FEM solution u_h. The researcher would choose a node P of interest and select a set of elements E_k, so that P is interior to $C = \cup_k E_k$. In addition, we require the number of nodes P_i, $i = 1, 2, \ldots, \delta$ in C distinct from P is no less than the dimension of V_e^h. In this case, we call C a *patch* for P. Select a basis for the polynomial space and use the nodes P_i to interpolate u_h across C. As there are more nodes than polynomial coefficients to estimate, then we will need to use least squares.

If we have superconvergence at the nodes, then we would expect that least squares would produce a better approximation of u in a neighborhood of P. We denote this function as $\hat{u}_P^h$. In turn, suppose that we were able to execute the least squares process using the actual values of u at the nodes. We denote this function by u_P^h. Further, we suppose that our family of partitions and refinements is selected so that once a point in D is a node in a partition then it remains a node in all subsequent refinements. We saw in Section 8.1 that such refinement processes exist for triangles and rectangles. Now, it is immediate that $\hat{u}_P^h \to u_P^h$ as $h \to 0$ and that the convergence is superconvergence.

Using Theorem 4.2.1, the difference between the two estimates for u in the neighborhood of P is an estimate of the FEM error. In Boroomand and Zienkiewicz (1997a, 1997b), the authors present extended examples of this process. The numerical evidence presented there is compelling.

Finally, the authors remark that there is no assurance that the least squares process will yield a well-posed problem. They do conjecture that if the points P_i are sufficiently irregular, then the linear system will be non-singular.

Bibliography

Ablowitz, M. J. and Fokas, A. S. (2003). *Complex Variables, Introduction and Applications*, 2nd edn. (Cambridge University Press).

Ahlfors, L. V. (1979). *Complex Analysis, an Introduction to the Theory of Analytic Functions of a Single Variable*, 3rd edn. (Mc Graw-Hill).

Atkinson, K. E. (1989). *An Introduction to Numerical Analysis*, 2nd edn. (J. Wiley).

Atkinson, K. E. and Han, W. M. (2009). *Theoretical Numerical Analysis, A Functional Analysis Framework*, 3rd edn. (Springer).

Aw, A. and Rascle, M. Resurrection of second order models for traffic flow. *SIAM J. Appl. Math.*, 60, 916–938.

Bellomo, N. and Dogbe, C. (2011). On the modeling of traffic and crowds: A survey of models, speculations and perspectives. *SIAM Review*, 53, 409–463.

Bellomo, N., Lods, B., Revelli, R. and Ridolfi, L. (2008). *Generalized Collocation Methods, Solutions to Nonlinear Problems* (Birkhauser).

Bialecki, B. (1998). Convergence Analysis of Orthogonal Spline Collocation for Elliptical Boundary Value Problems. *SIAM J. Numer. Anal.*, 35(2), 617–637.

Bialecki, B. and Fairweather, G. (2001). Orthogonal spline collocation methods for partial differential equations. *J. Comp. Appl. Math.*, 128(1–2), 55–83.

Boroomand, B. and Zienkiewicz, O. C. (1997a). Recovery by eqiulibrium patches (REP). *Int. J. Numer. Meth. Eng.*, 40, 137–154.

Boroomand, B. and Zienkiewicz, O. C. (1997b). An improved REP recovery and the effectivity robustness test. *Int. J. Numer. Meth. Eng.*, 40, 3247–3277.

Braithewaite, R. B. (1953). *Sceintific Explanation, A Study of the Function of Theory, Probability and Law in Science* (Harper Torchlight).

Brenner, S. C. and Scott, L. R. (2008). *The Mathematical Theory of Finite Elements Methods*, 3rd edn. (Springer).

Chen, S. Q., Wang, Y. W. and Wu, X. H. (2011). Rational spectral collocation method for a coupled system of singularly perturbed boundary values problems. *J. Comput. Math.*, 29(4), 458–473.

Chung, T. J. (2002). *Computational Fluid Dynamics* (Cambridge University Press, Cambridge).

Clough, R. W. (1980). The finite element method after twenty-five years: A personal view. *Comput. Structures*, 12, 361–370.

Douglas, J. and Dupont, T. (1974). *Collocation Methods for Parabolic Equations in a Single Space Variable* (Springer, New York).

Ekstrom, E. and Tysk, J. (2007). *Convexity Theory for the Term Structure Equation* (Univ. Uppsala D. M. Report 12).

Espinosa, A. and Jorgenson, J. (2003). *An Elementary Derivation of the Black–Scholes–Merton Formula* (available form the author).

Evans, L. C. (1998). Partial differential equations. *Amer. Math. Soc.*, 19, 662.

Fairweather, G. (2008). *Numerical Solutions to Two-point Boundary Value Problems* (available from the author).

Gerald, C. F. and Wheatley, P. O. (2003). *Applied Numerical Analysis*, 7th edn. (Pearson-Addison Wesley, Boston).

Gresho, P. M. and Sani, R. L. (1998). *Incompressible Flow and the Finite Element Method* (J. Wiley, USA).

Grcar, J. P. (2011). John von Neumann's analysis of Gaussian elimination and the origins of modern numerical analysis. *SIAM Rev.*, 53, 607–682.

Grindrod, P. (1988). Models of individual aggregation in single and multi-species communities. *J. Math. Biol.*, 26, 651–660.

Grindrod, P. (1991). *Patterns and Waves, The Theory and Applications of Reaction-Diffusion Equations* (Oxford University Press, UK).

Guazzelli, E. and Morris, J. F. (2012). *A Physical Introduction to Suspension Dynamics* (Cambridge, UK).

Gupta, V. and Kadalbajoom, M. K. (2011). A layer adaptive B-spline collocation method for singularly perturbed one-dimensional parabolic problem with a boundary turning point. *Numer. Methods Partial Differential Equations*, 27, 1143–1164.

Gustafsson, B., Kreiss, H. O. and Oliger, J. (2013). *Time-Dependent Problems and Difference Methods*, 2nd edn. (J. Wiley, USA).

Helburg, D. and Johansson, A. F. (2009). On the controversey around Daganzo's requium for and Aw–Rascle's resurrection of second-order traffic flow models. *Euro Phys. J.*, 69, 539–562.

Heubner, K. H., Dewhirst, D. L., Smith, D. E. and Byrom, T. G. (2001). *The Finite Element Method for Engineers*, 4th edn. (Wiley Interscience, USA).

Hesthaven, J. S. and Warburton, T. (2008). *Nodal Discontinuous Galerkin Methods, Algorithms, Analysis and Applications* (Springer, USA).

Hildebrand, F. B. (1974). *Introduction to Numerical Analysis*, 2nd edn. (Dover).

Jones, R. T. (1990). *Wing Theory* (Princeton University Press, USA).

Keller, E. F. (1980). Assessing the Keller Segel Model: How has it fared? *Lecture Notes in BioMathematics*, 38 (Springer, USA), 379–387.

Lax, P. D. and Richtmyer, R. D. (1956). Survey of the stability of linear finite difference equations. *Comm. Pure Appl. Math.*, 9, 267–293.

Li, Z.-C., Lu, T. T., Hu, H. Y. and Cheng, H. D. (2008). *Trefftz and Collocation Methods* (WIT Press, UK).

Loustau, J. and Bob-Egbe, B. (2010). Computing corresponding values of the Neumann and Dirichlet boundary values in for incompressible Stokes flow. *Involve J. Math.*, 3–4, 459–474.

Loustau, J. and Dillon, M. (1993). *Linear Geometry with Computer Graphics* (Marcel Dekker, USA).

Loustau, J., Lindorff, A., Irwin, S. and Svadlenka, J. (2013). Discontinuous polynomial collocation in two dimensions. *Risk and Decision Analysis*, 4, 47–57.

Lyness, J. N. and Jespersen, D. (1975). Moderate degree symmetric quadrature rules for the triangle. *J. Inst. Maths Applies*, 15, 19–32.

Mackenzie, D. (2013). If it smells like a traffic jam. *SIAM News*, 46.

Marsden, J. E. and Tromba, A. J. (2003). *Vector Calculus*, 5th edn. (W. H. Freeman, UK).

Munson, B. R., Young, D. F. and Okiishi, T. H. (2005). *Fundamentals of Fluid Mechanics*, 5th edn. (J. Wiley).

Murray, J. D. (2003). *Mathematdical Biology II: Saptial Models and Biomedical Applications* (Springer, USA).

Phillips, G. M. (2003). *Interpolation and Approximation by Polynomials* (*Canadian Math Soc.*, Springer).

Previte, J. P. and Hoffman, K. A. (2013). Period doubling cascades in a predator–prey model with a scavenger. *SIAM Rev.*, 55, 523–546.

Royden, H. L. (2011). *Real Analysis*, 4th edn. (Pearson, UK).

Rudin, W. (1976). *Elements of Real Analysis*, 3rd edn. (McGraw-Hill, New York).

Rudin, W. (1986). *Real and Complex Analysis*, 3rd edn. (McGraw-Hill, New York).

Shao, D., Levine, H. and Rappel, W. (2012). Coupling actin flow. Adhesion and morphology in a computational cell motility model. *Proc. Nat. Acad. Sci.*, 109, 6851–6856.

Shen, J., Tang, T. and Wang, L. L. (2011). *Spectral Methods, Algorithms, Analysis and Applications* (Springer, New York).

Simmons, G. F. and Robertson, J. S. (1991). *Differential Equations with Applications and Historical Notes*, 2nd edn. (McGraw-Hill, New York).

Strikwerda, J. C. (2004). *Finite Difference Schemes and Patial Differential Equations* (SIAM).

Strychalski, W. and Guy, R. D. (2013). A computational model of bleb formation. *Math. Med. Biol.*, 30, 115–130.

Theodorsen, T. (1932). Theory of wing sections of arbitrary shape. *NACA Technical Report*, 411 (NACA).

Theodorsen, T. and Garrick, I. E. (1934). General potential theory of arbitrary wing sections. *NACA Technical Report*, 452 (NACA).

Thomas, J. W. (1995). *Numerical Partial Differential Equations, Finite Difference Methods* (Springer, New York).

Thomas, J. W. (1999). *Numerical Partial Differential Equations, Conservation Laws and Elliptical Equations* (Springer).

Topper, J. (2005). *Financial Engineering with Finite Elements* (J. Wiley).

van der Berg, H. (2011). *Mathematical Models of Biological Systems* (Oxford University Press).

Wahlbin, L. B. (1995). *Superconvergence in Galerkin Finite Element Methods* (Springer).

Wodarz, D. and Komarova, N. L. (2015). *Dynamics of Cancer, Mathematical Foundations of Oncology* (World Scientific).

Xiu, D. B. (2010). *Numerical Methods for Stochastic Computations, A Spectral Method Approach* (Princeton Univ. Press).

Zienkiewicz, O. C., Taylor, R. L. and Nithiarasu, P. (2005). *The Finite Element Method for Fluid Dynamics*, 6th edn. (Elsevier North Holland).

Index

A

affine mapping, 31, 86, 239, 326

B

Banach space, 5
  L^1, 7
  L^∞, 7
  L^p, 7
  Lax–Richtmyer theorem, 195
  uniform boundedness
    principle (UBP), 197
Black–Scholes equation
  derived from Ito's lemma, 121
  derived from the heat
    equation, 131
boundary values
  Dirichlet boundary values, 81
  Neumann boundary values, 81

C

Céa's Theorem, 335
coercive form, 313
collocation, 122
  Chebyshev collocation points,
    142
  collocation inner product, 344
  collocation points, 122
  convergence theorem, 347
  degree of freedom, 124
  Dirichlet boundary values, 133
  Gaussian collocagtion, 125
  mesh parameter, 123, 344
  orthogonal spline collocation
    (OSC), 125
  reference element, 123
  spectral collocation, 140, 150
condensation, 270

D

degree of freedom, 75, 124–125,
  241–242, 247, 250–252, 265, 267,
  271, 274
density proof, 299
diffusion equation
  rendered with FDM, 62
divided difference, 18

E

elliptical PDE, 11, 16
  elliptical hypothesis, 313
  Helmholtz equation, 85
  Laplace equation, 41, 44
  Lax–Milgram theorem, 313
  uniformly elliptical, 311
error estimation, 105

F

Finite Difference Method (FDM)
Courant–Friedrichs–Lewy
theorem, 221
analytic domain of definition,
220
consistency for the Heat
equation, 198
consistent, 194
convergence (Poisson
equation), 225
convergent, 195
Courant number, 211
Crank–Nicolson, 54, 66
difference operator, 201,
205–207
Dirichlet boundary values, 53,
226, 275
displacement operator, 201,
205
explicit, 53, 66
finite difference scheme, 193
fractional step method, 64
implicit, 53, 66
Lax–Friedricks method, 215
Lax–Richtmyer theorem, 195
Lax–Wendroff, 213
leap frog, 64
mesh parameter, 193, 195,
198, 211, 223
Neumann stability, 57, 219
Neumann stability (hyperbolic
PDE), 212
Neumann stability (parabolic
PDE), 66
numerical domain of
definition, 220
Peaceman–Rachford, 64
stability, 56, 195, 219
trapezoid, 176
unconditionally unstable, 212

Finite Element Method (FEM),
84, 269
assembly, 79
condensation, 270
degree of freedom, 75, 125,
241–242, 247, 250–252, 265,
267, 271, 274
Dirichlet boundary values,
81–82, 88–89, 173, 175–176,
229, 274, 277, 306
element, 231
element diameter, 235
element patch, 351
error estimate, 351
FEM steps, 83
finite element partition, 231
local error estimate, 351
mesh parameter, 75, 160, 184,
236, 238, 265, 326, 329,
335–336
Neumann boundary values,
81, 90, 162, 168, 174, 176
nodes, 231
piecewise polynomial
interpolation, 78, 245
quasi-uniform, 237
reference element, 30, 86, 239,
244, 247, 255, 259, 326–327,
329–330
regular partition, 237, 331,
336
serendipity partition, 254, 261
Union Jack pattern, 233
weak form, 77
fluid
Bernoulli equation, 168
continuity equation, 168
incompressible, 39, 168
irrotational, 39
laminar, 38
Navier–Stokes equation, 167
Newtonian fluid, 166

normal stress, 165
Reynolds number, 15, 136,
  158, 167
shearing stress, 165
Stokes equation, 167
viscosity, 167
Fourier transform, 50
convoluted product, 50
discrete Fourier transform, 58
properties, 59

G

Gaussian quadratures, 342
Gelfand theorem, 218

H

Hölder continuous, 322
heat equation
  derived, 48
  Fourier transform solution,
    52
  rendered with FDM, 53, 62
Helmholtz equation
  derived from the wave
    equation, 85
  rendered with FEM, 88
Hermite interpolation, 338–339
Hermite polynomials, 123, 244,
  338, 344
Hermite quadrature, 340
Hilbert space, 291
  L^2, 6
  Lax–Milgram theorem, 313
hyperbolic PDE, 11, 16
  wave equation, 85

I

inner product, 6
collocation inner product, 344
Hermitian form, 7
Sobolev inner product, 291

inner product space, 6
Cauchy–Schwartz inqualiity, 6
Hilbert space, 6, 8, 291

J

Joukowski airfoil, 37

L

Lagrange polynomials, 19, 29, 78,
  141, 167, 172, 339, 341
Lax–Milgram theorem, 313
Lax–Richtmyer theorem, 195
Legendre polynomials, 343
Liapunov function
  definition, 100
  existence, 101
  stability, 100
linear transformation norm, 7
Lipschitz continuous, 232, 265
Lipschitz domain, 232

M

mesh parameter, 75, 123, 160, 184,
  193, 195, 198, 211, 223, 236, 238,
  265, 326, 329, 335–336, 344
metric, 5
metric space, 5
  complete, 5
multi-index, 241, 286

N

Navier–Stokes equation
  derivation, 167
norm, 4
  normed linear space, 5
  semi-norm, 5, 292, 327
  Sobolev norm, 291
normed linear space, 5
  Banach space, 5, 8, 195
  Hilbert space, 6, 8, 291

O

ODE
  asympotically stable, 97
  center, 97
  equilibrium point, 97–98, 100,
    101
  error estimation, 105
  Liapunov function, 100
  predator/prey population
    model, 95, 98–99, 113
  stability, 97
  stagnation point, 97
  Volterra model, 98
operator norm, 7

P

parabolic PDE, 11, 16
  diffusion equation, 62
  FDM consistency, 198
  heat equation, 48
  Laplace equation, 74, 77, 84
  Neumann stability, 66
PDE
  Black–Scholes equation, 121
  bond price equation, 134
  cell chemotaxis, 159
  coercive, 313
  continuity equation, 168
  diffusion equation, 62
  diffusion term, 310
  elliptical hypothesis, 313
  elliptical PDE, 11, 16
  forcing term, 310
  heat equation, 48
  Helmholtz equation, 85
  hyperbolic PDE, 11, 16
  Laplace equation, 41, 44, 74,
    77, 84
  linear, 10
  Navier–Stokes equation, 167
  reaction term, 310

  stochastic collocation, 112
  Stokes equation, 167
  transport diffusion equation,
    121
  transport term, 310
  uniformly elliptical, 311
  wave equation, 85
  weak form, 312
  well-posed, 192
polynomial interpolation
  Bramble–Hilbert lemma, 330
  convergence, 331
  divided diferences, 18
  error estimate (2 variable
    interpolation), 26
  error estimate (Hermite
    interpolation), 339
  Gaussian quadrature, 342
  Hermite interpolation,
    338–339
  Hermite polynomials, 123,
    244, 338, 344
  Hermite quadrature, 340
  Lagrange polynomials, 19, 29,
    78, 141, 167, 172, 339, 341
  Legendre polynomials, 343
  Newton form, 18
  piecewise polynomials, 78, 245
  triangular set of points, 21
  Vandermonde matrix, 27, 141
positive definite, 6
positive semi-definite, 6
predator/prey population model,
  95, 98–99, 113

R

reference element, 30, 86, 123, 239,
  244, 247, 255, 259, 326–327,
  329–330
Runge–Kutta method, 106

S

second-order PDE, 10
semi-norm, 5, 292, 327, 344
    positive semi-definite, 6
Sobolev space, 291
    $H^k(U)$, 291
    $H^{-1/2}$, 308
    H^{-1}, 307
    $H^{1/2}$, 308
    H_0^1, 307
    $H_0^{1/2}$, 308
    $W^{(k,p)}$, 318
    compact embedding, 324
    density argument, 299
    integration by parts, 299
    Sobolev conjugate, 318
    Sobolev density, 298–299
    Sobolev embedding, 323
    Sobolev inner product, 291
    Sobolev norm, 291
    Sobolev semi-norm, 292
    Sobolev trace, 302
    space of piecewise smooth
        functions, 294

Stokes equation
    intracellular fluid flow, 168
    rendered with FEM, 173
stress
    normal, 165
    shearing, 165
superconvergence, 350

T

triangulation
    Delauney triangulation,
        234–235
    Union Jack pattern, 88, 233

V

viscosity, 167
Volterra predator/prey population
    model, 97

W

weak derivative, 286
weak gradient, 286

Printed in the United States
By Bookmasters